SCIENCEWORLD

SW•7

> Explore
> Question
> Imagine

Peter Saffin
BSc, BEd, GD Env. Sci

Peter Stannard
BSc, DipEd

Ken Williamson
BSc (Hons), DipEd

Contributing authors: Hannah Fay Joanne Parker Melanie Lee

WESTERN AUSTRALIA

First published 2017 by
MACMILLAN EDUCATION AUSTRALIA PTY LTD
15–19 Claremont Street, South Yarra, VIC 3141

Visit our website at www.macmillan.com.au

Associated companies and representatives
throughout the world.

National Library of Australia
cataloguing in publication data

Creator:	Saffin, Peter, author
Title:	*ScienceWorld Western Australian curriculum 7* / Peter Saffin, Peter Stannard and Ken Williamson.
ISBN:	978 1 4202 3822 8 (paperback)
Notes:	Contributing authors: Hannah Fay, Joanne Parker and Melanie Lee
Target Audience:	For secondary school age
Subjects:	Science–Textbooks Science–Study and teaching (Secondary) Education–Curricula–Western Australia
Other creators/contributors:	Stannard, Peter, author Williamson, Ken, author

Publisher: Emma Cooper
Commissioning editor: Patrick O'Duffy
Project editor: Barbara Delissen
Proofreader: Ronél Redman
Illustrators: Brent Hagen, Chris Dent, Guy Holt, Nives Porcellato and Andrew Craig
Cover and text designer: Dim Frangoulis
Production control: Janine Biderman
Photo research and permissions clearance: Alison Hayles and Katy Murenu
Typeset in Merriweather Light 9.5/13.5 by Polar Design Pty Ltd
Cover image: Shutterstock/Victor Maschek
Indexer: Karen Gillen

Printed in China

Warning: It is recommended that Aboriginal and Torres Strait Islander peoples exercise caution
when viewing this publication as it may contain images of deceased persons.

Contents

Using *ScienceWorld* iv
Foreword vii
Links to the Western Australian Curriculum viii
Online resources 1

CH•1 **Introduction to the lab 2**
1.1 Laboratory equipment 4
1.2 Safety in the laboratory 9
1.3 Using a Bunsen burner 12
1.4 Science is investigating 14

CH•2 **Working scientifically 26**
2.1 Inferring and predicting 28
2.2 Measuring 33
2.3 Using graphs 40
2.4 Experimenting 42

CH•3 **Forces 48**
3.1 Forces around you 50
3.2 Frictional forces 59
3.3 Gravitational forces 65

CH•4 **Simple machine technology 72**
4.1 Simple machines 74
4.2 Pulleys and gears 80
4.3 How things fly 88

CH•5 **Classifying living things 98**
5.1 Classifying things 100
5.2 The five kingdoms 107
5.3 Animals and plants 115

Doing a project 128

CH•6 **Ecosystems 132**
6.1 Living in a food web 134
6.2 Ecosystems 144
6.3 Ecosystems under threat 150

CH•7 **Earth, moon and sun 160**
7.1 How the Earth moves 162
7.2 Phases, eclipses and tides 169
7.3 Discovering space 175

CH•8 **Separating mixtures 184**
8.1 What's a mixture? 186
8.2 Solutions 187
8.3 Separating mixtures 193

CH•9 **Sustainable Earth 208**
9.1 Water as a resource 210
9.2 Sustainable resources 216
9.3 Minerals and energy 224

Answers to Reviews 236
Glossary 246
Index 250
Acknowledgements 253

Using ScienceWorld

ScienceWorld takes a constructivist approach to learning, helping students explore what they already know, then building on that knowledge as they progress. This guide will show the key features of *ScienceWorld*, and help you understand how best to use it in the classroom.

Each chapter starts with a **Chapter opening page** containing an engaging photo for discussion. It shows the main content and skills to be covered in that chapter.

Students' prior knowledge is explored with a **Get started** activity, each with an **Explore**, **Question** or **Imagine** task that introduces the chapter topic.

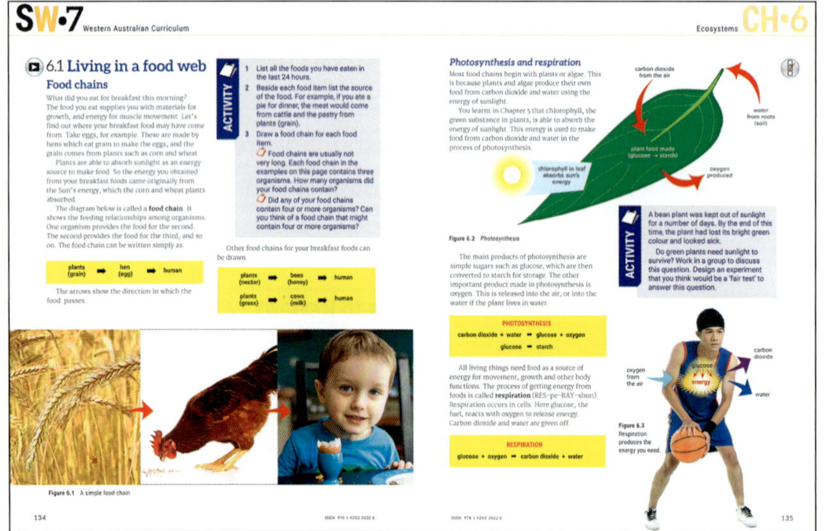

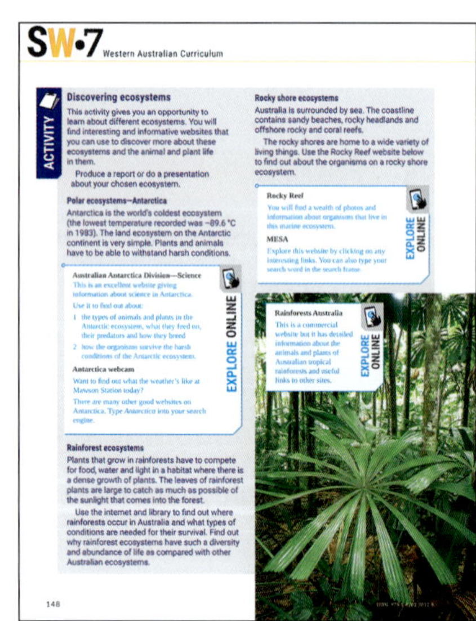

Each chapter is broken into sections focusing on certain aspects of theory and skills covered. *ScienceWorld* uses engaging photos, illustrations, cartoons and contexts to make science accessible to all students.

Activities are dispersed throughout each section to reinforce concepts and provide hands-on learning opportunities for each lesson.

Explore online boxes provide opportunity for wider research.

ISBN 978 1 4202 3822 8

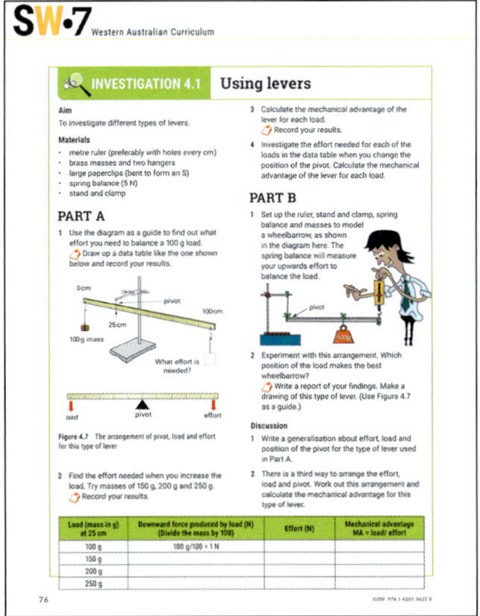
SW•7 Western Australian Curriculum

INVESTIGATION 4.1 Using levers

Investigations provide opportunities for students to apply Science Inquiry skills, while exploring key concepts.

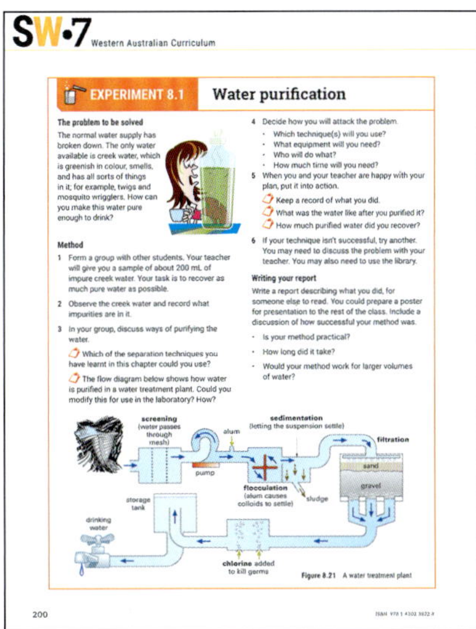
SW•7 Western Australian Curriculum

EXPERIMENT 8.1 Water purification

Experiments allow students to design their own experiments and inventions. Students explore and apply science skills and method while solving problems. This allows students to discover and engage with science concepts for themselves.

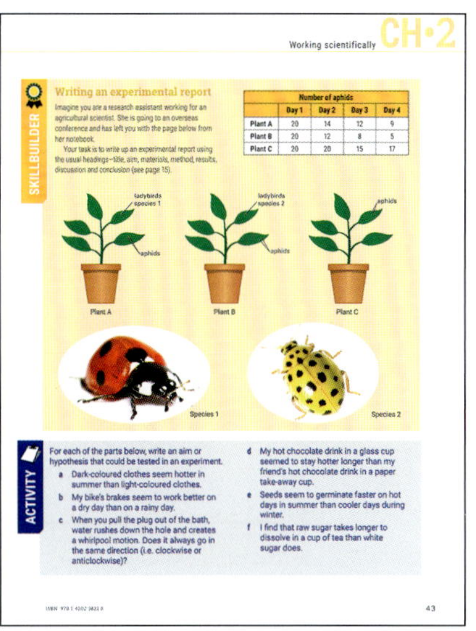
Working scientifically CH•2

SKILLBUILDER

Writing an experimental report

ACTIVITY

Skillbuilders teach key skills explicitly, supporting a clear progression of skill development throughout the book.

Separating mixtures CH•8

SCIENCE AS A HUMAN ENDEAVOUR

Desalination

Science as a human endeavour features bring science to life; putting science in context historically, for today, and for the future. These features include activities that allow students to understand the nature and context of science and to imagine the future.

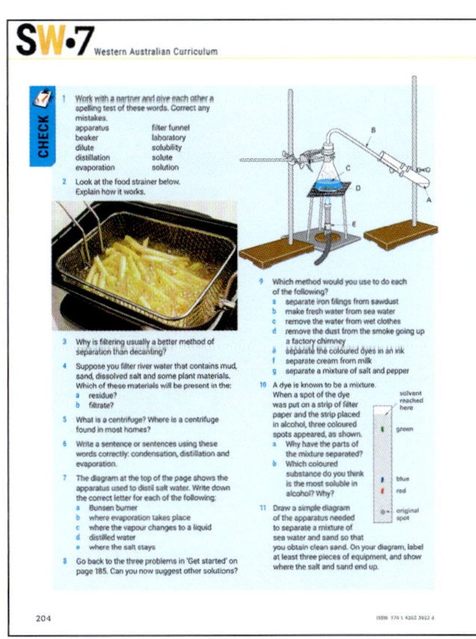
SW•7 Western Australian Curriculum

CHECK

Check questions review students' understanding of concepts for each section.

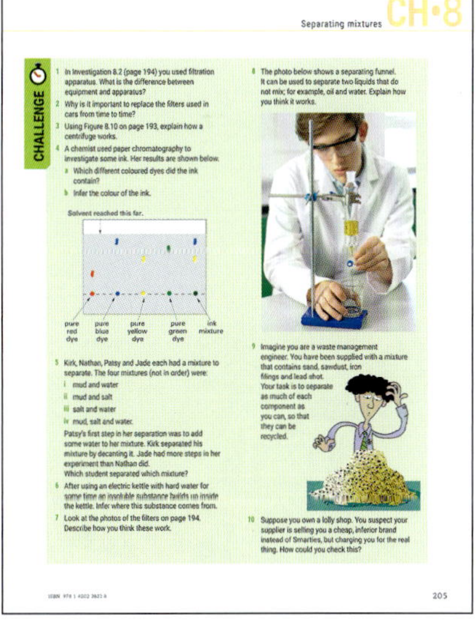
Separating mixtures CH•8

CHALLENGE

Challenge questions provide an opportunity for students to apply their knowledge and skills in context, and challenge students to think at a higher level.

ISBN 978 1 4202 3822 8

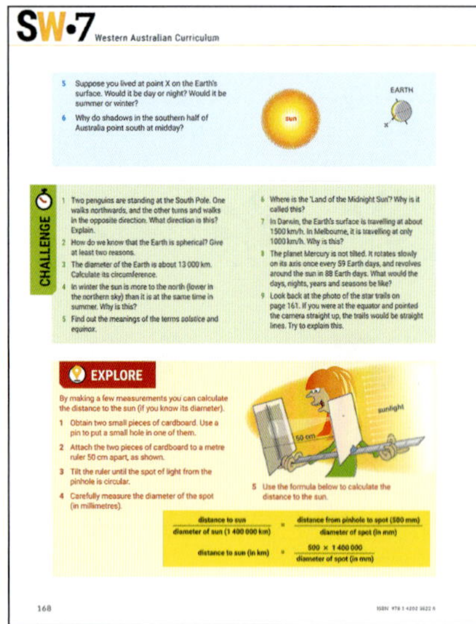

Explore sections allow students to create, invent and inquire into concepts learnt throughout the chapters.

Each chapter ends with a **Main ideas** cloze exercise to test students' understanding of the key chapter concepts through comprehension.

The **Chapter review** then provides an opportunity for students to revise key science knowledge and skills developed through each chapter.

This edition contains more information about the latest emerging technology, and introduces the concept of **STEM**. This addition to the *ScienceWorld* approach will help keep your students more informed about future trends in science and technology.

ISBN 978 1 4202 3822 8

Foreword

As you probably know, I'm mad about science. Every day I learn something new about the world around me—dark matter and dark energy, living creatures of all shapes and sizes, the amazing irrationalities and untapped abilities of our human brain etc. My *Great Moments in Science* radio series/podcast is one way in which I explore these things and try to make them easier for people to understand. In this book you will explore all these things and learn how to think scientifically, asking questions about the world and imagining new solutions to these questions.

Doing science can lead to so many different and fascinating careers where you can design intelligent robots, use giant telescopes in space, produce food in a world where global warming is real, or go as an astronaut to planet Mars! It's fun to apply your knowledge of science to the real world and let your imagination run riot. After all, it's not the answer that gets you the Nobel Prize, it's the question. You could be the next Elizabeth Blackburn who won a Nobel Prize for her work on the telomere—a previously unexplored section of the human chromosome that gives us new and deep insights into aging.

And remember the words of Richard Feynman, 'Science is a way of trying not to fool yourself' ...

Dr Karl

ISBN 978 1 4202 3822 8

Links to the Western Australian Curriculum

This scope and sequence provides an overview of how *ScienceWorld* 7 covers the Western Australian Curriculum. The focus is on the Science Understanding strand, although only some of the Science as a Human Endeavour content and elaborations are covered in this version of the scope and sequence. Included online in the teacher support are curriculum scope and sequence guides that detail how *ScienceWorld* covers the Western Australian Curriculum content descriptions across all four books, and these also include a full mapping of the Science as a Human Endeavour and Science Inquiry Skills strands.

Abbreviations:
SHE: Science as a Human Endeavour
BS: Biological Sciences
CS: Chemical Sciences
ESS: Earth and Space Sciences
PS: Physical Sciences

ScienceWorld 7

Chapter & Unit titles	Science Understanding	Elaborations
1 Introduction to the lab		
1.1 Laboratory equipment	Science Inquiry Skills	
1.2 Safety in the laboratory	Science Inquiry Skills	
1.3 Using a Bunsen burner	Science Inquiry Skills	
1.4 Science is investigating	Science Inquiry Skills	
2 Working scientifically		
2.1 Inferring and predicting	Science Inquiry Skills	
2.2 Measuring	Science Inquiry Skills	
2.3 Using graphs	Science Inquiry Skills	
2.4 Experimenting	Science Inquiry Skills	
3 Forces		
3.1 Forces around you	PS: Change to an object's motion is caused by unbalanced forces, including Earth's gravitational attraction, acting on an object (ACSSU117)	• investigating the effects of applying different forces to familiar objects • investigating common situations where forces are balanced, such as stationary objects, and unbalanced, such as falling objects
3.2 Frictional forces	PS: Change to an object's motion is caused by unbalanced forces, including Earth's gravitational attraction, acting on an object (ACSSU117)	• investigating the effects of applying different forces to familiar objects • investigating common situations where forces are balanced, such as stationary objects, and unbalanced, such as falling objects
3.3 Gravitational forces	PS: Change to an object's motion is caused by unbalanced forces, including Earth's gravitational attraction, acting on an object (ACSSU117)	• exploring how gravity affects objects on the surface of Earth

ISBN 978 1 4202 3822 8

4 Simple machine technology		
4.1 Simple machines	PS: Change to an object's motion is caused by unbalanced forces, including Earth's gravitational attraction, acting on an object (ACSSU117)	• investigating a simple machine such as a lever or a pulley system
4.2 Pulleys and gears	PS: Change to an object's motion is caused by unbalanced forces, including Earth's gravitational attraction, acting on an object (ACSSU117)	• investigating a simple machine such as a lever or a pulley system
4.3 How things fly	PS: Change to an object's motion is caused by unbalanced forces, including Earth's gravitational attraction, acting on an object (ACSSU117)	• investigating a simple machine such as a lever or a pulley system • investigating common situations where forces are balanced, such as stationary objects, and unbalanced, such as falling objects
5 Classifying living things		
5.1 Classifying things	BS: Classification helps organise the diverse group of organisms (ACSSU111)	• considering the reasons for classifying such as identification and communication • grouping a variety of organisms on the basis of similarities and differences in particular features • using provided keys to identify organisms surveyed in a local habitat
5.2 The five kingdoms	BS: Classification helps organise the diverse group of organisms (ACSSU111)	• classifying using hierarchical systems such as kingdom, phylum, class, order, family, genus, species • using provided keys to identify organisms surveyed in a local habitat
5.3 Animals and plants	BS: Classification helps organise the diverse group of organisms (ACSSU111)	• grouping a variety of organisms on the basis of similarities and differences in particular features • using scientific conventions for naming species • using provided keys to identify organisms surveyed in a local habitat
Doing a project	Science Inquiry Skills	
6 Ecosystems		
6.1 Living in a food web	BS: Interactions between organisms can be described in terms of food chains and food webs; human activity can affect these interactions (ACSSU112)	• constructing and interpreting food webs to show relationships between organisms in an environment • recognising the role of microorganisms within food chains and food webs
6.2 Ecosystems	BS: Interactions between organisms can be described in terms of food chains and food webs; human activity can affect these interactions (ACSSU112)	• constructing and interpreting food webs to show relationships between organisms in an environment

6.3 Ecosystems under threat	BS: Interactions between organisms can be described in terms of food chains and food webs; human activity can affect these interactions (ACSSU112)	• exploring how living things can cause changes to their environment and impact other living things, such as the effect of cane toads
7 Earth, moon and sun		
7.1 How the Earth moves	ESS: Predictable phenomena on Earth, including seasons and eclipses, are caused by the relative positions of the sun, Earth and the moon (ACSSU115)	• comparing times for the rotation of Earth, the sun and moon, and comparing the times for the orbits of Earth and the moon • explaining why different regions of the Earth experience different seasonal conditions
7.2 Phases, eclipses and tides	ESS: Predictable phenomena on Earth, including seasons and eclipses, are caused by the relative positions of the sun, Earth and the moon (ACSSU115)	• investigating natural phenomena such as lunar and solar eclipses, seasons and phases of the moon • modelling the relative movements of the Earth, sun and moon and how natural phenomena such as solar and lunar eclipses and phases of the moon occur
7.3 Discovering space	SHE: Scientific knowledge has changed peoples' understanding of the world and is refined as new evidence becomes available (ACSHE119)	• investigating how advances in telescopes and space probes have provided new evidence about space
8 Separating mixtures		
8.1 What's a mixture?	CS: Mixtures, including solutions, contain a combination of pure substances that can be separated using a range of techniques (ACSSU113)	• recognising the differences between pure substances and mixtures and identifying examples of each
8.2 Solutions	CS: Mixtures, including solutions, contain a combination of pure substances that can be separated using a range of techniques (ACSSU113)	• recognising the differences between pure substances and mixtures and identifying examples of each • identifying the solvent and solute in solutions
8.3 Separating mixtures	CS: Mixtures, including solutions, contain a combination of pure substances that can be separated using a range of techniques (ACSSU113)	• investigating and using a range of physical separation techniques such as filtration, decantation, evaporation, crystallisation, chromatography and distillation • exploring and comparing separation methods used in the home
9 Sustainable Earth		
9.1 Water as a resource	ESS: Water is an important resource that cycles through the environment (ACSSU222)	• considering the water cycle in terms of changes of state of water • investigating factors that influence the water cycle in nature • exploring how human management of water impacts on the water cycle
9.2 Sustainable resources	ESS: Some of Earth's resources are renewable but others are non-renewable (ACSSU116)	• considering what is meant by the term 'renewable' in relation to the Earth's resources • considering timescales for regeneration of resources
9.3 Minerals and energy	ESS: Some of Earth's resources are renewable but others are non-renewable (ACSSU116)	• considering what is meant by the term 'renewable' in relation to the Earth's resources • considering timescales for regeneration of resources

ISBN 978 1 4202 3822 8

Online resources

Throughout this book you will find links to activities and video or audio files. The activities are for students to practise key skills, or to reinforce learning on key concepts. Activities vary in type and include crosswords, matching, drag and drop, labelling, multiple choice, true and false, and sequencing activities. Students can repeat these activities as revision, and practise them at any time.

Each activity is scored and the teacher can review student progress in the digital mark book. In *ScienceWorld* 7 there are approximately 80 activities.

When one or more activities are available, you will find an icon on the page of the book where it is most relevant to learning.

We hope you enjoy using these activities to improve learning outcomes.

Quokka

The cover of this book shows a quokka, a small marsupial that lives on Rottnest Island, located off the coast of Western Australia near Perth. Marsupials have an important characteristic: they carry their young in a pouch. Most marsupials are native to Australia, with around 70% of all known marsupials being found on the mainland and Tasmania.

More information about quokkas can be found on page 120.

ISBN 978 1 4202 3822 8

IN THIS CHAPTER

Science Inquiry Skills

- identify and correctly name laboratory equipment
- list safety rules for the science laboratory
- demonstrate the correct use of a Bunsen burner
- do a risk assessment for a science experiment
- write a report of a science experiment, using the correct headings
- make accurate observations and record them

CH•1 Introduction to the lab

ISBN 978 1 4202 3822 8

GET STARTED: *EXPLORE*

A science **laboratory** (la-BOR-a-tree) is a specially designed room where you can carry out experiments safely.

- How is a science laboratory different from other classrooms in the school? Discuss this in a group.
- Draw a large floor plan of your laboratory. Show the position of each of the following items, labelling them clearly:
  - workbenches
  - teacher's bench
  - gas taps and emergency shut-off tap
  - water taps and sink
  - power points and emergency trip switch
  - preparation room
  - doors (including emergency exit)
  - fume cupboard
  - heating equipment cupboard
  - glassware cupboard
  - rubbish bin and broken glass bin
  - fire extinguisher
  - fire blanket
  - sand bucket
  - first aid kit
  - safety shower and eye bath.

Make sure you know why your laboratory contains each of these items.

Risk assessment and planning

It is important to ensure the safety of yourself, other students and teachers while working in a laboratory. All activities you undertake may have some risk of injury occurring. For that reason, you need to carry out a risk assessment before doing any Activity, Investigation or Experiment. A **risk assessment** is how you check that what you are going to do will be as safe as possible. A risk assessment may involve checking the following things to understand if they are safe, or what risk there may be to using them: chemicals, equipment, and the method to be used or the steps to be followed.

After assessing the risk it is then important to look at how to reduce the risk. Some things that can reduce a risk may include:

- tying back hair, wearing a lab coat and safety glasses
- swapping an unsafe chemical or piece of equipment for a safer one
- clearing enough space around you so others are not too close
- following special rules.

If you are not sure if something is safe, ask your teacher.

ISBN 978 1 4202 3822 8

1.1 Laboratory equipment

In the science laboratory you will find many different pieces of equipment. Before you can begin experimenting, you need to be able to identify these items and know what they are used for. You also need to be able to spell their names correctly, and to draw them when you write reports of experiments.

Containers and other useful items

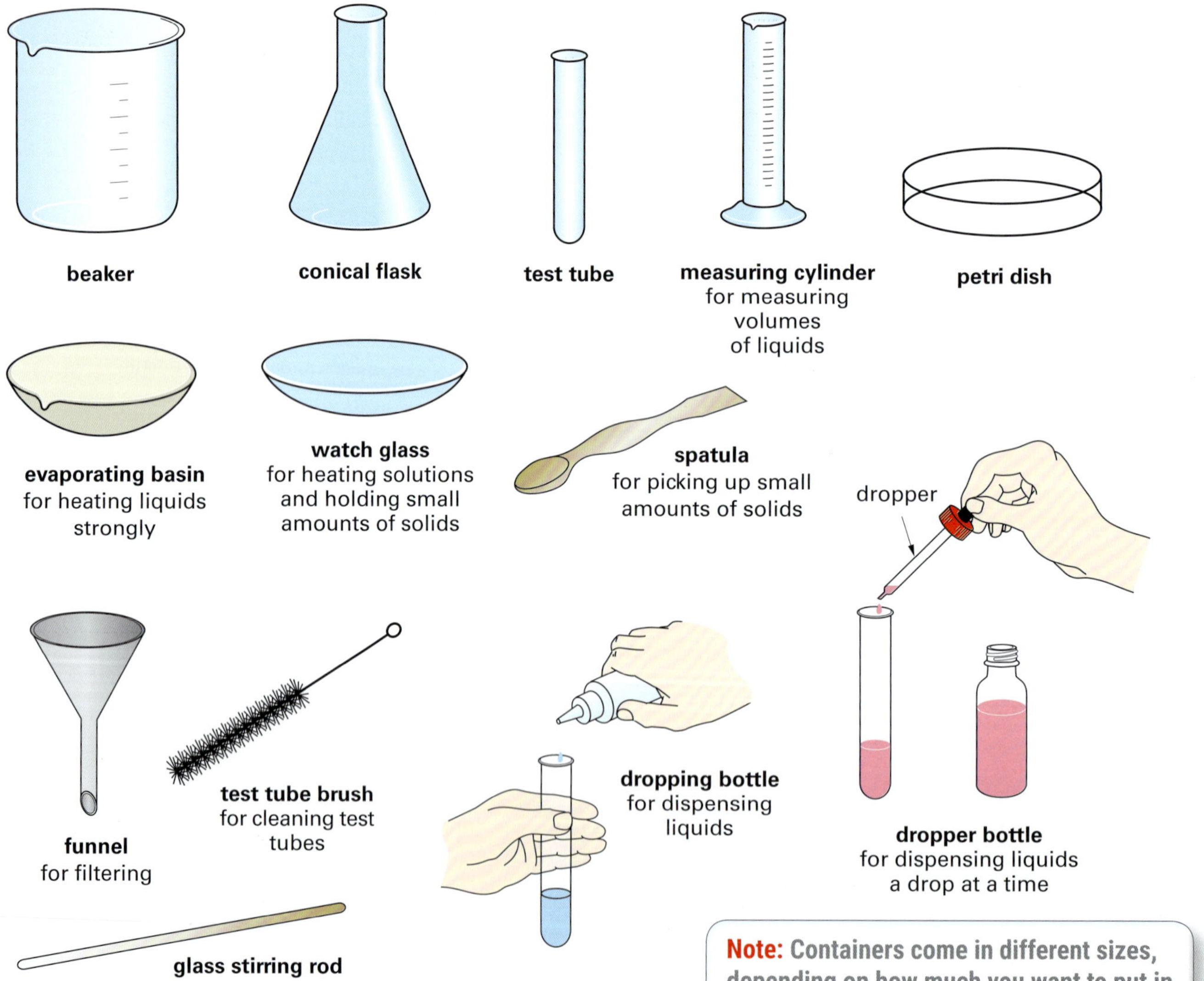

beaker

conical flask

test tube

measuring cylinder for measuring volumes of liquids

petri dish

evaporating basin for heating liquids strongly

watch glass for heating solutions and holding small amounts of solids

spatula for picking up small amounts of solids

dropper

funnel for filtering

test tube brush for cleaning test tubes

dropping bottle for dispensing liquids

dropper bottle for dispensing liquids a drop at a time

glass stirring rod

Figure 1.1 Laboratory equipment

> **Note:** Containers come in different sizes, depending on how much you want to put in them. Some, such as beakers, can be made from glass or plastic.

ISBN 978 1 4202 3822 8

Holding things

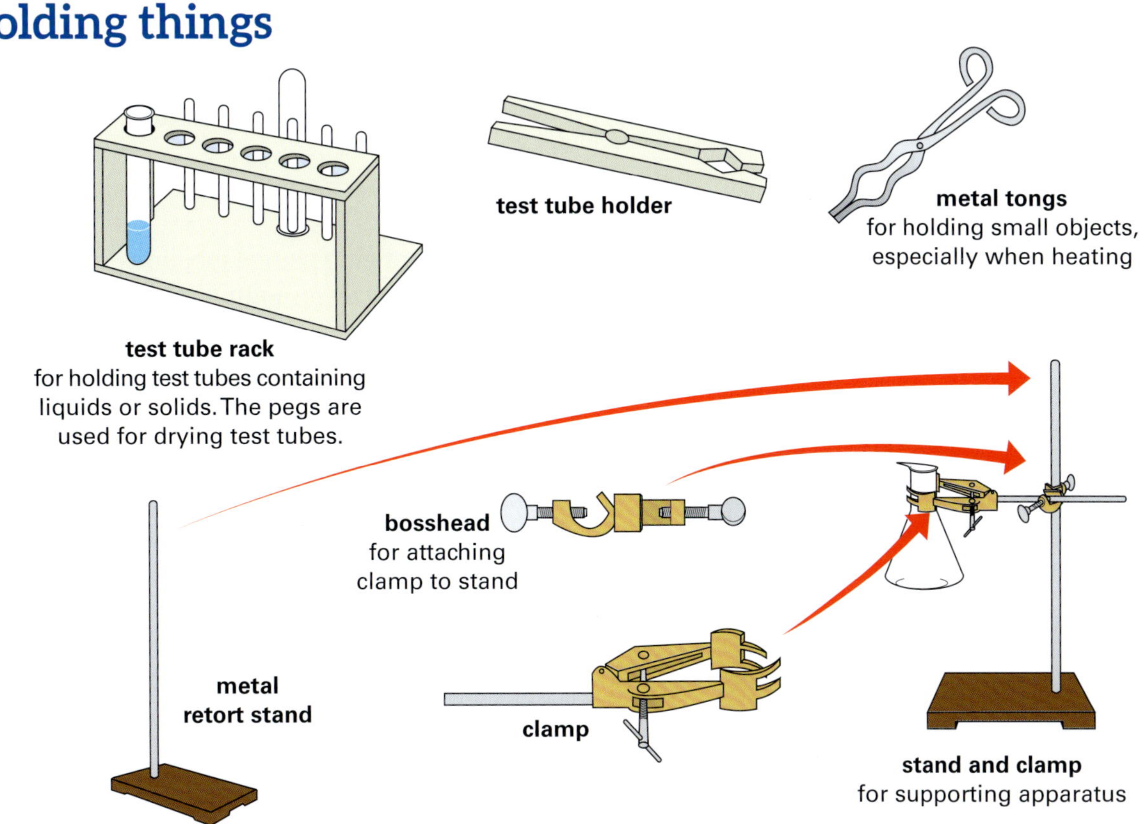

Figure 1.2 Apparatus for handling and securing equipment

Heating apparatus

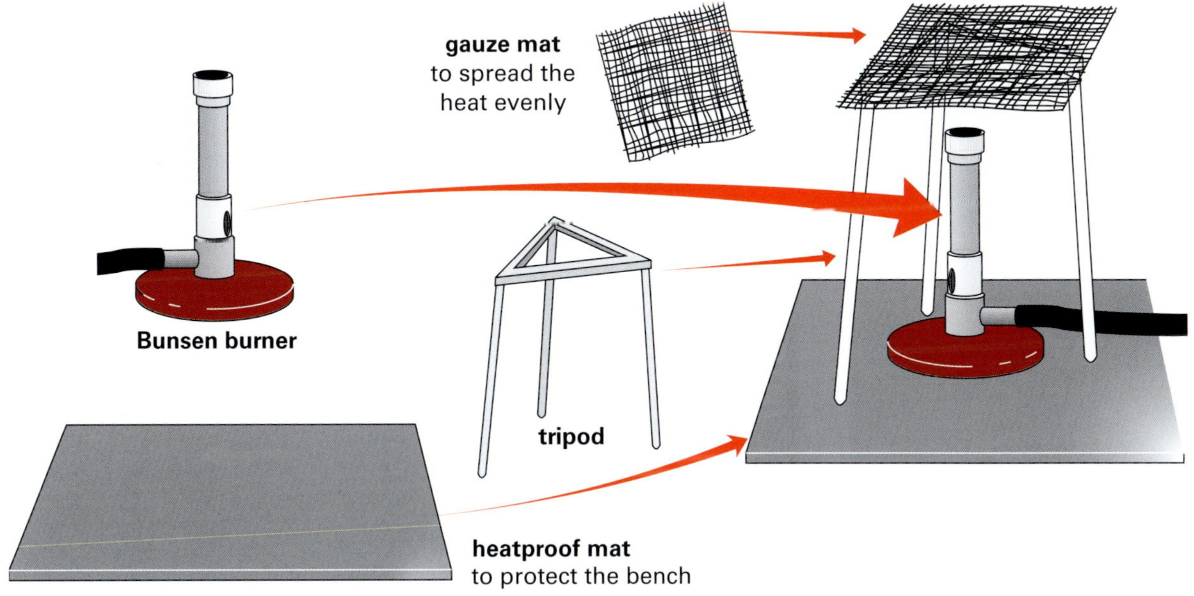

Figure 1.3 Equipment for heating things

ISBN 978 1 4202 3822 8

Drawing science equipment

It is best to keep drawings of science equipment simple. Look at the figures on this page. Figure 1.4 shows a photograph of apparatus, while Figure 1.5 shows the same apparatus as a two-dimensional diagram. In Figure 1.6, the drawings on the left are three-dimensional views, while the drawings on the right show them in a two-dimensional form. This is how you should draw equipment for your science investigations. Note how much simpler the right-hand drawings are. For example, there is no line across the mouth of the test tube, beaker or flask.

When science equipment is put together for a purpose, such as heating water in a flask, it is called **apparatus** (see the photo below). Notice how much simpler the diagram below it is. For example, the tripod has been drawn with only two legs.

When you are drawing apparatus like this, you should:

- use a pencil, for ease of correction if you make a mistake
- label the drawing using label lines
- use a ruler for all straight lines
- not use shading or colouring.

Note: There are plastic templates available for drawing scientific apparatus.

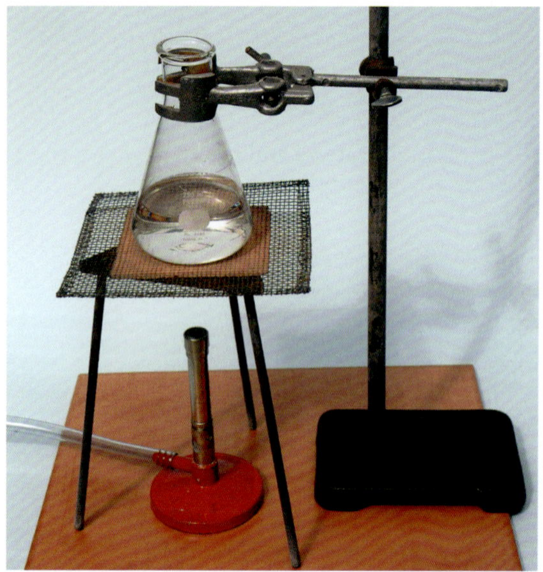

Figure 1.4 Apparatus for heating a flask of water

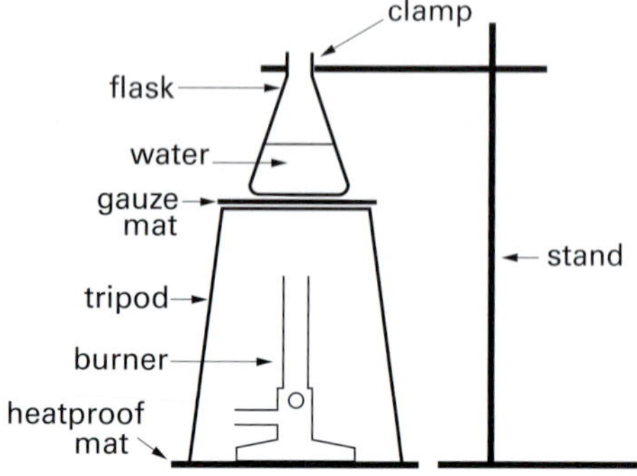

Figure 1.5 Two-dimensional diagram of apparatus for heating a flask of water

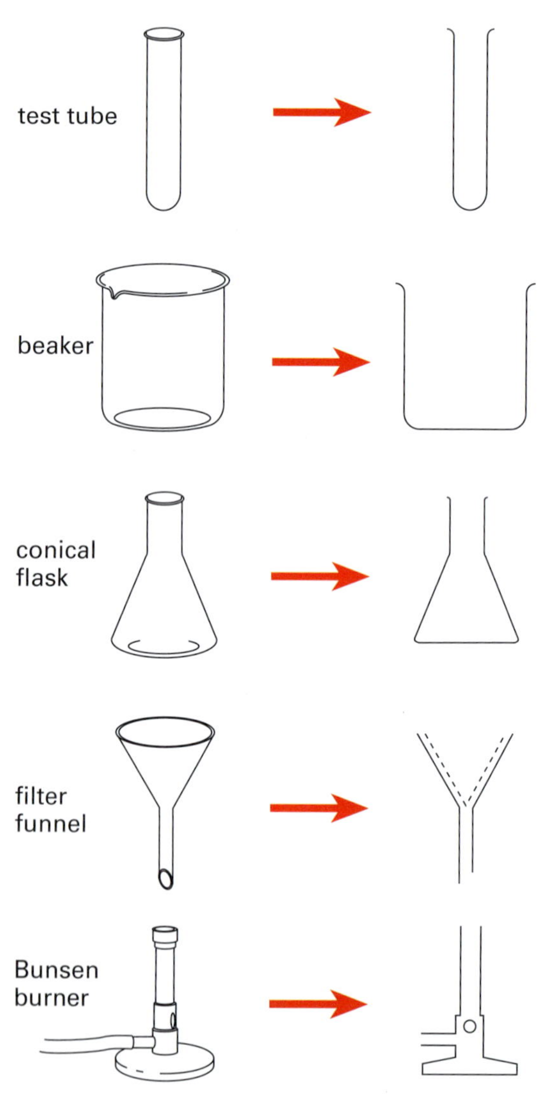

Figure 1.6 Three-dimensional and two-dimensional diagrams of apparatus. Use the two-dimensional diagrams on the right for your science investigations.

ISBN 978 1 4202 3822 8

SKILLBUILDER

Data tables

In the next activity you are going to look at a number of pieces of laboratory equipment. You will record the name of each piece and what it is used for. A way to record information so that it is easy to read and understand is by drawing a table.

In other investigations you will have to record results that include numbers or measurements. These numerical results are called **data**. You record data in a **data table**.

For example, suppose you counted the number of people in your class with blue eyes, brown eyes and green eyes. The results are easy to read if they are in a data table like the one shown.

Eye colour	Number of people
blue	8
brown	11
green	9

Exercise

Sam and Amanda were investigating how long it takes bean seeds to germinate at different temperatures. They recorded their results in a data table.

At 10 °C the bean seeds took on average 8 days to germinate. At 20 °C they took 6 days. At 5 °C they took 12 days, and at 30 °C they took 4 days. None germinated when the potting mix was at 50 °C.

Draw up a data table for the results.

ACTIVITY

Your teacher will place about eight pieces of numbered equipment on your table.

Your group's task is to identify the equipment by using the information on the previous pages.

For each piece of equipment, record its number, name, size (if it is a container) and what it is used for. Record this information in a data table like the one below.

Equipment number	Equipment name	Size (if container)	What is it used for?	Diagram of equipment
1				
2				
3				
...				

ISBN 978 1 4202 3822 8

SCIENCE AS A HUMAN ENDEAVOUR

Josh's laboratory

Dr Josh Mylne is a plant biologist working on how plants know that winter has passed. Winter makes plants flower faster, and many crops wait until after winter before they flower and make seeds.

Most of Josh's work is done in his laboratory using some of the equipment you have seen and identified. His plants are grown on plastic petri dishes in growth rooms or in soil in glasshouses.

Look at the equipment in the photo of Josh's laboratory. Make a list of as many pieces of equipment as you can recognise.

Figure 1.7 Dr Josh Mylne, plant biologist, in his lab

CHECK

1 Match each item of equipment listed below with its use. Write the correct pairs in your notebook in a table as shown.

Apparatus	Use

tripod	a general purpose glass container for small amounts of material
gauze mat	for holding hot objects
spatula	placed on top of a tripod to spread the heat
test tube	you stand equipment on this when heating things
beaker	for picking up small amounts of solids
Bunsen burner	a general purpose glass container with a pouring lip
stand and clamp	for heating things
metal tongs	for holding equipment in place

2 Draw simple two-dimensional drawings of the following pieces of equipment:

- round-bottom flask
- measuring cylinder
- tripod
- stand and clamp
- filter funnel
- evaporating basin

Check with your teacher that you have drawn them correctly.

3 Which equipment would you need to:

- measure out exactly 20 mL of water?
- heat a small volume of liquid?
- heat a small amount of solid strongly?
- take a small amount of powder from a bottle and dissolve it in water?
- test whether a small piece of fabric will burn?
- boil about 200 mL of water?
- add 2 drops of liquid A to about 5 mL of liquid B?

4 Look at the equipment in Figure 1.4 on page 6. What other container could you use to heat water? What is the advantage of the flask?

ISBN 978 1 4202 3822 8

1.2 Safety in the laboratory

A laboratory is a place for doing things. You will enjoy working there. However, to make the laboratory a safe place for everyone, there are two main rules to follow.

1 Know what you are doing in the laboratory—read the instructions carefully before you start.
2 Always think of others and behave sensibly.

Follow the safety rules on the right and accidents should not happen. Many accidents can be avoided by keeping alert and using common sense. Read about the types of accidents that can occur and how to avoid them. If an accident does occur, report it to your teacher.

Eye injuries can be caused by liquids splashing into your eyes during investigations.

- Always wear safety glasses whenever there is a chance of liquid splashing into your eyes, especially when heating things.
- Always wear safety glasses when you see the safety glasses symbol on Investigation pages.
- Never point a test tube towards yourself or anyone else. If you get a chemical in your eye, wash your eye immediately with lots of water, and tell your teacher. Your laboratory may have a special eye bath to make this easier.

Safety rules

1 Do not enter the laboratory unless you are with a teacher.
2 Never touch equipment in the laboratory unless you are told to use it.
3 Do not eat or drink in the laboratory.
4 Always walk—never run.
5 Wear protective clothing—a laboratory coat or apron—and, when appropriate, safety glasses.
6 Never taste anything.
7 Do not use paper to light Bunsen burners, and do not put burning or hot things in the rubbish bin.
8 Keep books, paper and clothing away from flames.
9 Tie up long hair.
10 Always point test tubes away from people.
11 Check with your teacher on how to dispose of waste liquids and solids. Broken glass should be cleaned up using gloves, a brush and dustpan, and placed in a special bin.
12 If you spill something on your skin or clothes, wash it immediately with lots of water. Tell your teacher.
13 Report all accidents and breakages to your teacher.
14 After heating equipment, let it cool on a heatproof mat before picking it up. This will avoid burns.
15 Clean all equipment after use and put it back where you got it from. Clean and dry your work bench.
16 If you see something that looks unsafe report it to your teacher.

ISBN 978 1 4202 3822 8

Poisoning can be caused by breathing in fumes during an investigation, or by tasting chemicals or spilling them on your skin.

- Never taste anything, and never bring food or drink into the laboratory.
- Check the labels on chemicals before you use them.

ACTIVITY

In a small group, discuss what each of these labels or symbols means.

Give examples of where you would find each of these.

toxic

flammable

corrosive

radioactive

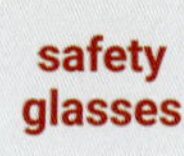

safety glasses

Cuts are caused mainly by broken glass.

- Use gloves and a brush and shovel or dustpan to clean up any broken glass and put it into the special bin.

Burns can be caused by touching hot equipment, or by spilling hot liquid.

- Treat these types of burns with cold running water for about 10 minutes.
- More serious burns can occur when using a Bunsen burner. If this happens, tell your teacher immediately.

Fires are always possible when using burners.

- Do not use paper to light a burner, and never place burning things in rubbish bins.
- If you have long hair, it is essential you tie it back whenever you are using a burner.
- If there is a fire, stay calm and call for help.
- If a person's hair or clothing catches fire, remember the three rules: stop, drop and roll. The person must stop moving around, drop to the floor and roll. While the person is rolling, a fire blanket should be quickly wrapped around them to smother the flames.

Damage to clothing and skin can occur when chemicals, especially corrosive liquids such as acids and alkalis, are spilt.

- Wear a lab coat or other protective clothing when doing investigations.
- If there is a spill, wash the area immediately with lots of water and send someone to tell the teacher.
- In serious cases it may be necessary to use the safety shower.

ACTIVITY

1 At the start of the year your teacher may have given you a copy of the laboratory safety rules for your school.

Work in a small group and compare the rules in the list on page 9 with your school's rules.

- Which rules are the same?
- Which are different? Suggest why these rules are different. Your teacher may want to discuss this with the whole class.

2 Without looking at the information on this page, make a dot-point list of the things you would do in the following situations:

- You have spilt a chemical on your skin.
- Your friend has picked up a very hot piece of glassware.
- You have splashed a liquid in your eye.

ISBN 978 1 4202 3822 8

Disposal of chemicals

To protect our environment it is essential to dispose of chemicals properly. You should never put leftover solids down the sink. Some liquids can safely be poured down the sink, but others cannot, so always follow your teacher's instructions.

At home, you must also be careful how you dispose of chemicals. You should not put oil or petrol or corrosive substances such as caustic soda down the sink. These substances may pass through sewage treatment plants into creeks and rivers and harm the plants and animals that live there. Councils usually provide places at the local dump where you can take liquids such as used oil and mineral turpentine. Industries must also take special care in the disposal of chemicals, and there are laws to enforce this.

Figure 1.8 Correct disposal of waste ensures the environment is protected.

CHECK

1. Select three of the safety rules from the list on page 9 and explain why each is important.
2. Design a poster to illustrate one of the safety rules.
3. Look at the cartoon showing Ian and Penny in the laboratory.
   a What has happened to Ian?
   b What was he trying to do?
   c What did he do wrong?
   d What should Penny tell him to do?

4. Explain two things that are unsafe in the following situation.

5. A school has the following rules for its science laboratory:
   - Leave bags outside.
   - Wear shoes with leather uppers.
   - Clean and put away all equipment when you have finished with it.

   Explain why you think each of these rules is necessary.

ISBN 978 1 4202 3822 8

1.3 Using a Bunsen burner

In the laboratory a Bunsen burner is used to heat things. The burner is named after a German chemist called Robert Wilhelm Bunsen. He found that he could get a cleaner and hotter flame if he allowed air to mix with the gas before it was burnt. This piece of technology quickly became essential in science laboratories, and led to improvements in gas burners used in everyday life.

How a burner works

As the gas flows through the gas jet, air is drawn in through the air hole. This mixture of gas and air then burns at the top of the barrel to produce a flame.

You can control the temperature of the flame by turning the metal ring or collar. As you rotate this collar, you either open or close the air hole, which then changes the amount of air mixed with the gas. The more open the air hole is, the more air is drawn in, and the hotter the flame.

Air hole closed—yellow safety flame

The yellow safety flame is easy to see. Use it if you want to leave the burner on for a short time without using it for heating.

Air hole open—blue heating flame

The blue flame is very hot and hard to see, and can cause serious burns. The hottest part of the flame is about 1500 °C—higher than the temperature needed to melt silver and gold!

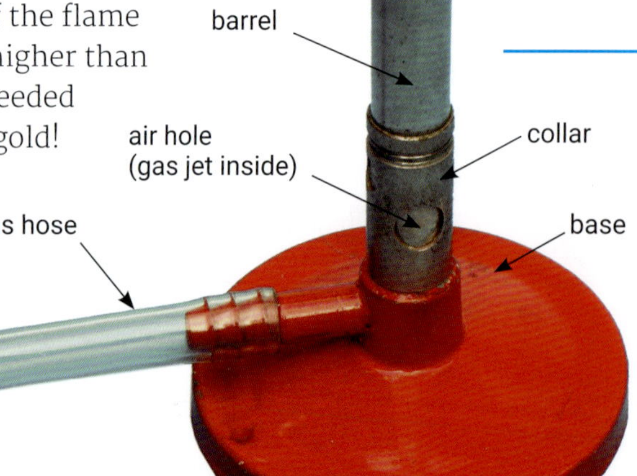

Figure 1.9 The parts of a Bunsen burner

ACTIVITY

Before you start an investigation, you need to understand what you are doing. You should read through the steps, look at the diagrams and prepare data tables where necessary.

To help you make the most of the investigations, use a 'Risk assessment and planning' box. Read the one on the next page.

You can avoid most laboratory accidents if you are aware of any risks to your safety before you start. You can do this by completing a risk assessment.

Read through Investigation 1.1: Using a Bunsen burner.

1 In a group, list the risks involved in this investigation, using what you have learnt on pages 9 and 10.
2 For each risk, discuss which safety precautions you will need to take.

ROBERT BUNSEN

Robert Bunsen spent most of his life working in a laboratory. He nearly killed himself with arsenic poisoning and he lost one eye in an explosion when a small piece of glass went into his eye.

Find out more about this interesting scientist by searching the internet for **Robert Bunsen**.

You will find a number of websites.

What did you find most interesting about him?

ISBN 978 1 4202 3822 8

INVESTIGATION 1.1 Using a Bunsen burner

Aim

To use a Bunsen burner correctly.

Materials

- Bunsen burner
- heatproof mat
- tripod
- gauze mat
- matches
- piece of copper wire
- metal tongs
- 250 mL beaker

Rules for safe use

1. Keep the burner away from books, and away from the edge of the bench.
2. Use a heatproof mat under the burner.
3. Always light the burner with the air hole closed.
4. Switch to a yellow safety flame when not heating.
5. The barrel of the burner gets very hot. If you have to move the burner, turn it off first. Move it by holding the base or the gas hose.
6. Check that the gas is off properly when you have finished.

Risk assessment and planning

Read through the experiment carefully before you do it, then answer these questions.

- Why should you light the match before you turn on the gas?
- What does the sign mean?
- What is the purpose of Part C?

PART A Lighting the burner

1. Place the Bunsen burner on a heatproof mat. Connect the gas hose to a gas tap.
2. Rotate the collar so that the air hole is closed.
3. Light a match, turn on the gas, and bring the match close to but not over the top of the barrel, as shown. The gas should ignite.

   If a hissing noise comes from the burner or a flame burns at the jet, immediately turn off the gas. The burner is said to be 'burning back' and needs cleaning. Report this to your teacher.

Figure 1.10 Lighting a Bunsen burner

4. Observe the flame with the air hole closed. This flickering yellow safety flame is not very hot, and is very sooty.
5. Gradually open the air hole, noting carefully what happens to the flame. This roaring bluish flame is the one you use for heating. You can change the size of the flame by adjusting the gas tap.

   In your science notebook, draw diagrams of the yellow safety flame and the blue heating flame. Use coloured pencils.

Note: Whenever you see this symbol it means that there is something to record in your science notebook.

PART B How hot is the flame?

1. Turn the gas on fully and open the air hole.
2. Use metal tongs to hold a piece of copper wire in the flame. Note how long it takes for the wire to become red hot.
3. Let the wire cool down and then repeat step 2 in other parts of the flame.

   On your diagram of the blue flame, mark in the hottest part.

ISBN 978 1 4202 3822 8

PART C Comparing flames

For Part C you will need to work with another group. One group will use a yellow flame and the other a blue flame.

1 Add 50 mL of water to a beaker. (Use the graduations on the side of the beaker.)

2 Set up the heating apparatus as shown below.

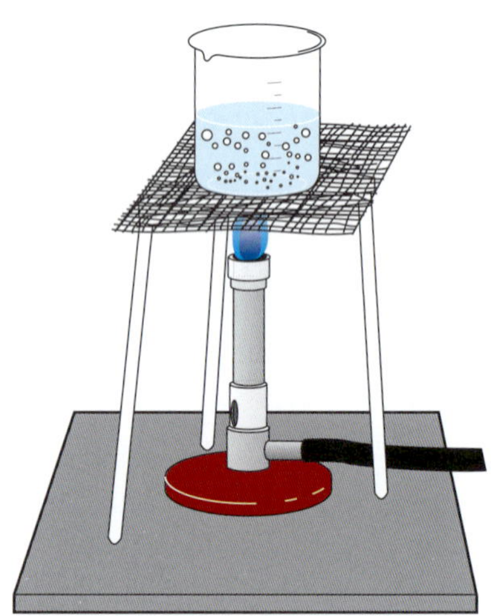

3 Light both burners. Leave one with the air hole closed (yellow flame) and the other with the air hole open (blue flame).

4 Start heating the beakers at the same time, and time how long the water takes to boil for each. Which flame boiled the water first?

Discussion

1 Which flame was hotter?

2 When the beakers are cool, lift each one and check underneath. What do you observe?

3 Which flame was easier to see?

4 When you are not using a burner you should always leave the air hole closed. Why?

5 Suggest why it is important to light the burner with the air hole closed.

6 What is the purpose of the gauze mat?

7 What is the purpose of the heatproof mat?

1.4 Science is investigating

Scientists like Dr Josh Mylne carry out their investigations in the laboratory and in the field to answer questions, such as why plants need cold weather to flower and fruit.

Josh plans his investigations carefully and makes many observations. An **observation** is something you can find out with your senses. We mainly use our sense of sight, but you can also feel the texture of an object or whether it is hot or cold.

Scientists also take measurements during investigations and record them in data tables. Here Josh is counting and marking the leaves of plants that have been in the coldroom. Notice the data table in his recording book.

Figure 1.11 Dr Josh Mylne making observations

ISBN 978 1 4202 3822 8

Writing reports

As well as planning his investigations carefully, Josh plans how he is going to record his observations and write how he conducted his investigations. He does this in a report. A report is important because other people can find out what he did and what he discovered.

A report is organised using seven headings.

TITLE the name of the investigation, your name and the date.

AIM you say why you did the investigation. Sometimes this is a question.

MATERIALS a list of equipment and chemicals you used in the investigation.

METHOD you say what you did in the investigation in numbered steps. Whenever possible, include a large, neat diagram of the apparatus.

RESULTS you record the data. Data includes qualitative observations (words) and measurements (numbers). Usually these are recorded in a data table. This makes the data easier to read.

DISCUSSION you try to explain your results, and list any problems that you experienced. You might also explain how you could improve the investigation.

CONCLUSION you answer the question posed in the aim.

Sometimes in your conclusion, you can write a general statement or **generalisation**—one that seems true in most cases. For example, a student investigating the stopping distances of toy trucks concluded: The heavier the truck is, the longer it takes to stop.

You will not always be able to make a generalisation like this, and in some cases it may not be possible to make a conclusion at all.

In Investigation 1.2 on the next page, you can practise writing your own report.

SKILLBUILDER

Filtering

In Investigation 1.2 you are going to filter a solid from the liquid in a beaker. To do this, you will need to know how to fold a filter paper and set up the apparatus for filtering.

The diagram below shows how to fold a filter paper. Your teacher will give you a filter paper to fold.

Your teacher will give you some cold tea with lots of tea leaves in it. Your job is to filter the tea.

A quick way to do this is to *decant* the tea (pour off most of the tea, leaving the tea leaves behind), then filter the remainder.

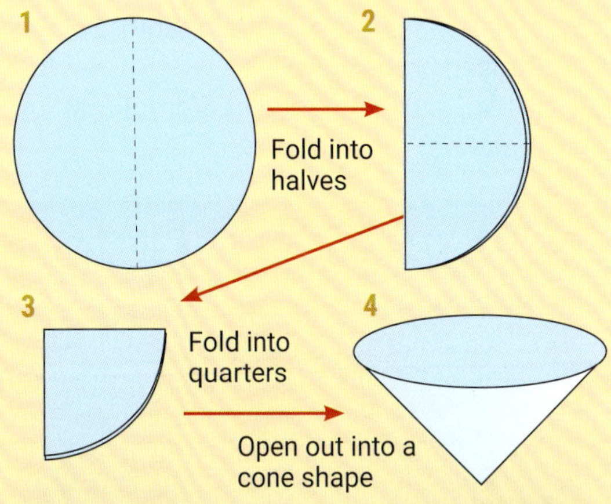

Figure 1.12 The apparatus for filtering

ISBN 978 1 4202 3822 8

INVESTIGATION 1.2 Making milk glue

SAFETY GLASSES MUST BE WORN

Aim

To find out if glue can be made from milk.

Materials

- skim milk (about 100 mL)
- white vinegar (about 25 mL)
- baking soda (about 5 g)
- two 250 mL beakers
- spatula
- stirring rod
- filter funnel and paper
- stand and clamp (or filter stand)
- Bunsen burner
- tripod, gauze mat and heatproof mat
- matches

Risk assessment and planning

- Carefully read through the method and list the safety precautions you will have to take. Discuss this with your teacher before you start.
- Why should you wear safety glasses during this investigation?

Method

1 About one-third fill a 250 mL beaker with skim milk.

2 Add 25 mL of vinegar to the milk.

3 Set up a tripod, gauze mat, Bunsen burner and heatproof mat.

4 Heat the mixture slowly, stirring all the time with the stirring rod. When you see small white clumps (called curds) forming, turn off the burner. The curds will fall to the bottom of the beaker. You have made cottage cheese!

5 While the mixture settles and cools, set up the filtering apparatus.

6 Decant the clear liquid (called the whey) into another beaker, and try not to lose any of the curds. Pour the liquid down the sink.

7 Fold the piece of filter paper, and put it in the funnel.

8 Pour the curds into the filter paper. When all the liquid has filtered through, gently scrape the curds into a beaker.

9 Add 20 mL of water and one spatula of baking soda to the curds.

10 Stir to make a paste. This is the milk glue.

11 Test your glue by sticking paper or ice-cream sticks together.
Record your observations when the glue dries.

Writing your report

> Write a full report of the experiment, using the headings from the previous page.
> In this case the TITLE and AIM have been written for you.
> Under METHOD you should write, in your words, what you did. Include diagrams to help your description.
> Under RESULTS record your observations of what happened during the investigation.
> In the DISCUSSION try to explain your results and list any improvements you would make to the method.
> In the CONCLUSION write down an answer to the question in the aim.

Note: Compare your report with the one on the next page, but do not look at it until you have written your own.

ISBN 978 1 4202 3822 8

Investigation 1.2 8 March

Making Milk Glue

Aim:
To find out if glue can be made from milk.

Materials:
100 mL skim milk, 25 mL white vinegar, 5 g baking soda, spatula, stirring rod, filter funnel and paper, stand and clamp, Bunsen burner, tripod, gauze mat, heatproof mat and matches

Method:
1. About 100 mL of milk was added to a 250 mL beaker, then about 25 mL of vinegar was added.
2. The mixture was heated slowly with a burner while stirring.
3. When white curds were noticed, the burner was turned off.
4. The mixture was set aside to cool, while the filtering apparatus was set up.
5. The liquid was decanted from the mixture, and the curds were filtered.
6. The curds were scraped into a beaker.
7. About 20 mL of water and a spatula of baking soda were added to the curds.
8. The mixture was stirred to make a paste.

stirring rod
beaker
milk and vinegar
tripod
Bunsen burner

Results:
The glue paste was tested with paper and with wooden ice-cream sticks. We found that the glue stuck paper together really well. But the wooden ice-cream sticks came apart with a little force.

Discussion:
We think milk glue works really well with paper. However, we found out on the internet that the curds are actually made from casein which is a protein in milk. This casein glue was used by the Egyptians as a wood glue.
We don't know why our glue didn't work well on wood. But we think we should have dried the curds better before we added the baking soda.

Conclusion:
A glue can be made from milk.

Skills for investigating

In this chapter you have learnt how scientists make careful observations of their investigations and then record them in a report. Making accurate observations and recording them are important skills in science.

These skills are also used in other fields. If you have read mystery stories or watched them on TV, you will know that many crimes are solved because somebody has made careful observations and written accurate reports.

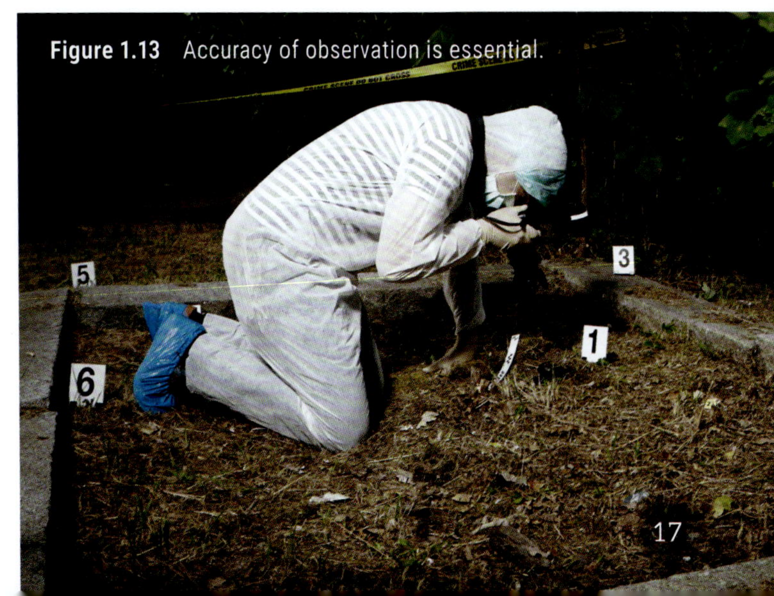

Figure 1.13 Accuracy of observation is essential.

ISBN 978 1 4202 3822 8

SKILLBUILDER

Observing and recording

Making accurate observations and recording them are important skills in everyday life—not just for a scientist like Josh Mylne or a crime scene investigator. Whatever job you have later in life, it will be an advantage if you can make detailed observations and record them for other people to read.

Imagine you are a Martian visiting Earth for the first time. You discover two strange objects:

- a sugar cube (there is no sugar on Mars)
- a burning candle (things don't burn on Mars).

Your job is to note as many observations as possible to record in the spaceship's log.

Use as many senses as you can—sight, hearing, touch, smell and taste. It may be a good idea to take measurements, and you may want to do something to the object; for example, put the sugar cube in water.

You should also observe the candle before, during and after burning.

Work in a group of two or three, and record detailed observations using complete sentences (not in note form). A sentence must contain a verb and, of course, should always begin with a capital letter and end with a full stop.

When you have finished, join with another group to compare your observations. How many observations did you make of the burning candle? A good observer should be able to make about 50!

ISBN 978 1 4202 3822 8

Observing and recording

Aim

To practise the skills of observing and recording.

Risk assessment and planning

Carefully read the instructions for each of the four parts. Look at Parts A, B and C. For each part say why you need to wear safety glasses.

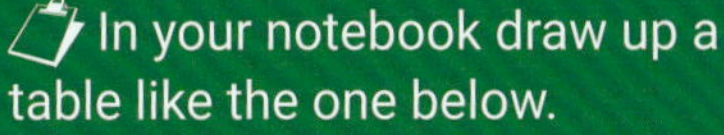

In your notebook draw up a table like the one below.

Use as many senses as possible—sight, touch, smell and hearing—but do not taste any of the chemicals.

Part	Observations
A	
B	

PART A

Materials

- limewater
- drinking straw
- flask (e.g. 250 mL)

Method

Pour about 50 mL of limewater into the flask.
Blow through a drinking straw into the limewater.

Record what happens.

PART B

Materials

- test tube
- small piece of zinc
- dilute **hydrochloric acid** (1 M) in dropping bottle

Method

Place a small piece of zinc in the clean test tube.
Cover the zinc with dilute hydrochloric acid.

Record as many observations as you can.

PART C

Materials

- test tube
- spatula
- sodium thiosulfate crystals (hypo)
- dilute **hydrochloric acid** (1 M)

Method

Use the spatula to add about a teaspoon of sodium thiosulfate to the test tube. One-third fill the test tube with water and shake to dissolve the crystals.

What do you notice when you feel the test tube?

Add about 10 drops of dilute hydrochloric acid to the test tube.

What happens now?

ISBN 978 1 4202 3822 8

PART D

Materials

- ink pad
- methylated spirits and paper towel (for cleaning up)
- hand lens

Method

1 Examine your fingers using the hand lens. Can you see the fingerprint patterns?

2 Prepare a fingerprint chart on white card or paper, as shown. Each box needs to be at least 3 cm × 3 cm.

	Thumb	Index	Middle	Ring	Little
Right hand					
Left hand					

3 Place the ink pad and your fingerprint chart on the edge of the bench. Roll your right thumb on the ink pad, then carefully roll it over the right thumb spot on the chart as shown below.

4 Repeat this procedure for all fingers, on both hands.

5 Use the methylated spirits and paper towel to clean your fingers.

6 Use a hand lens to examine your fingerprints. Are any of them the same?

Use the photographs below to try to classify your prints as arches, loops or whorls. (A composite print has several of these patterns joined together.)

Collect data from the whole class on the numbers of arches, loops, whorls and composite fingerprints. Which type is most common?

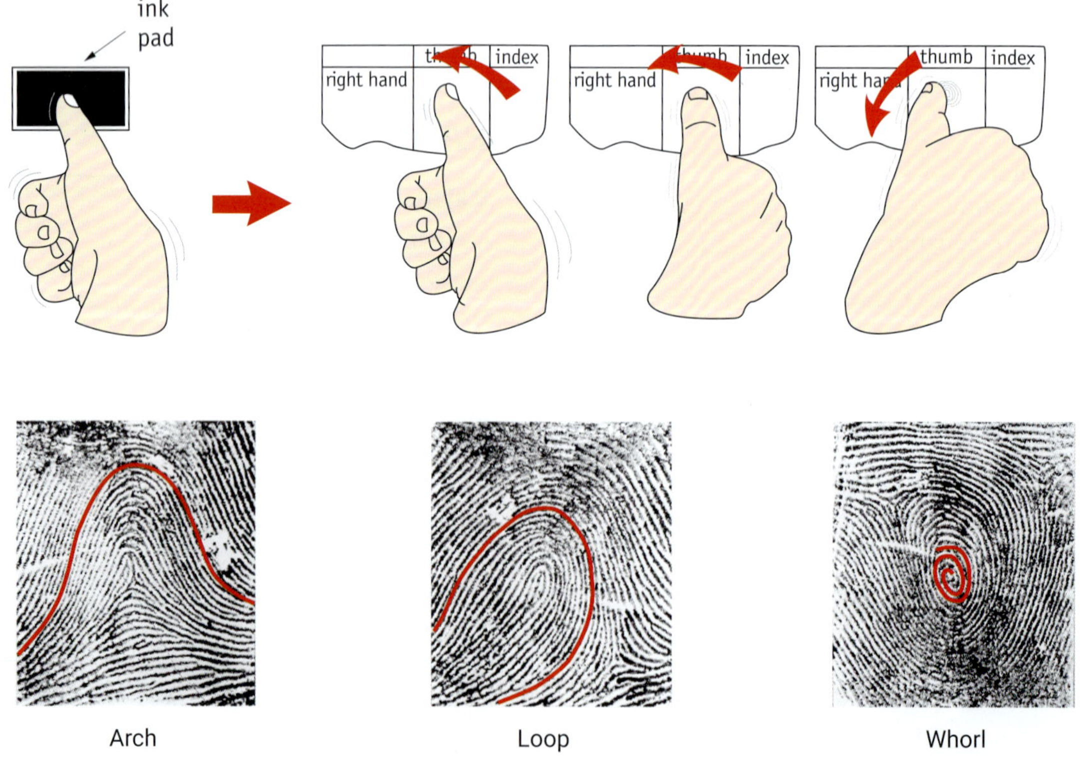

Arch Loop Whorl

ISBN 978 1 4202 3822 8

Fingerprints

Look closely at your fingers and you will see a pattern of ridges on your skin. In Investigation 1.3, when you put ink on your fingers, these ridges formed fingerprints when you placed your fingers on paper.

Your fingers are normally covered by small amounts of sweat and fats that have been given off by the tiny glands found in your skin. When you touch something, you leave behind traces of sweat and fats that were on the ridges of your fingers. These prints are usually invisible, but when they are dusted with powder that sticks to the sweat and fats, they can be seen clearly.

You may have found out in Investigation 1.3 that everyone's fingerprints are different. Scientific investigators use this knowledge to test for fingerprints at crime scenes to help identify suspects.

EXPLORE ONLINE

To learn more about fingerprints, and how to dust for fingerprints around school or home, search for '**Connections Academy fingerprints**'.

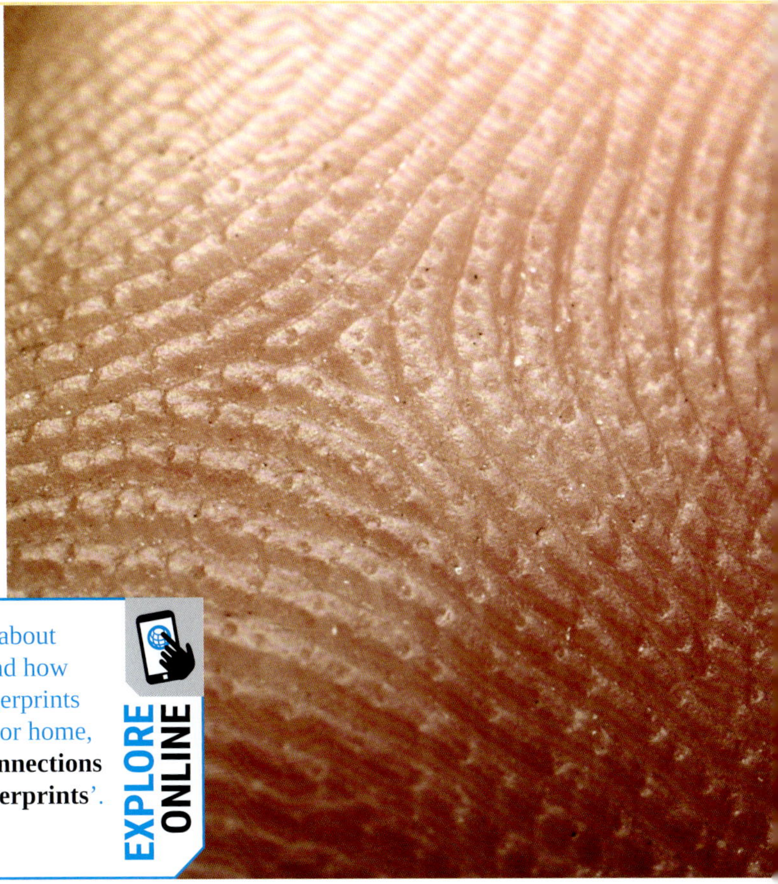

Figure 1.14 Close-up of a fingerprint

CHECK

1 Draw a Bunsen burner in your notebook. Label the five major parts shown in the diagram.

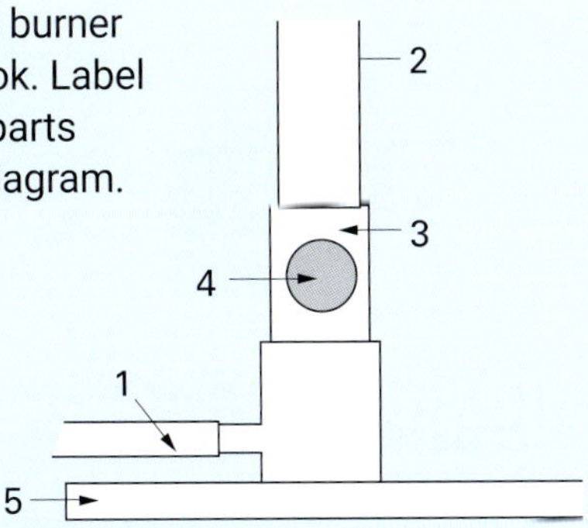

2 Give two reasons why the blue flame is a hazard. Why is the yellow flame called a safety flame?

3 Alistair's teacher asked him to write in point form the steps in lighting a Bunsen burner so that other students could follow them.

Alistair wrote the first step:

1 Connect the burner hose to the gas tap.

Complete Alistair's task.

4 Why is it important to turn off the gas if a burner flame goes out?

5 Copy and complete this table.

	Air hole	
	open	**closed**
What colour is the flame?		
Is the flame easy to see?		
Does the flame make a noise?		

6 You are heating a beaker of water when you have to get some extra equipment. What should you do with the Bunsen burner? Explain.

7 List the seven headings you use when writing a report. Briefly describe what each heading means.

ISBN 978 1 4202 3822 8

8 Josh Mylne, like other scientists, writes full reports of his investigations. He details what he does in the investigations and lists all the results. He posts his investigations on the internet and often publishes them in science journals. Suggest why scientists do all of these things.

9 Look at the diagram below. Write down how you would set up this apparatus, by putting the sentences below in the correct order.

A Put a gauze mat on the tripod.
B Put the burner under the gauze mat.
C Half fill the beaker with water.
D Adjust the air hole so it is open.
E Put the beaker on the tripod and gauze.
F Put the tripod on the heatproof mat.
G Light the burner.

10 You want to add a few crystals of copper sulfate to about 3 teaspoons of water, and heat the water to dissolve the crystals. Make a list of the equipment you think is most suitable for this task. Beside each item of equipment explain why you chose it.

11 Compare the skills used by a scientist doing an investigation to those of a detective trying to solve a crime.

12 Look at the drawing below of an insect for 10 seconds, then cover it up. Now try to draw the shape as you remember it. Finally, compare your drawing with the one here.

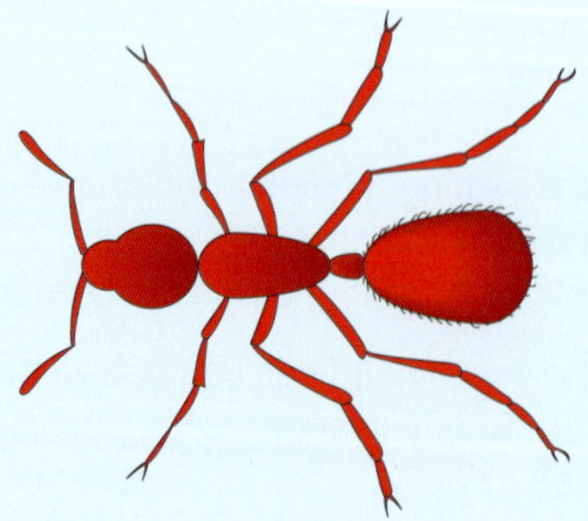

13 When making observations, you often compare and contrast what you are observing with something else. You do this by looking for similarities and differences. Here is an example:

Mars is similar to Earth in size and gravity. It is different from Earth in that it is much cooler.

For each pair of objects below, write a sentence using 'similar to' and a sentence using 'different from'. In each case, explain how the objects are similar or different.

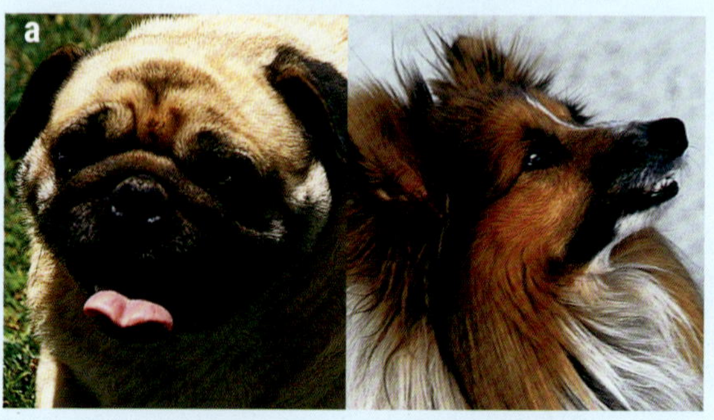

b

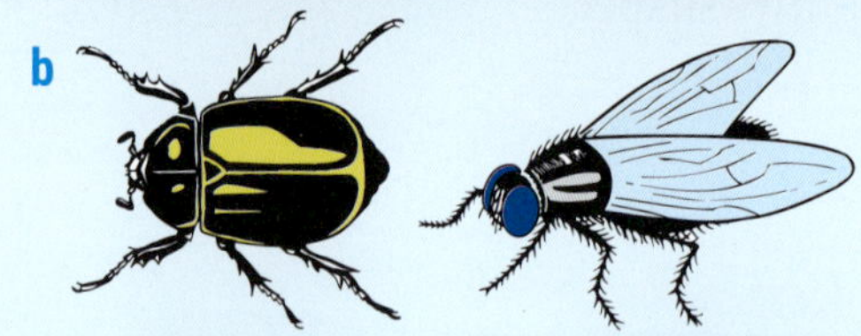

c

14 The cartoon shows two ways of smelling a gas. Which one is safer? Why?

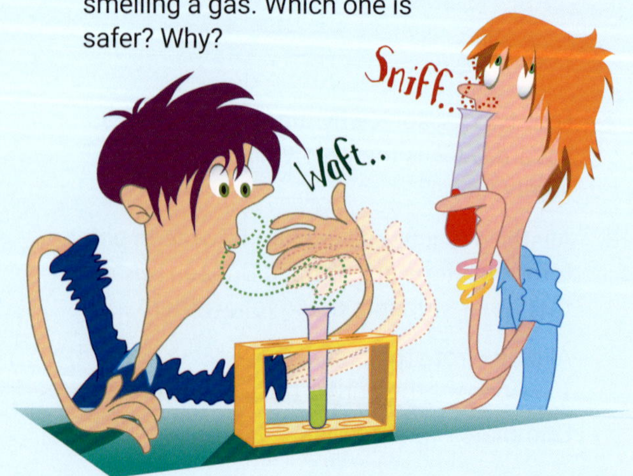

ISBN 978 1 4202 3822 8

15 How good are you at observing? Can you spot the six differences between the two cartoons below?

16 People who witness a crime often have to record their observations of a suspect they saw at the crime scene. Study the photo below for 15 seconds. Then shut the book and try to record as much information as possible to help police identify the suspect.

Figure 1.15 Could you be a reliable witness at the scene of a crime?

CHALLENGE

1 A gas stove burner needs to give a clean flame so that it doesn't make saucepans sooty. The flame must also be easily controlled. The diagram here shows the design of a typical burner.

In what ways is this gas stove burner similar to a Bunsen burner? In what ways is it different?

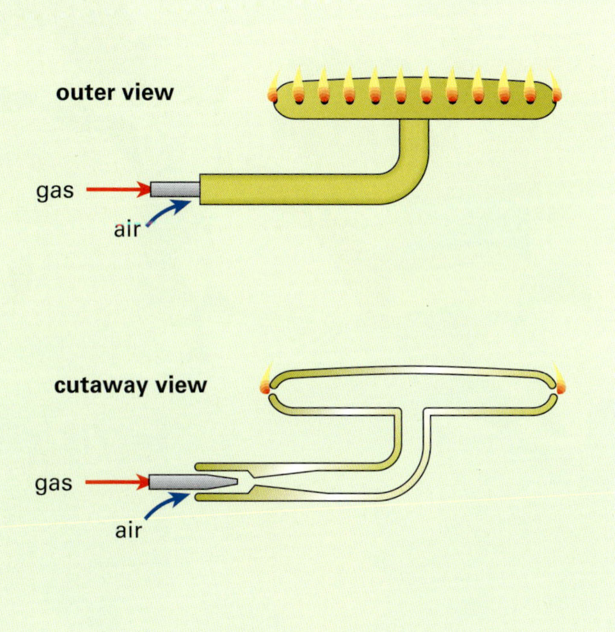

2 Somebody gives you a clear, colourless liquid in a bottle. What could it be? What observations could you make to try to find out what it is?

3 Collect five similar items; for example, five leaves, five insects, five shells or five pieces of laboratory glassware. Label them 1 to 5. Choose one of the items and write a detailed description of it, without naming it.

Pass your description to another student and ask them to pick which one of the five items you described. If they cannot tell which it is, you need to make your description clearer or more complete.

ISBN 978 1 4202 3822 8

MAIN IDEAS

Copy and complete these statements to make a summary of this chapter. The missing words are on the right.

1 You must be able to correctly identify the equipment in a science _____.

2 There is a standard way to draw scientific _____.

3 You must obey the _____ rules for the science laboratory.

4 You need to know how to use a _____ burner correctly.

5 It is important to wear safety _____ whenever there is a chance of anything getting into your eyes.

6 You must take special care in the handling and _____ of chemicals.

7 Accurate _____ and recording are essential skills in science.

8 A good report of an investigation usually has the following headings: title, _____, method, results, discussion and _____.

9 A _____ assessment is always completed before doing an experiment.

aim
glasses
safety
Bunsen
observations
disposal
apparatus
laboratory
risk
conclusion

CH•1 REVIEW

1 Look at Figure 1.16 below. Make a list of the laboratory rules that are being broken.

Figure 1.16 Breaking the rules of the laboratory

ISBN 978 1 4202 3822 8

2 Write the correct terms for the following. Make sure your spelling is correct.
- a a room with special equipment for conducting science experiments
- b equipment put together for a science experiment
- c a device used for heating in the laboratory
- d a tool used for holding small objects, especially when heating
- e an item used for stirring
- f a piece of equipment that spreads the heat evenly from a burner
- g the purpose of an experiment

3 What should you do if:
- a you burn your finger on a hot tripod?
- b you drop a test tube of hot acid on the floor?
- c you splash some liquid in your eye?
- d your sleeve catches fire when you are using a burner?

4 Why are safety glasses such an important safety item in the science laboratory? Give examples of when it is important to use them.

5 Which of the following statements are true and which are false? Rewrite the false ones to make them correct.
- a Test tubes hold less than flasks.
- b A spatula is used for stirring.
- c Immediately treat burns with cold running water.
- d If you spill acid on yourself, wipe it off with a cloth.
- e When lighting a burner, turn on the gas before striking the match.
- f You should always put a gauze mat under a Bunsen burner.
- g The hottest part of a Bunsen flame is near the top of the barrel.
- h To heat a test tube gently, you should use a small blue flame.

6 Draw a labelled diagram of the apparatus you would need to boil water in a beaker.

7 Many homes have gas stoves for cooking. The flame is produced by the Bunsen burner method. Do you think the air hole will be open or closed? Explain your answer.

8 Suggest a safe way of disposing of each of the following chemicals.
- a dilute hydrochloric acid
- b mineral turpentine (used with oil-based paints)
- c copper sulfate crystals

Questions 9 and 10 refer to the animals shown below.

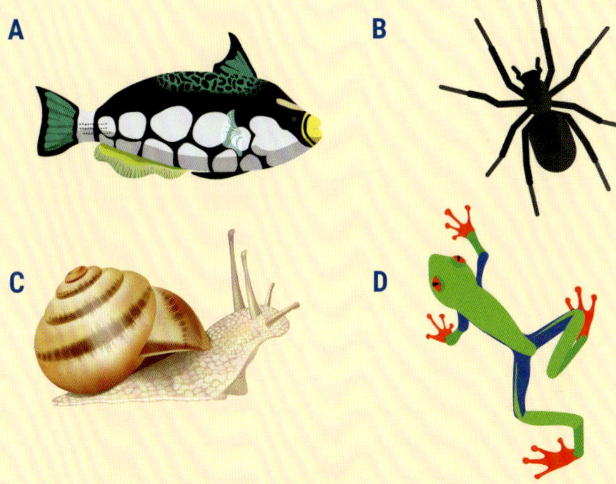

9 Match each of the following observations with the animal it best describes.
- a This animal has one sharp spine on its back.
- b This animal has a coiled shell.
- c This animal has a round body and eight legs.

10 List at least three features of animal D.

LAB REVIEW

Do this test in pairs. Your partner will watch what you do and note any errors you make. They will discuss these with you when you have finished. Then swap jobs and check your partner's skills. Here is the task.

Light a Bunsen burner. Then one-third fill a test tube with water and boil it using the burner. Heat the tube carefully so that the water does not splash out of the tube.

Check your answers on page 236.

ISBN 978 1 4202 3822 8

IN THIS CHAPTER

Science Inquiry Skills

- make inferences and predictions, based on observations
- accurately read the scale on various measuring instruments
- display data in bar graphs and line graphs
- design an experiment to answer a question or solve a problem
- write an experiment report based on second-hand data

CH•2 Working scientifically

ISBN 978 1 4202 3822 8

GET STARTED: *IMAGINE*

You and your group time travel thousands of years back in time.

Each night you look up in the sky and notice a white shining object that people call the moon. You observe that it changes shape from a round object to a thin crescent then back to a round object.

- Work in your group and write as many inferences as you can to explain your observations.
- Is there any way to test your inferences?
- On a particular night you notice that the moon has a very thin crescent shape. Predict the shape of the moon five days later. How did you arrive at your answer?

ISBN 978 1 4202 3822 8

2.1 Inferring and predicting

Look at Figures 2.1 and 2.2. Both are photos of the same structure on the surface of the planet Mars. Figure 2.1 was taken in the late afternoon in 1976. Figure 2.2 was taken 25 years later in 2001.

In 1976, scientists wondered about the origin of the 'face'. People thought the massive structure might have been carved by an ancient Martian civilisation. Or maybe it was formed from erosion of the Martian surface and the face is an optical illusion.

These two statements are called **inferences.** An inference is an explanation of an observation. In 1976, these two inferences were made to explain the face, and one of them was almost certainly wrong. The photo taken in 2001 showed that the 'face' was simply a geological structure, more than likely formed by erosion. So the first inference was wrong.

Inferring is an important skill in science, and you need to remember three things about it.

Making inferences

1 You can usually make several different inferences from the same observation.
2 Observations are correct, provided the observer has been careful and honest in reporting the observations. However, inferences made from these observations can be incorrect. They can be tested by further observations.
3 It is important not to confuse observations and inferences. Otherwise you might think something is a 'fact' when it is only an 'educated guess'.

Figure 2.1 The 'face' on Mars taken by the Viking spacecraft in 1976. The dots correspond to areas where data was lost during transmission from Viking to Earth.

Figure 2.2 The 'face' on Mars taken by the Mars Global Surveyor in 2001. If you half close your eyes, you can still make out the 'eyes' and 'nose' of the 'face'.

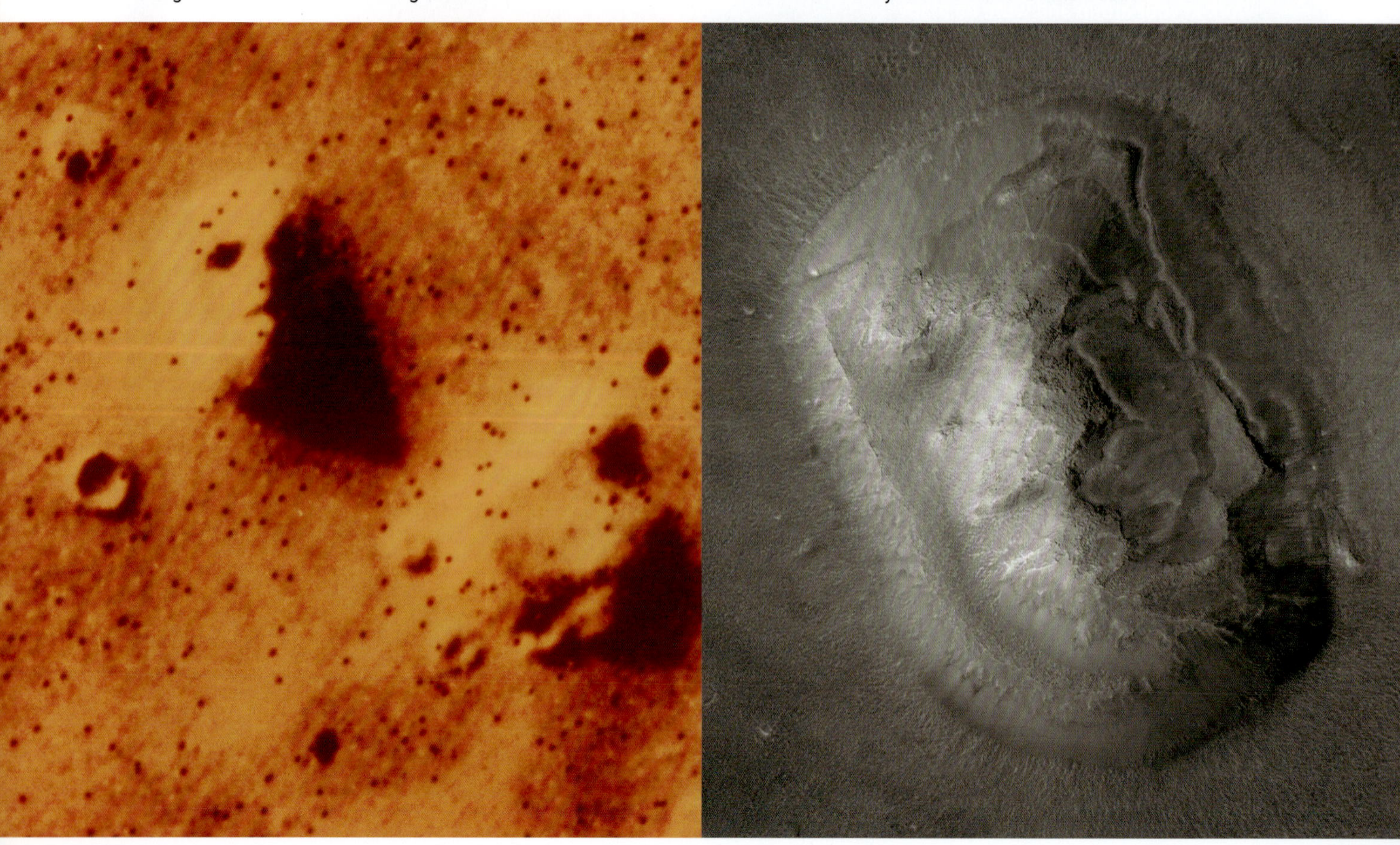

ISBN 978 1 4202 3822 8

Figure 2.3 Different inferences from the same observation

Figure 2.4 Inferences can be wrong.

Making predictions

Another important skill is **predicting**. This is making a forecast of what a future observation might be.

Predictions are based on your observations and what you already know. For example, if you have been observing the moon for a number of nights, you can confidently predict whether there will be a full moon tonight. Otherwise you can only guess, and you will probably be wrong.

SKILLBUILDER

Averaging

In Investigation 2.1 you are going to repeat measurements and average your results. Repeating measurements improves the accuracy of observations.

For example, suppose you timed how long a model car took to go down a ramp four times: 5 seconds, 7 seconds, 8 seconds and 8 seconds. You only want one measurement, so you calculate the average.

To do this you:

- add all the measurements together, and
- divide this total by the number of measurements.

$$5 + 7 + 8 + 8 = 28 \qquad \text{average} = \frac{28}{4} = 7$$

The average is more accurate than if you had taken the first measurement of 5 seconds.

SCIENCE AS A HUMAN ENDEAVOUR

Predicting Pluto

In the last eight years of his life, astronomer Percival Lowell (born in Boston 1855, died 1916) searched the night skies for Planet X—the planet he predicted had to be beyond Neptune.

Lowell based his predictions on calculations he had made observing the positions of the two planets Neptune and Uranus. He reasoned that the movement of the two planets was influenced by another planet. He called the predicted planet Planet X, since the planet couldn't be named until it was discovered.

Pluto was finally discovered by Clyde Tombaugh at the Lowell Observatory in Arizona on 18 February 1930. Although the International Astronomical Union declared Pluto a proper planet in 1999, it was reclassified as a dwarf planet by the Union in 2006.

ISBN 978 1 4202 3822 8

INVESTIGATION 2.1 Pendulum predictions

Aim

Does the distance of swing or the length affect the time it takes for a pendulum to make one swing?

Materials

- stand and clamp
- string (about 50 cm)
- large paperclip
- blank piece of A4 paper and adhesive tape
- a weight; for example, a steel nut
- ruler
- stopwatch

Risk assessment and planning

You need to work in a small group of at least three people. Read through the method and design data tables for your results. The data tables are a very important part of this investigation. Don't start without them!

PART A Changing the distance of swing

Method

1 Set up the apparatus as shown in the diagram.

2 Draw a line on the paper directly under the hanging pendulum. This is the *zero mark*.

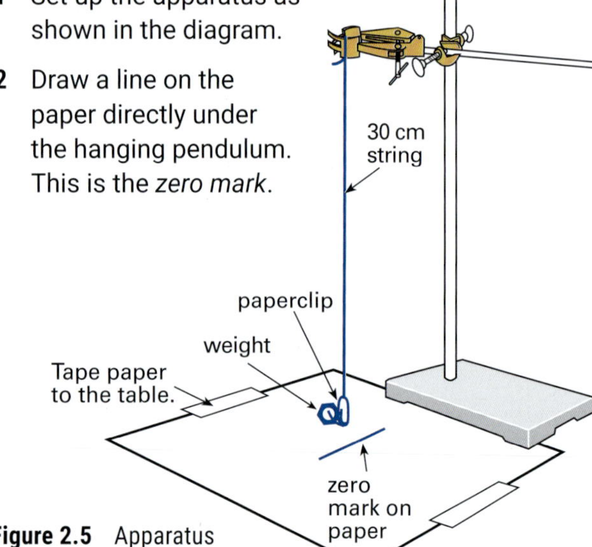

Figure 2.5 Apparatus for measuring swing

3 Measure 5 cm from the zero mark and draw a line parallel with it. Hold the ruler vertically on this line.

4 Now pull the weight out so that it touches the ruler. Let the weight go and pull the ruler away. Have another person time how long it takes to swing out and back to the 5 cm mark. Record the time in your data table.

5 Repeat this another three times. (You can get more accurate results if you time four swings and divide by 4 to get the average time.)

6 Repeat steps 3 to 5, but this time pull the weight out 10 cm. Record your results and find the average time for a swing. Predict what will happen if you pull the weight out 15 cm or 20 cm. Test your prediction.

Discussion

1 Suggest why you used the zero mark on the paper.

2 Why did you hold the ruler vertically just before you let the weight go?

3 Write a report of your investigation using the headings on page 15.

4 Why did you time *two* distances of swing before you made your prediction? Why not make a prediction after the first distance?

PART B Changing the length

In this part of the investigation, keep the swing distance the same (say 5 cm) and the mass the same, but change the length of the pendulum.

Shorten the string length by 5 cm. Time the swings. Then shorten the length by another 5 cm. Time the swings.

Use the results to predict what will happen if you shorten the string by another 5 cm.

ISBN 978 1 4202 3822 8

1 Look at the cartoon below.
 a What was the boy's inference?
 b On what observations did he base his inference?
 c Was his inference correct?

2 Explain, in your own words, the difference between an inference and a prediction.

3 Which of the following are observations, and which are inferences?
 a The leaves of this plant are drooping.
 b I think this is a sugar solution.
 c The inside of the Earth is molten rock.
 d The temperature of the water is 23 °C.
 e This toy must have a magnet in it.

4 When Sven saw the burning candle below, he made the inference that the candle must be in a draught. What observations made him think this? Use his inference to predict what might happen to the candle if it keeps burning.

5 For each situation below (a–d), decide which statement is an observation, which is an inference, and which is a prediction.

a

b i The left-hand end of the see-saw is lower than the right-hand end.
 ii If the person on the left-hand end gets off, the right-hand end will fall.
 iii One person is heavier than the other.

c i In 2 seconds he will drop the bar.
 ii The bar is very heavy.
 iii He is lifting the bar.

d i The minute hand is on the 6 and the hour hand is between 8 and 9.
 ii The time is 8.30 am.
 iii In half an hour the bell will ring.

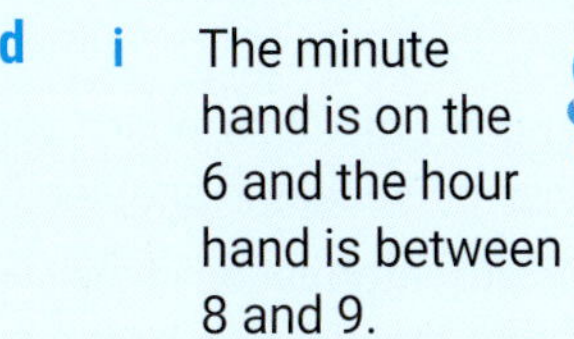

ISBN 978 1 4202 3822 8

6 Look at the cartoon below.
 a What observations has the adult made?
 b What two inferences could he make?

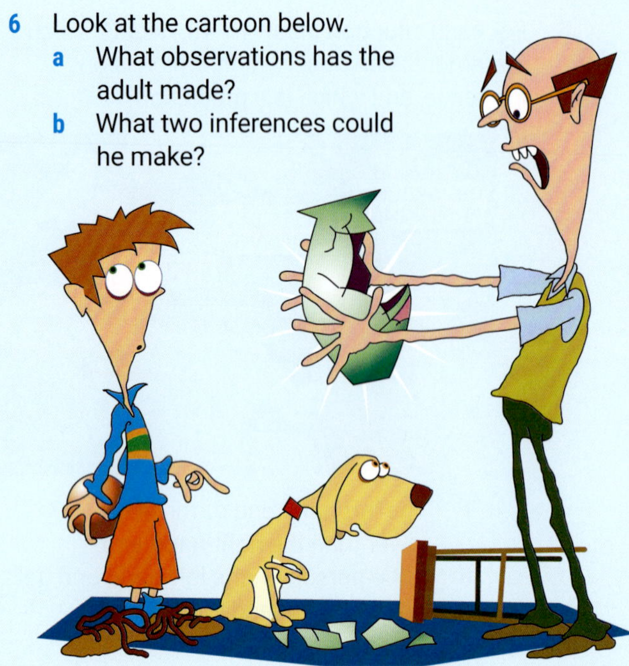

7 Cameron has a mouse in a cage. The mouse has an exercise wheel with a counter on it. Cameron wrote down the counter reading each morning, but his little brother tore the corner off his data table.
 a Predict what the counter reading for day 4 should be (approximately).
 b Explain how you made your prediction.

	Counter
Day 1	49
Day 2	100
Day 3	152
Day 4	

CHALLENGE

1 Make up your own example (as in Check 5 on page 31) to show the differences between an observation, an inference and a prediction.

2 The weather bureau's predictions about the week's weather are sometimes wrong. Suggest a reason for this.

3 Five students were discussing the results of an experiment.
Duncan: The plants in pot C are the tallest.
Rohan: Yes, that's because we watered them more often than the others.
Cameron: No, pot C must have better soil, because we gave all the pots the same amount of water.
Jess: That could be, but I noticed that pot C was closer to the window than the others.
Gavin: Anyway, the plants in pot D certainly didn't grow very well.
Draw up a table and put the students' observations in one column and their inferences in another.

4 You look up your star sign information in the newspaper. Today it says that you will meet a dark-haired person, you will travel overseas soon, and your lucky numbers are 2, 5, 11 and 21.
Would you class this information as scientific predictions? Give a reason for your answer.

5 Look at these tracks made on the beach.

 a Infer what made the tracks.
 b Infer the order in which the tracks were made. Discuss your answers with others.

6 a What observations can you make about the surface of Mars on page 28?
 b Is the sun shining from the left or the right of the photo taken in 1976? Is your answer an observation or an inference?
 c Which inference about the face do you think is correct? Why?

ISBN 978 1 4202 3822 8

2.2 Measuring

There are two different types of observations. One is a description in words, such as the colour of a car or the smell of a flower. These observations are said to be **qualitative** (KWAL-i-tate-ive). The other type of observation involves measurements; for example, a 70 kg person or the 9 cm tail of a mouse. These measurements involve numbers, and are said to be **quantitative** (KWON-ti-tate-ive).

Note that measurements are made up of a number and a unit. For example, your height might be 155 cm. Centimetres are the units used. Without the units the number has no meaning. A friend may tell you she has 1000 in the bank. You might think she is rich, until she says it is 1000 cents. So the unit is just as important as the number.

Some measuring instruments have digital readouts; for example, digital watches. Other instruments have a scale with numbers on it and a pointer that moves along the scale. To read these instruments, you must estimate the position of the pointer against the scale.

Reading a scale is simple if you follow the five steps in the Skillbuilder below.

Quantity	Instrument	Common units
length	metre rule or tape measure	millimetre mm (1/1000 m) centimetre cm (1/100 m) metre m kilometre km (1000 m)
mass	balance	gram g (1/1000 kg) kilogram kg tonne t (1000 kg)
time	watch or clock	second s minute min hour h
temperature	thermometer	degree Celsius °C
volume (liquids)	measuring cylinder	millilitre mL (1/1000 L) litre L

Notes on the table

1 Use these prefixes for smaller and larger units:
micro (μ) = one-millionth
milli (m) = one-thousandth
centi (c) = one-hundredth
kilo (k) = a thousand
mega (M) = a million

2 The volume of solids is more commonly measured in cubic centimetres (cm^3) or cubic metres (m^3). One cubic centimetre is the volume of a cube with a side of 1 cm.

3 When measuring the volumes of liquids and gases, 1 cm^3 is the same as 1 mL.

SKILLBUILDER

How to read a scale

1 Decide which way the scale reads—up, down, or left to right.
2 Work out what each division on the scale stands for.
3 Find the closest numbered division before the pointer.
4 Count the number of divisions from the numbered division to the pointer. Calculate their value.
5 Add the value of these divisions to the numbered division.

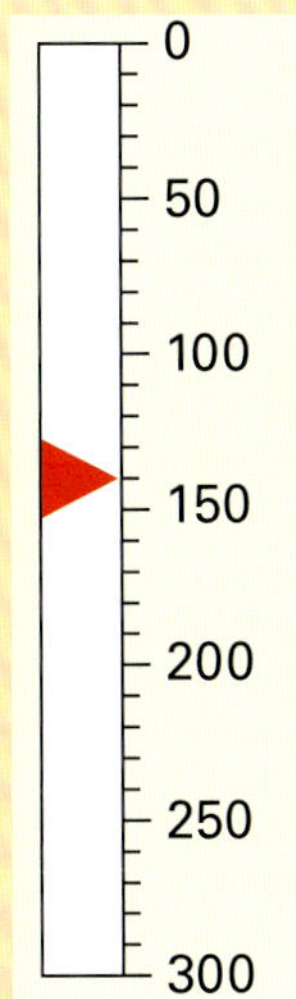

Example

1 The scale reads from top to bottom.
2 In between 0 and 50 there are five divisions, so each division represents 10 units.
3 The closest numbered division before the pointer is 100.
4 There are four extra divisions after the 100. Each is 10 units, which gives an extra 40 units.
5 So the scale reading is: 100 + 40 = 140.

ISBN 978 1 4202 3822 8

Estimating readings

When reading a scale, you will often find that the pointer lies between two lines. In these cases you have to estimate the reading. For example, on the scale below, the pointer is between the 0.60 and the 0.70 position, but not exactly in the middle. The reading is more than 0.65 but less than 0.70. It can be estimated as 0.67.

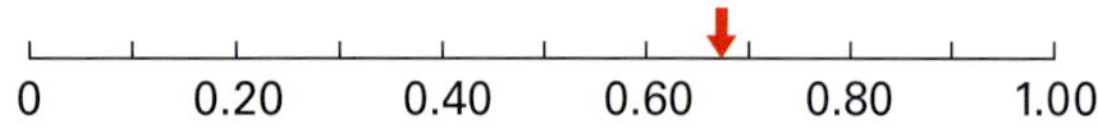

Accuracy

Remember—you cannot get a better measurement than your measuring instrument allows. All measuring instruments are accurate only within limits. Scales used on any instrument are marked off into smaller and smaller divisions. The smallest division determines the accuracy of the instrument. For example, in Figure 2.6, thermometer B can measure the temperature of the water more accurately than thermometer A.

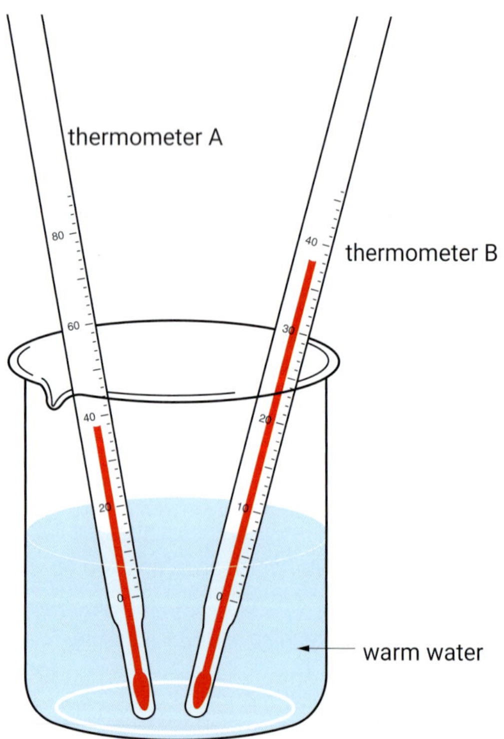

Figure 2.6 Why does thermometer B give a more accurate reading than thermometer A?

Errors

It is difficult to say any measurement is exact. Mistakes or errors occur in all measurements. These errors can occur when you make a mistake reading a scale or writing down the measurement. They can occur because an instrument is not working properly or because you are not using it correctly.

Parallax error

Parallax error occurs when you don't look straight over the pointer. You need to look square on to a measuring instrument.

The student on the left will be able to make an accurate measurement, but the student on the right will have parallax error in his measurement.

Reading the bottom of the meniscus

To avoid errors when measuring liquids in measuring cylinders, always read the bottom of the **meniscus** (me-NIS-kus)—the curved water surface. Keep your eye level with the meniscus. The volume of water below is 87 mL, not 88 mL.

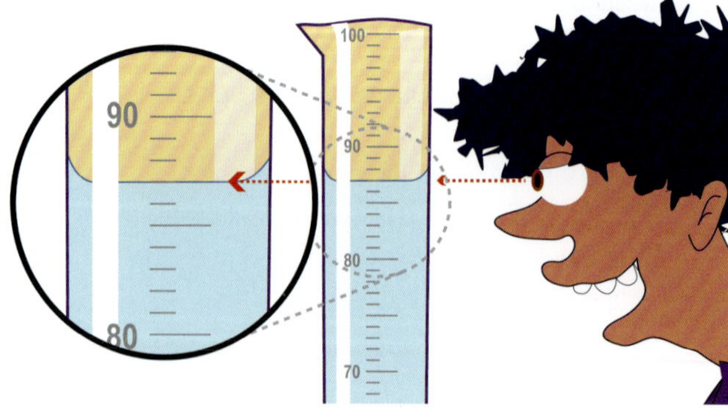

Figure 2.7 Always read the bottom of the meniscus.

ISBN 978 1 4202 3822 8

Measuring things

Aim

To measure length, temperature, mass, time and volume accurately.

Risk assessment and planning

Make a list of the equipment you will need for each part of the investigation. You will need to design data tables for some parts.

PART A Length

Is the length of the index finger and the hand span of your right hand the same as for your left hand?

Your data table should look something like this:

	Length of index finger	Hand span
left hand		
right hand		

Suggest how you could improve the accuracy of your results.

Are your hands smaller, larger or about the same size as the average for the hands of other people in the class? (Give measurements.)

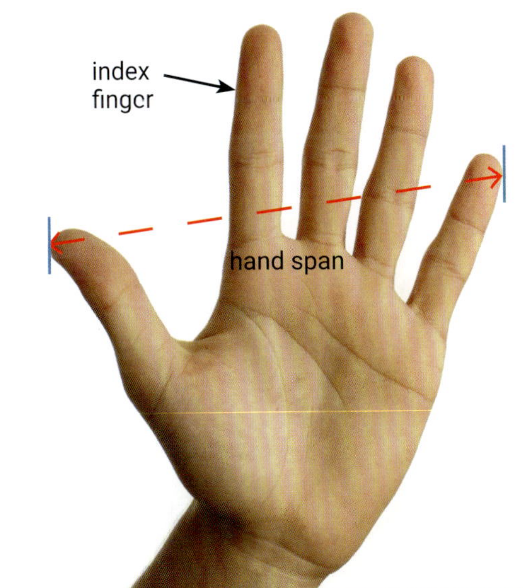

PART B Temperature

What is the temperature of the following?

- **a** the air—in the shade and in the sun
- **b** water—from cold and hot water taps
- **c** crushed ice
- **d** crushed ice with several spoonfuls of salt mixed with it
- **e** inside a refrigerator and inside a freezer
- **f** under your armpit

Record your measurements in a data table like the one below.

Location/object	Temperature (°C)

Compare your measurements with those made by other students. If there are any differences, it may be wise to check for errors.

Hints for using thermometers

1. Thermometers break easily, so take special care with them. Do not roll them or drop them. When putting them away, place them in their special tube or box. If you break a thermometer, report it to your teacher immediately.
2. The bulb of the thermometer has to be put into the substance so you can measure the temperature, and left there while you read the scale.
3. Don't hold the bulb of the thermometer, as this changes the temperature.
4. You can't get a reading straight away. Wait until the red alcohol (or silver mercury) column stops moving.
5. Before reading the thermometer, turn it so that the scale is facing you.

ISBN 978 1 4202 3822 8

PART C Volume of a liquid

How can you accurately measure the volume of a liquid?

1 Use a beaker with graduations on the side to measure exactly 50 mL of water.
2 Pour the water into a 100 mL measuring cylinder.
Record the volume as accurately as you can. (Remember to read the volume at the bottom of the meniscus.)
Are the two readings the same? Suggest reasons for any differences.
Why are your volume measurements more accurate when using a measuring cylinder than when using a graduated beaker?

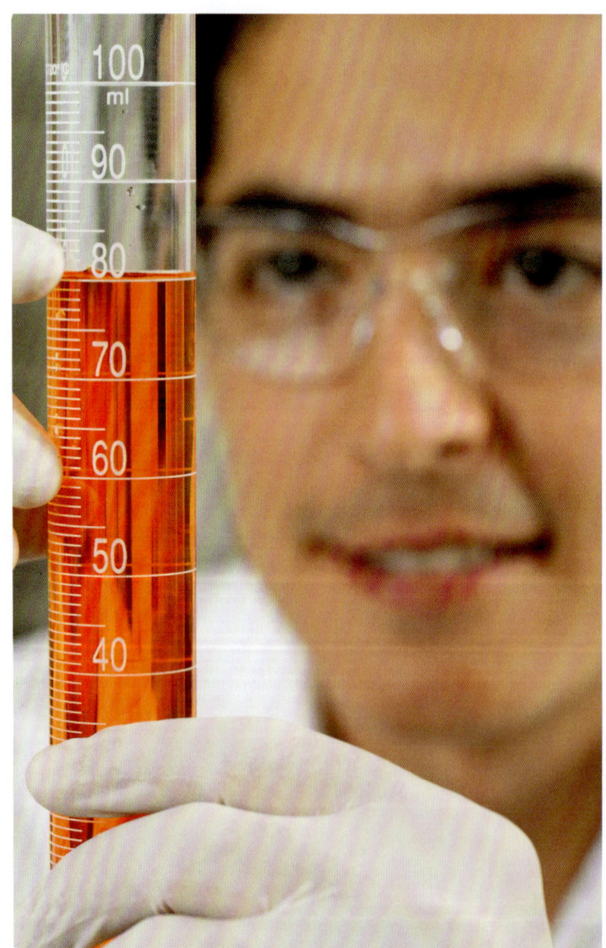

Figure 2.8 Use a measuring cylinder for more accurate readings.

PART D Volume of a solid

How can you measure the volume of an irregular solid?

For regular solids such as cubes and spheres, you can find the volume by measuring the sides or diameter and using a maths formula.
For irregular solids such as a stone, you can find the volume by a method called *displacement*. The diagrams below show you how to do it.

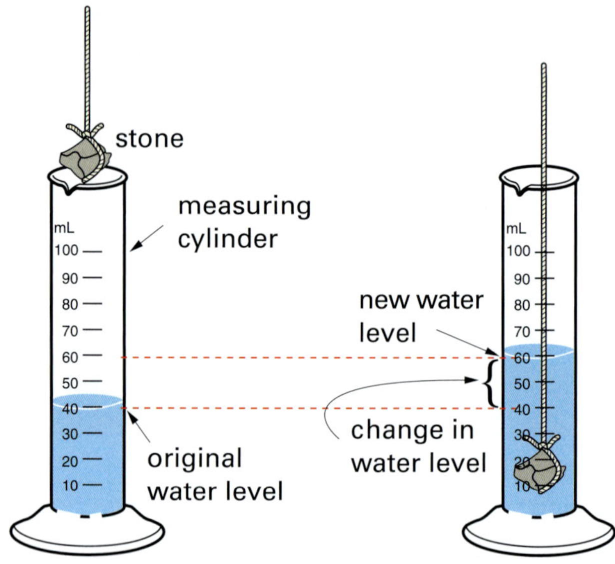

Figure 2.9 The volume of an object can be measured by submerging it in water in a measuring cylinder. What is the volume of this stone?

Find a stone or other irregular solid and determine its volume. Try to solve these two problems:

- How would you find the volume of a large stone that will not fit into a measuring cylinder?
- How would you find the volume of an irregular piece of styrofoam? (Styrofoam floats in water!)

ISBN 978 1 4202 3822 8

PART E Mass of objects

How can you measure the mass of an object?
The most common balance in school laboratories is the electronic balance with a digital readout.

Method

1 Make sure the balance is on a firm surface such as a lab bench. The reading can be affected if the balance is unsteady.
2 Turn on the balance and wait until the screen reads 0.00 g. If it doesn't, press the zero key.
3 Place an object on the pan and wait until the reading is steady.
Record the mass of the object.
4 Weigh two or three other objects.
Record the mass each time.
5 Ask another group to weigh your objects.
Compare their results to yours. Suggest reasons for any differences.

When measuring the masses of laboratory chemicals, you should always use a container. Knowing this, solve the following measurement problems.

1 How can you use a beaker and measuring cylinder to find the mass of 100 mL of water?
2 How can you measure out exactly 1 gram of a white powder?

PART F Mass of liquids and solids

When working in the laboratory you need to be able to accurately measure the masses of solids and liquids. This requires the use of a container to hold the liquid or solid.

1 Plan a method for solving the following two problems and check it with your teacher before you start.
2 Make sure you record a description of your method and any calculations you use in finding the masses.
    - Find the mass of 100 mL of water.
    - Measure out exactly 1 gram of salt. Use a watch glass as your container.

PART G Time and temperature

How hot does water get when you boil it?
Collect the equipment you need to set up the apparatus shown. Add 100 mL of water to the beaker.
Light the burner. Heat the water and measure the temperature every minute.

- Record your results in the data table you designed in the 'Risk assessment and planning' box.
- What pattern can you see in the results?
- Keep your results. You will use them on page 41.

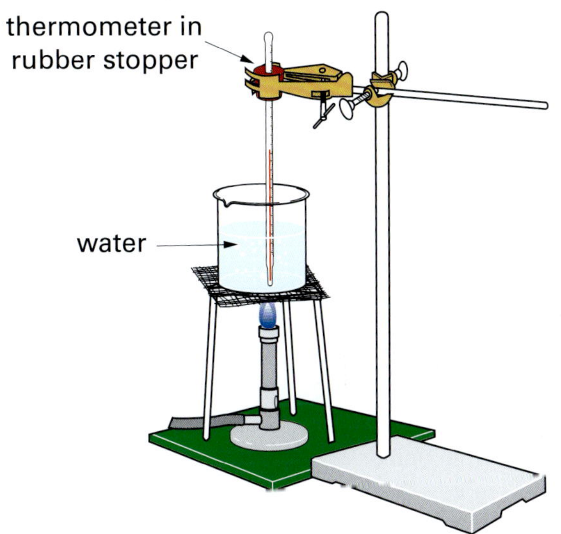

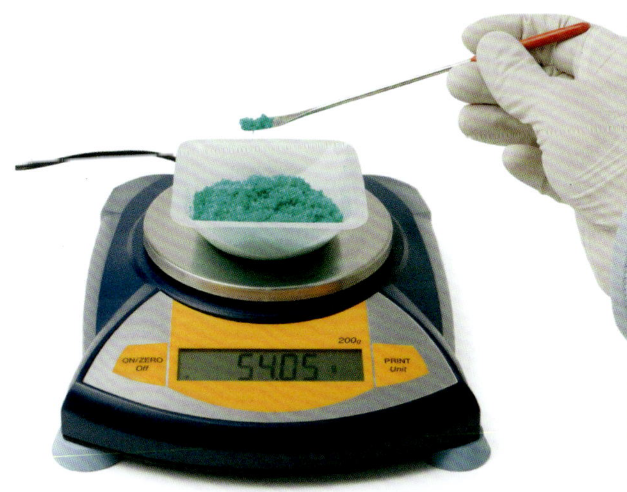

Figure 2.10 Measuring chemicals requires the use of a container.

ISBN 978 1 4202 3822 8

1 Copy and complete the following sentences.

a Every quantity has a _____ of measurement.

b An instrument for measuring mass is called a _____.

c The curved surface of a liquid is called a _____.

d For measuring volumes, a measuring cylinder is more _____ than a beaker.

e Measurements made during an investigation should be recorded in a _____ table.

f An observation that does not contain a measurement is said to be _____.

g Parallax _____ occurs when you don't look straight over the pointer on a measuring instrument.

2 Write five correct sentences by joining words from each of the columns in the table; for example: 'Large lengths are measured in kilometres.'

Very large	lengths	are measured in	mega	metres
Large			kilo	
Small	volumes		milli	litres
Very small			micro	

3 What unit does each of these symbols represent?

a mm

b m

c s

d mL

e cm^3

f kL

4 Which of the following observations are quantitative?

A The gas has an odour like rotten eggs.

B This soft-drink can contains 375 mL.

C The ant has six legs.

D This rock contains large pink crystals.

E The magnet did not attract the nails.

F The maximum temperature was 28 °C.

G The water froze overnight.

H The fizzing lasted about 2 minutes.

5 Reading a scale is a skill that requires a lot of practice. Read the scales on this page. Some you will need to estimate.

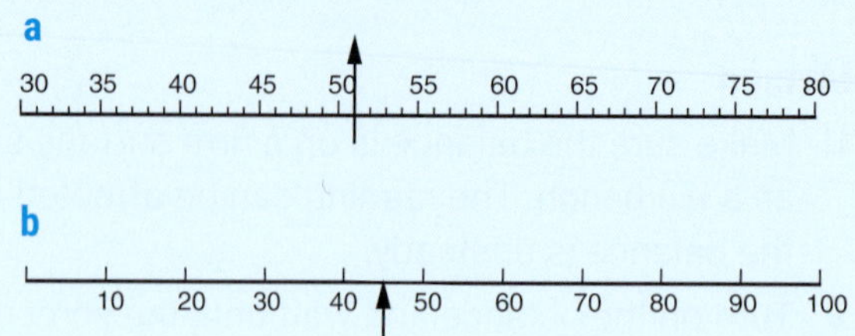

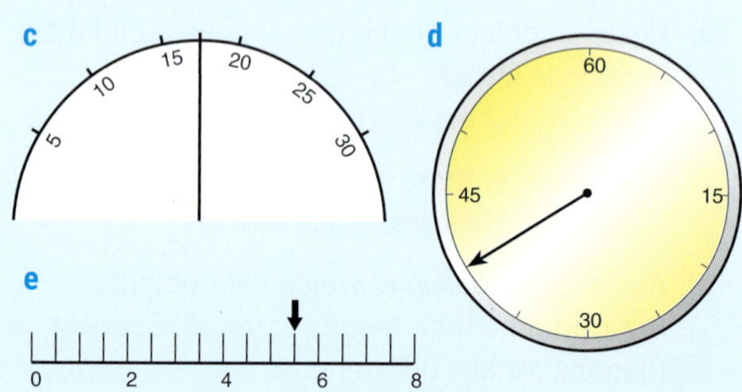

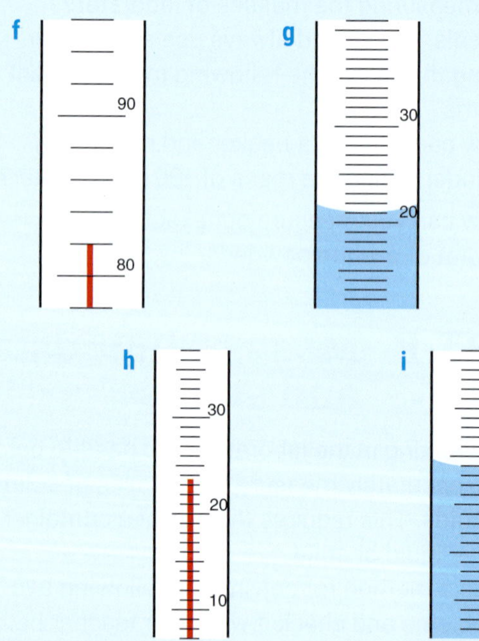

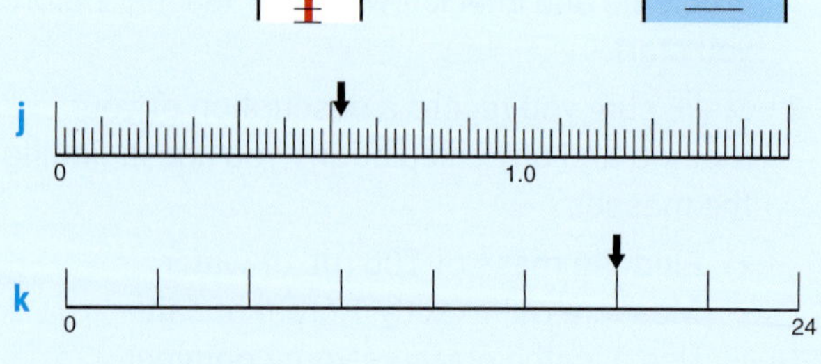

ISBN 978 1 4202 3822 8

6 A measuring cylinder is filled with water to the level shown in a. A small object is then dropped into the cylinder and the water level rises to that in b. What is the volume of the object?

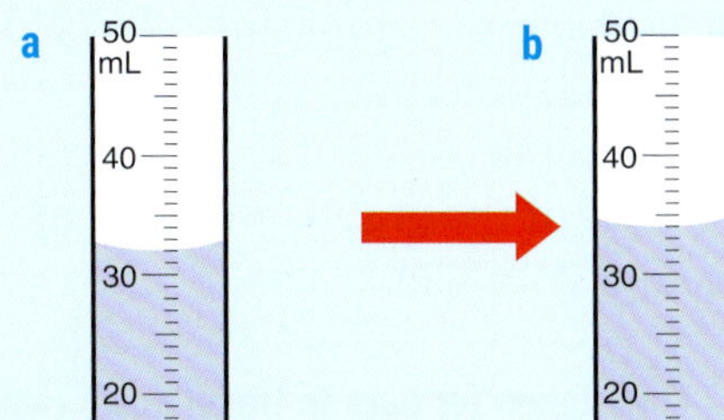

7 The two thermometers below were used to measure the temperature of tap water. Which one would give a more accurate measurement? Why?

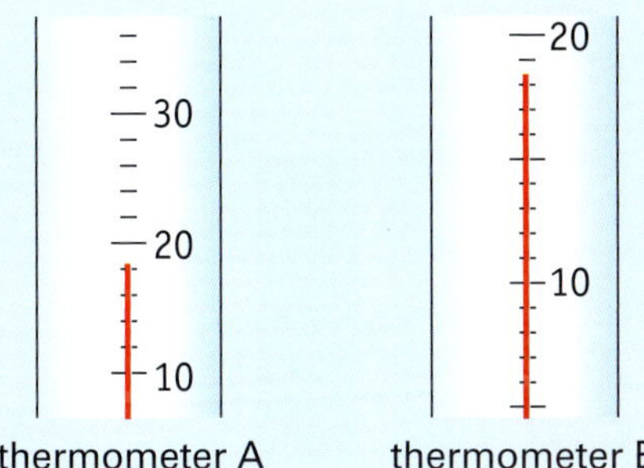

8 Scott found the mass of a beaker of water. He then tipped out the water and found the mass of the empty dry beaker. Here are his results:

full beaker	418 grams
empty beaker	112 grams

- a What was the mass of water in the beaker?
- b What would the total mass be if Scott added 100 mL of water to the empty beaker? (Hint: 1 mL of water has a mass of 1 gram.)

9 Convert the following from one unit to the other.

- a 250 cm to m
- b 3.5 L to mL
- c 3600 m to km
- d 0.245 m to cm
- e 3 h to s
- f 430 s to min
- g 3.5 kW to W
- h 0.00002 s to μs

10 Look at the cartoons below. For each case say what error in measurement the student has made.

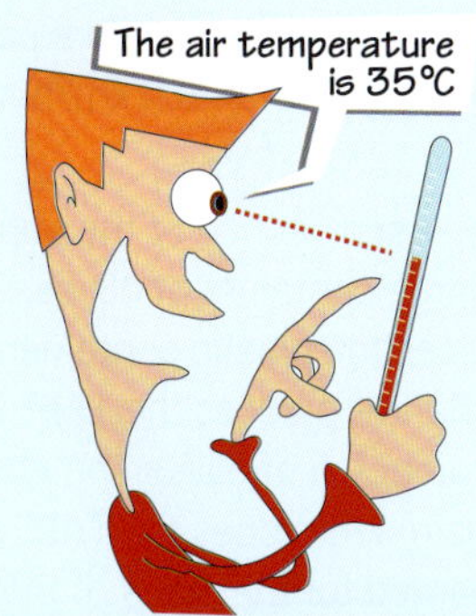

ISBN 978 1 4202 3822 8

1 Ann and Peter are driving to the coast. Ann is driving and Peter (in the passenger's seat) says she should slow down because she is doing more than 100 km/h. Ann says he has made a parallax error and she is doing less than 100 km/h. What does she mean?

2 At the delicatessen section of the supermarket, you order a large container of olives. The sales assistant places the plastic container on the electronic balance, presses the TARE button, fills the container full of olives and weighs the container again. What does the TARE button do? Were you charged for the mass of the container?

3 A swimming pool is 50 metres long, 20 metres wide, 1 metre deep at the shallow end and 2 metres deep at the deep end. What volume of water is needed to fill it? (Give your answer in megalitres, where 1 ML = 1000 m^3.)

4 Suggest why some medicine glasses are this shape.

2.3 Using graphs

So far in this book you have learnt and practised a number of science skills. For example, you have:

- made and recorded observations
- devised data tables for these observations
- made inferences from observations
- made predictions from observations, inferences and your experience.

As well as these skills, you have learnt how to use laboratory equipment, read a scale, calculate averages and reduce errors.

Another important part of an investigation is displaying your data in a graph, diagram or chart.

Bar graphs

Suppose you were investigating how long it took different model cars to travel down a wooden ramp. Here are the results.

Model car	A	B	C	D	E
Average time to travel down ramp (seconds)	7	9	4	6	5

A very useful way of comparing data is to draw a bar graph (sometimes called a bar chart). In this case, the time (in seconds) is on the vertical or *y*-axis of the graph, and the type of car is on the horizontal or *x*-axis.

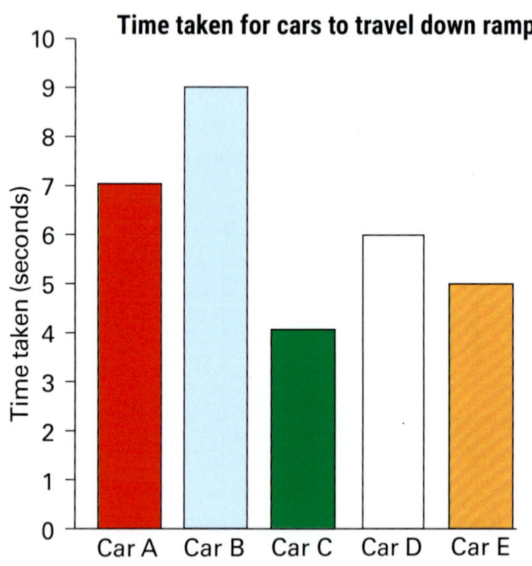

Line graphs

Sometimes you want to show the *relationship* between two things being measured. In this case you would draw a line graph of the data. For example, a group of students was investigating the growth of seedlings. They measured the height of the seedlings every day. Here are their results.

Time (days)	Height (cm)
0	0
1	1.0
2	2.1
3	2.8
4	3.8
5	5.0
6	5.8

ISBN 978 1 4202 3822 8

Before you start on your line graph, you have to decide which measurement goes on which axis. On a line graph the *independent* measurement goes on the horizontal axis. The *dependent* measurement goes on the vertical axis.

In this case, time is the independent measurement, and height is the dependent measurement. Height is dependent because the height the seedlings grow *depends* on how many days (time) you let them grow.

With this experiment the points don't lie on a straight line, therefore you will need to draw *a line of best fit*. A line of best fit is where the line doesn't need to pass through all the points, as in the graph shown. Refer to the Skillbuilder on page 56.

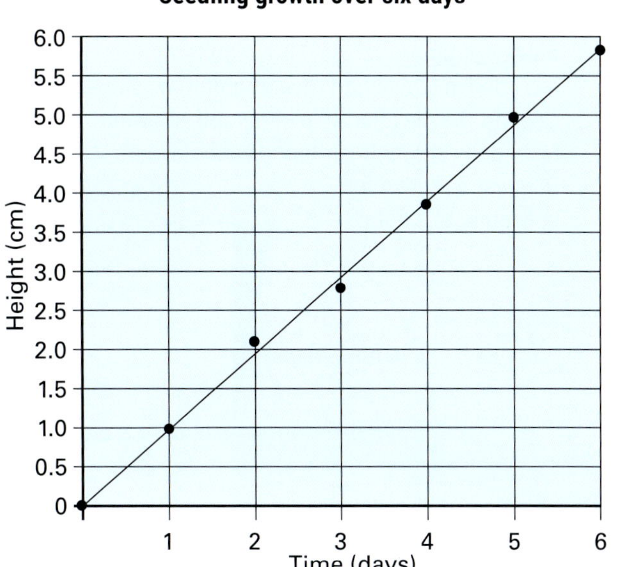

CHECK

1 The results of an investigation are shown in the data table below.

a Write down an aim for the investigation.

Solid	Melting point (°C)
ice	0
mothballs	80
wax	44
hypo (photographic fixer)	48

b Say how you think the investigation was done.

c Write a conclusion for the investigation.

2 Philippa wrote this in her science notebook:

We measured the mass of five mice. The first mouse had a mass of 155 grams. The second mouse measured 163 grams. The third mouse measured 180 grams; the fourth 135 grams; and the fifth mouse had a mass of 149 grams.

Record this information in a data table.

3 Adam and Duncan observed the motorbikes passing the school in one hour. They saw 10 Suzukis, 8 Hondas, 5 Yamahas, 9 Kawasakis, and 2 they could not identify. Draw a bar graph of this data.

4 Use the data table to answer the questions below.

Student	Height in cm	Student	Height in cm
Belinda	133	Emma	136
Robert	138	Paul	142
Duncan	157	Vanessa	156
Darryl	160	Chadi	135
Katherine	140	Annalissa	139

a What is the average height of the girls?

b What is the average height of the boys?

c What is the average height of the 10 students?

d Which students are taller than the average?

e What conclusion can you make from these results?

5 Ella recorded data from an investigation. She is unsure whether to draw a bar graph or a line graph. Explain what she should do and why.

6 Look at your results in Investigation 2.2 Part G.

a Use your results to plot a line graph.

b What did you purposely change in this experiment?

ISBN 978 1 4202 3822 8

CHALLENGE

1 Fifty students were asked 'What is your favourite flavour of potato crisps?' The results were recorded in a pie chart (sector graph) as shown. Convert the pie chart to a bar graph, showing how many students preferred each flavour. Suggest a reason for collecting this data.

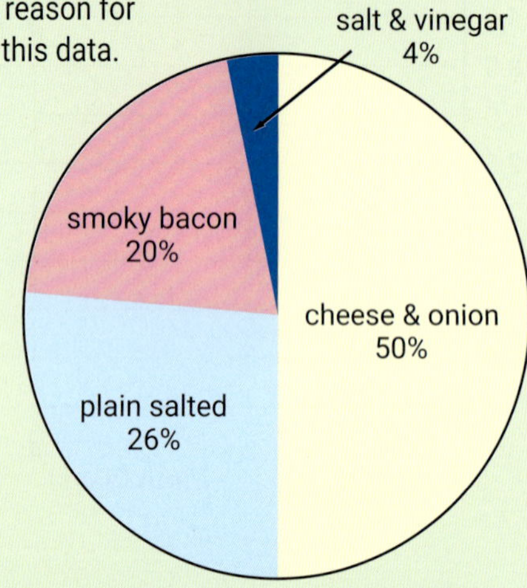

2 The table below shows the times of high tide at a beach.

	Mon	Tues	Wed	Thurs	Fri
am	6:10	7:00	7:50	8:40	
pm	6:35	7:25	8:15	9:05	

a What pattern can you see linking the times of the high tides?

b Predict the times of the two high tides on Friday.

2.4 Experimenting

You have probably heard about scientists doing experiments and then wondered what the difference is between an experiment and an investigation. These terms mean much the same thing—scientists carefully planning laboratory or field work to show that something is true (or not true).

An **experiment** always involves designing tests to answer a question or solve a problem. For example, when you cut an apple and leave it for a few hours, the white flesh inside starts to turn brown. You have an idea that it is something in the air that causes the apple to go brown. So your aim might be: *To test the idea that if a cut apple is covered to exclude air, it won't go brown*. You then design tests to show whether this is true or untrue.

Designing experiments

The important thing to remember about designing experiments is that your aim must be a statement or question that is able to be tested. Such a statement is called a **hypothesis** (high-POTH-e-sis). For example, the hypothesis *Plants grow better in white light than blue light*, is easy to design tests for.

When planning experiments and writing reports, you use the same headings as you did for the report on page 15. In other words, you start with a *title*. Then write an *aim*, list the *materials* you will need and write the *method* so that others can follow it. You then collect *results*, write your *discussion* and finally your *conclusion*.

On the next page you can practise designing an experiment.

Figure 2.11 Scientists use experiments to solve problems.

ISBN 978 1 4202 3822 8

SKILLBUILDER

Writing an experimental report

Imagine you are a research assistant working for an agricultural scientist. She is going to an overseas conference and has left you with the page below from her notebook.

Your task is to write up an experimental report using the usual headings—title, aim, materials, method, results, discussion and conclusion (see page 15).

Number of aphids				
	Day 1	**Day 2**	**Day 3**	**Day 4**
Plant A	20	14	12	9
Plant B	20	12	8	5
Plant C	20	20	15	17

ACTIVITY

For each of the parts below, write an aim or hypothesis that could be tested in an experiment.

a Dark-coloured clothes seem hotter in summer than light-coloured clothes.

b My bike's brakes seem to work better on a dry day than on a rainy day.

c When you pull the plug out of the bath, water rushes down the hole and creates a whirlpool motion. Does it always go in the same direction (i.e. clockwise or anticlockwise)?

d My hot chocolate drink in a glass cup seemed to stay hotter longer than my friend's hot chocolate drink in a paper take-away cup.

e Seeds seem to germinate faster on hot days in summer than cooler days during winter.

f I find that raw sugar takes longer to dissolve in a cup of tea than white sugar does.

ISBN 978 1 4202 3822 8

EXPERIMENT 2.1 Testing reaction time

The problem to be solved

How quickly can you catch a falling ruler?

When a person holds a ruler between their thumb and fingers and lets it go, they see it fall and then their fingers quickly try to catch it. You can measure a person's reaction time by finding out how far the ruler fell before it was caught.

The problem in this experiment is to find out which of your senses—sight, hearing or touch—reacts fastest to catch a falling ruler.

Designing your experiment

1 **Setting up the test**
Have one person hold a metre ruler (a 50 cm one can be used) between another person's thumb and fingers. Make sure the 0 cm mark is level with the top of their thumb and fingers.

2 **Testing the senses**
Drop the ruler and measure how far it fell before it was caught. Do this four times and average the results.
Repeat the test but this time blindfold the person or ask them to shut their eyes. They have to catch the ruler after they hear you shout NOW or GO. To test a person's touch reaction, tap them on their arm or hand.

3 **Designing tests**
Work in a small group and design the tests to solve the problem. Make sure you write a draft of your method, list the materials needed and discuss safety issues. Show this to your teacher before you start.

4 **Plotting graphs**
Include bar graphs of the results in the results section of your report.

5 **Writing your report**
Write a full report of the experiment. Include diagrams where appropriate. You could also take a photo of the test to include in your report. Consider typing your report and drawing data tables and graphs using graphing software.

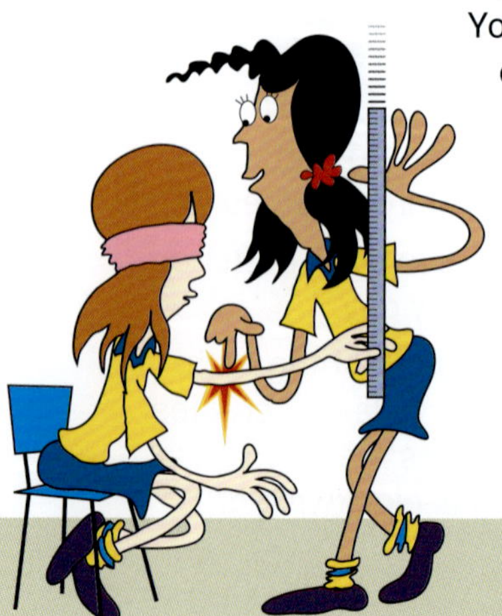

ISBN 978 1 4202 3822 8

ACTIVITY

Where can science take me?

You have now learnt a little about the laboratory, how experimenting works and what scientists do. You have probably also done some science activities at primary school and maybe even at home. Have you thought about what sorts of jobs you could get that need an understanding of science? There are many jobs and careers that you could think about.

Did you know that many of the jobs that will exist in 10 to 15 years' time have not even been invented yet? Use your imagination to think about how fast robotics and artificial intelligence (AI) are changing and moving forwards. Can you think of any robotics jobs that may be needed in the future?

In science we often talk about the areas or disciplines of science, including:

- biological science
- chemical science
- physical science
- Earth and environmental science
- space science.

During your study you will learn about each of these areas in more detail. Each of these disciplines has many areas within it too, and scientists from each area often work together to solve problems, invent things and work on scientific projects.

Your task is to conduct research to select a career in science that you find interesting and would like to know more about. You then need to construct a presentation in the format of your choice on this career.

1 Select your career and get your teacher to check it is suitable before going on.

2 Research your chosen career and answer the following questions:

   a Which areas of science are especially important to know about for this career?

   b Describe the work a person in this job would do. Think about what a day or a week in this job might be like.

   c What do you think are the most interesting things about this job?

   d How would this job benefit society?

   e How important do you think this job is for the future? Do you think it will still be needed in 15 years' time, or will it become different or be replaced?

   f Which subjects would you need to study at school and/or university to get into this career?

EXPLORE ONLINE

To get you started, check out some of these websites: **Science Buddies**, **ABC Science careers** and **Splash ABC**, where you can start your search for various scientific careers.

ISBN 978 1 4202 3822 8

Copy and complete these statements to make a summary of this chapter. The missing words are on the right.

1 An inference is an _____ of an observation. The inference may or may not be correct.

2 It is important not to confuse observations and _____.

3 _____ is making a forecast of what a future observation will be, based on past observations.

4 Measurements are _____ observations, and are made up of a number and a _____.

5 The results of an experiment can be recorded in a _____, and displayed in graphs.

6 An _____ is an investigation that involves designing tests to answer a question or solve a problem.

7 A good report of an experiment has the following headings: _____, aim, method, results, discussion and _____.

conclusion
experiment
data table
quantitative
inferences
predicting
explanation
title
unit

CH•2 REVIEW

1 Maria walked into her bedroom. What she saw is shown on the right.

Which of the following are observations, and which are inferences?

a My new coat has been ripped.
b The vase has been broken.
c That awful dog has done this.
d There are marks on my bedspread!

2 For each statement below, decide whether it is an aim, a conclusion, an observation or a prediction.

a Which magnet is stronger?
b Magnet B will pick up more iron filings than magnet A because it is bigger.
c Magnet A picked up 10 paperclips, and magnet B picked up 15.
d Magnet B is stronger than magnet A.

ISBN 978 1 4202 3822 8

3 The temperature was measured at 9 am and at 3 pm, as shown.

a What was the temperature at 9 am?

b What was the temperature at 3 pm?

c What was the difference in temperature between 9 am and 3 pm?

d What was the average of the two temperatures?

4 Six students each measured how long it took a tablet to dissolve in water. They used water at different temperatures. The table below shows their results.

Student	Temperature of water (°C)	Dissolving time (seconds)
Cory	30	22
Katherine	25	24
Darryl	35	20
Duncan	40	18
Emma	20	27
Belinda	45	16

a Sort out the data, and see if you can recognise a pattern in it. Write a generalisation (see page 15) for the results.

b Draw a graph of their results. Why is it better to draw a line graph for these results?

5 Heidi has dropped a ball from two different heights, and measured how high it bounced each time. She recorded her results in a bar graph.

a Predict how high the ball will bounce if she drops it from 75 cm onto the same surface.

b Predict the bounce height for a drop height of 200 cm.

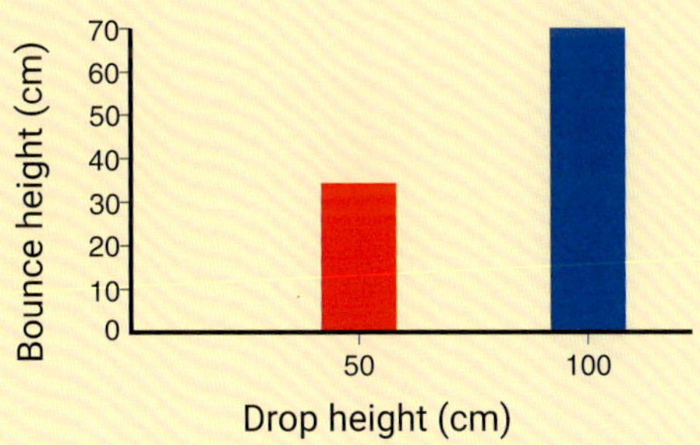

6 Brent did an experiment and made these sketches to show his results.

a Write a report of Brent's experiment. Make sure you include the title, aim, materials, method, results (including a data table), discussion and conclusion.

b Would it be better to draw a bar graph or a line graph of the results? Explain the reasons for your choice.

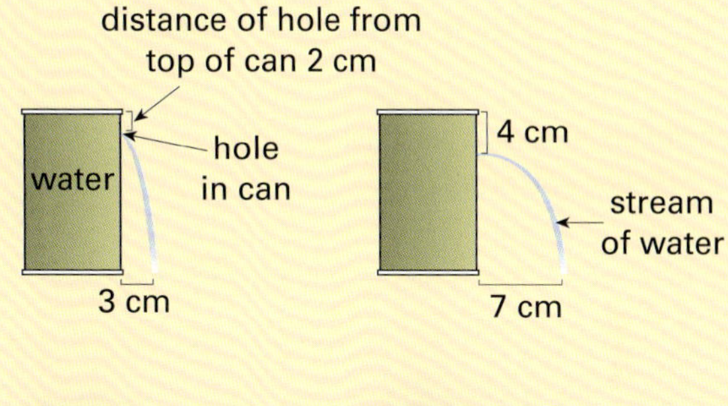

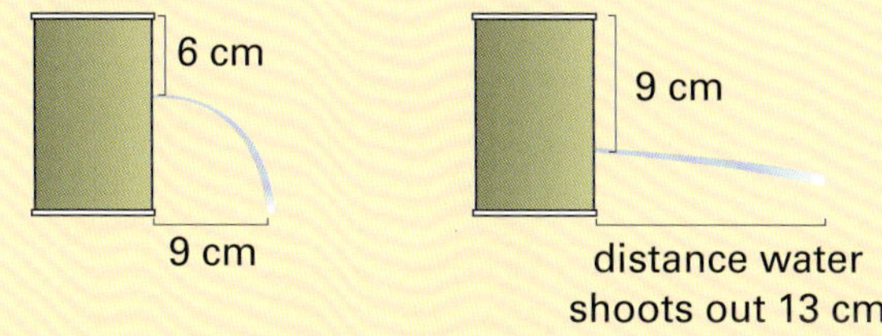

c Predict what would happen to the stream of water if the hole was 12 cm from the top of the can.

LAB REVIEW

Do this in pairs, checking your results. Discuss reasons for any differences in the measurements, and repeat them if necessary.

1 Measure the temperature inside and outside the laboratory. What is the difference?

2 Work out an accurate way of measuring the volume of a drop of water from a dropper. Then do it.

Check your answers on page 237.

ISBN 978 1 4202 3822 8

IN THIS CHAPTER

Science Understanding

- investigate the effects of applying different forces to familiar objects
- investigate common situations where forces are balanced and unbalanced
- explore how gravity affects objects on the surface of the Earth
- consider how gravity keeps planets and moons in orbit
- understand the importance of the work of Isaac Newton

Science Inquiry Skills

- design and carry out controlled experiments to answer questions about friction and gravity
- draw a line of best fit to show a pattern in the results of an experiment

CH•3 Forces

ISBN 978 1 4202 3822 8

GET STARTED: *IMAGINE*

Work in a small group to solve one or more of the following problems.

> Imagine a large asteroid has been detected on a collision course with Earth. What could be done to prevent the collision?

> You want to find out which type of adhesive tape is best for sticking things to the bench. Design an experiment to find out.

> Imagine that you want to design and build a billycart that will beat all others in a 100-metre downhill race. Discuss the features you would include in your design.

ISBN 978 1 4202 3822 8

3.1 Forces around you

When you lie in bed at night everything is so still it seems that nothing moves. But your chest moves up and down as you breathe, and your heart is pumping blood throughout your body. While this is happening, the Earth is hurtling through space at 100 000 kilometres per hour! All this motion is caused by forces, which affect every moment of your life.

Forces are pushes or pulls. They can start objects moving, and they can stop, speed up, slow down or change the direction of moving objects. They can lift things, or cause them to turn, bend or twist. They can also prevent motion; for example, a handbrake on a car stops it from rolling down a hill.

Look at the cartoon below.
Find examples of:

- **a** pushing forces. (What is doing the pushing?)
- **b** pulling forces. (What is doing the pulling?).

Which of the examples are:

- forces caused by gravity?
- lifting forces?
- forces where something bends?
- muscular forces?
- electrical forces?
- magnetic forces?

Figure 3.1 Forces are all around you.

ISBN 978 1 4202 3822 8

INVESTIGATION 3.1 Everyday forces

Aim

To experience a range of different forces.

> **Risk assessment and planning**
>
> This experiment has five parts. Discuss with your teacher whether you will do all parts or whether different groups will do different parts. Make sure you read the instructions before you start.

PART A Pushing forces

Materials

- table tennis ball
- drinking straws

Method

1 Use the table tennis ball and a straw to answer these questions.
   - **a** What happens if you blow on the ball while it is moving towards you?
   - **b** What happens if you blow on it while it is moving away from you?
   - **c** What happens if you blow on it while it is moving across in front of you?

   Discuss your answers with others.

2 Play a game of blowball. This game is played in groups of four, with two in each team. Use a straw each, and set up two sets of goal posts. See who can score the most goals.

PART B Gravitational forces

Materials

- 20 cent coin
- scissors and paper

Method

1 Cut out a piece of paper the size of the coin.

2 Hold the piece of paper in one hand and the coin in the other, at the same height. Drop them both at the same time.

   Write an inference to explain what happened. Name the forces that were acting on the coin and the paper, and in which direction they were acting.

3 Roll the paper into a ball and drop it and the coin again.

   Write an inference to explain what happened.

PART C Electrostatic forces

Materials

- piece of paper
- plastic pen or ruler

Method

Tear a piece of paper into small bits. Rub a plastic pen briskly on your clothes. (A woollen jumper works well.) What happens when you bring the pen near the pieces of paper?

ISBN 978 1 4202 3822 8

The force acting here is due to an electrostatic charge on the pen. Does the pen need to be touching the paper, or can the electrostatic force act over a distance?

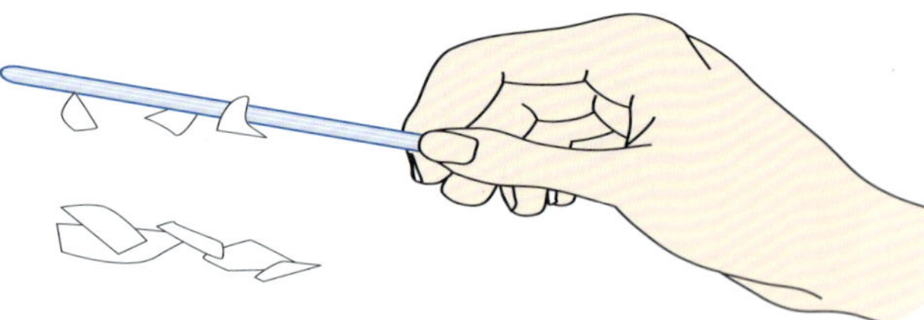

PART D Magnetic forces

Materials

- 2 bar magnets with their poles marked

Method

1 Hold one magnet in each hand. Feel what happens as you slowly bring the end of one magnet close to the end of the other.

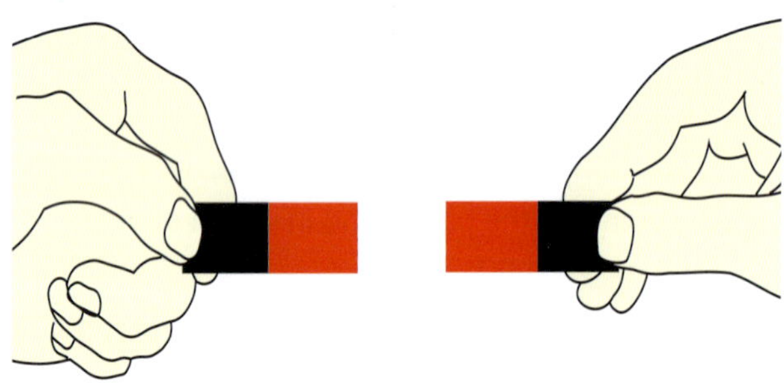

2 Repeat the test, but use the other end of one of the magnets.
Write a generalisation describing the forces acting between the two magnets.

PART E Buoyancy forces

Materials

- bucket of water
- balls of different types, e.g. golf, table tennis, rubber, styrofoam

Method

1 Put a table tennis ball into a bucket of water. Push it to the bottom of the bucket and let it go.

2 Do the same with the other balls.
What forces act on a ball in water? In which direction do they act?
Try to write a generalisation about what happens to different balls in water.

ISBN 978 1 4202 3822 8

What do forces do to objects?

You have seen so far that forces can:

- start motion
- stop motion
- speed up motion
- slow down motion
- change the direction of motion
- change the shape of an object.

Some forces act by contact and are called *contact forces*. For example, when you push something by hand, or pull it with a rope, you are using contact forces. Other examples are the wind blowing the trees and ocean waves crashing on rocks.

Some forces do not need contact, and can act at a distance. These are *non-contact* forces. For example, two magnets exert a force on each other without even touching. Other examples of non-contact forces are gravitational and electrostatic forces.

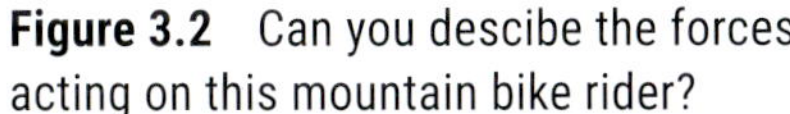

Figure 3.2 Can you descibe the forces acting on this mountain bike rider?

INVESTIGATION 3.2 Demonstrating forces

Aim

This is a 'design-it-yourself' investigation.

You have to design ways to show the six things that forces do to objects (see the list at the top of the page). Name the forces that are contact forces and the forces that are non-contact forces.

Materials

You will be given a rubber band, a tennis ball, a piece of playdough, a magnet and some paperclips. You can use other materials, but you will have to discuss the use of these with your teacher.

Planning

1. In a small group, discuss how you will show each of the six things forces do to objects. For example, for 'starting motion' you might flick a tennis ball with a ruler.
2. Think of ways to demonstrate both contact forces and non-contact forces.
3. Show your plan to your teacher before you start.

Writing your report

Draw up a large table with *What we did, What we observed, What the force did* and *Contact or non-contact force* as column headings.

ISBN 978 1 4202 3822 8

Balanced and unbalanced forces

In a tug-of-war, there are two equal forces acting in opposite directions. There is no motion until one force becomes greater than the other. You can use arrows to show the direction and strength of the forces.

Figure 3.3 Balanced and unbalanced forces

Bicycle forces

To start off riding your bike, you use your muscles to push on the pedals. This force then turns the back wheel, which pushes on the road, causing the bike to start moving. There are also frictional forces that tend to slow you down.

However, your pushing force is greater than the frictional forces and the bike speeds up. The forces are *unbalanced*, causing an increase in speed.

When you reach a constant speed, the forces are *balanced*. If you stop pushing on the pedals, the forces are again unbalanced, and the bike soon stops (unless you are going down a hill!).

unbalanced forces—
slowing down

Figure 3.4 Forces on a bike change as you speed up and slow down.

ISBN 978 1 4202 3822 8

Measuring forces

A spring stretches when a pulling force acts on it, and is squashed or compressed when a pushing force acts on it. The bigger the force, the more the spring is stretched or compressed. For this reason, a spring can be used to measure the strength of forces. A pointer attached to the spring moves as the spring changes length, and the force can be read on a scale. To measure larger forces, you use a stronger spring. Spring balances measure pulls, and kitchen or bathroom scales measure pushes.

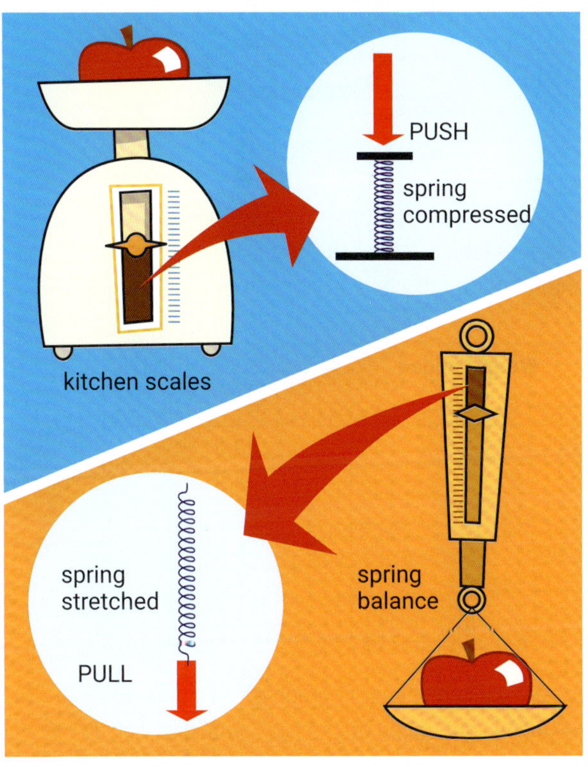

Figure 3.5 Measuring forces

The unit used to measure force is the **newton** (N), named after Sir Isaac Newton. The table below gives you some idea of the approximate sizes of some forces.

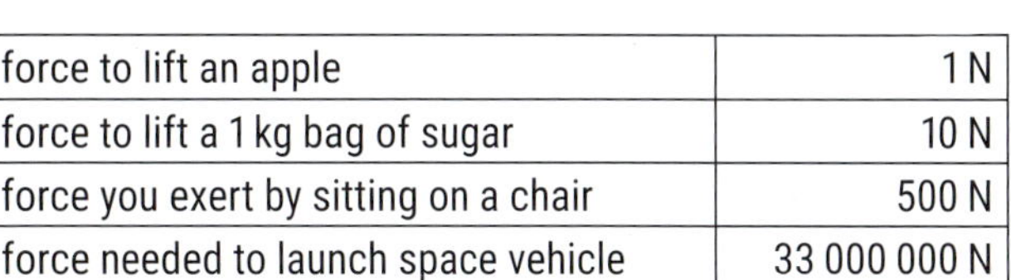

force to lift an apple	1 N
force to lift a 1 kg bag of sugar	10 N
force you exert by sitting on a chair	500 N
force needed to launch space vehicle	33 000 000 N

You can also use force measurers in a horizontal position. For example, you can use a spring balance to measure the force needed to pull a door open.

Figure 3.6 This force measurement laboratory can apply forces to objects to test their strength and breaking point.

Obtain several different spring balances, e.g. 5 N, 20 N and 100 N. If possible, obtain some kitchen scales, e.g. 2 kg (20 N) and some bathroom scales, e.g. 120 kg (1200 N).

If the scale is in kilograms, multiply by 10 to get newtons. If the scale is in grams, divide by 100 to get newtons.

Measure a range of pulls and pushes. For pulls use a spring balance. For small pushes use kitchen scales, and for large pushes use bathroom scales.

You could measure the force needed to:

- push or pull a sliding door
- open a can of soft drink
- pull sticky tape off the bench
- break a piece of fishing line
- turn on a light switch
- lift a shot-put
- push a button on an electrical appliance.

Record all your measurements in a data table.

ISBN 978 1 4202 3822 8

SKILLBUILDER

Drawing and interpreting graphs

Experiment 1

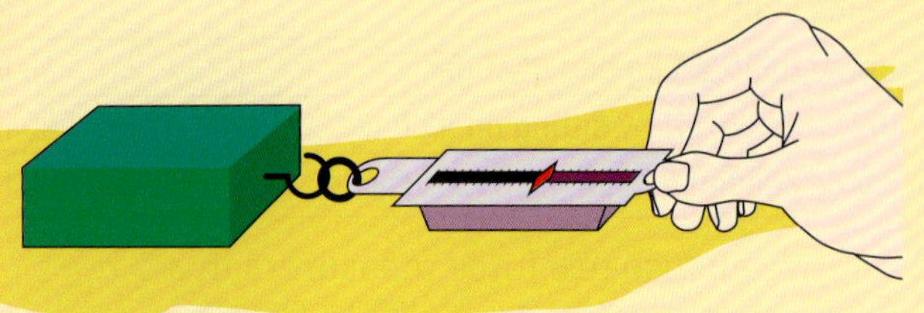

Jack decided to check whether friction is greater for heavy objects than for light objects. He used a spring balance to pull a wooden block across the desk. Then he pulled two blocks, one on top of the other, then three and four. Here are his results.

Number of blocks	Force needed to pull blocks (N)		
	Trial 1	Trial 2	Trial 3
1	3.1	3.3	3.2
2	6.7	6.4	6.4
3	9.8	9.5	10.0
4	12.8	12.6	13.1

1 What was the aim of Jack's experiment?
2 Why do you think Jack did each test three times?
3 Calculate the *average* force for each test.
4 Use graph paper to draw a *line of best fit* graph of Jack's data, with the number of blocks on the horizontal axis. (See the Skillbuilder on page 67.)
5 What pattern can you see in Jack's results?
6 What conclusion can you draw from the results? Try to write a general statement or *generalisation*—one that seems true in most cases.

Experiment 2

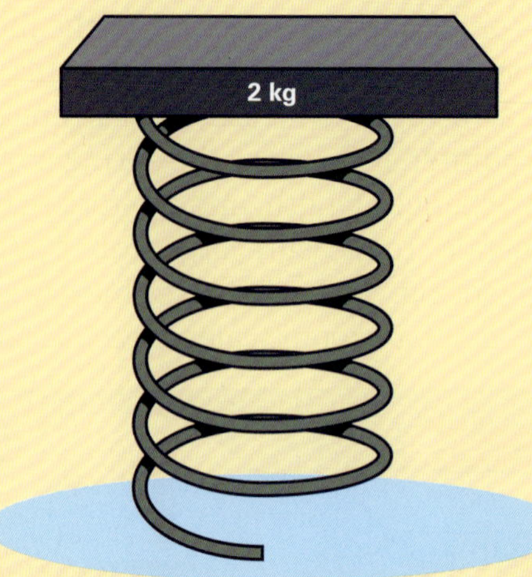

Lia put a 2 kg weight on top of a bed spring and measured the height of the spring. She then added more and more weights, and recorded the results, as shown.

Weight added (kg)	Height of spring (cm)
0	10.0
2	8.8
4	7.6
6	6.4
8	5.2
10	4.0

1 Draw a line of best fit graph of Lia's data, with the weight added on the horizontal axis.
2 Write a generalisation linking the height of the spring to the weight added.
3 Suppose you add a 5 kg weight to the spring. *Predict* the height of the spring.
4 If the spring is compressed to 5 cm, predict the weight on it.
5 If more than 10 kg is added to the spring, the graph is no longer a straight line. Suggest a reason for this.

ISBN 978 1 4202 3822 8

1 Copy and complete the sentences below. Choose from these words:

direction pull move push change

a A force is a _____ or a _____ .
b When you open a door, you _____ .
c When you lift something, you _____ .
d A force can also _____ the shape of an object.
e A force can make things _____ .
f A force can also make moving things change _____ .

2 The diagrams show some forces in action. The forces are shown with arrows.
a For each picture name the object that the force acts on. For example, in **A** the force acts on the ball.
b Choose from the list below to say what the force is doing in each picture.
- starting an object moving
- slowing down an object that is moving
- changing the direction of movement
- balancing another force, and preventing movement
- bending an object .

3 Some forces can act over a distance, rather than by contact.
a What does this statement mean?
b Name three types of forces that can do this.

4 What forces cause a bike to slow down when you stop pedalling?

ISBN 978 1 4202 3822 8

5 Look at the spring balance, kitchen scales and bathroom scales below.
   a Write down the reading on each.
   b Where necessary, convert the reading to newtons.
   c Which shows the largest force? Is it a push or a pull?

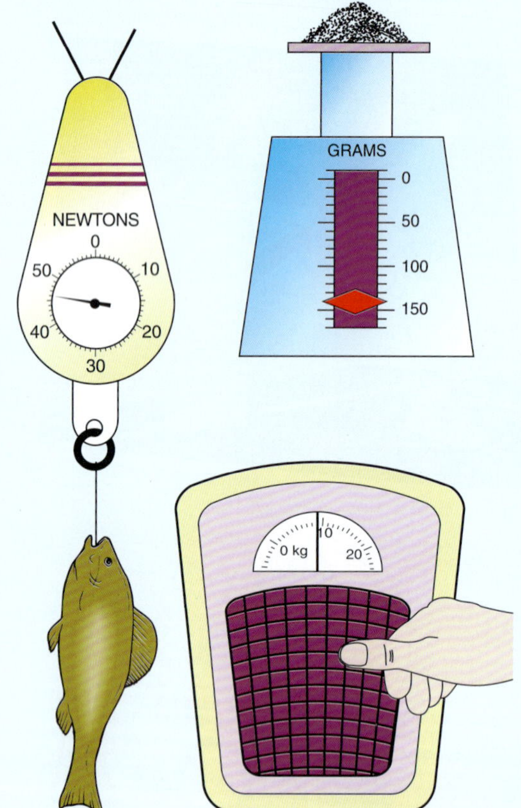

6 How do the bathroom scales work? Draw a diagram.

7 Sketch a person parachuting from a plane. Draw arrows to represent the forces acting on the person. When are the forces balanced?

8 You are sitting on a chair. There are two balanced forces.
   a What are the forces?
   b What would happen if these forces were not balanced?

9 Imagine you are abseiling down a cliff. What are the two forces acting on you? Are they balanced?

EXPLORE

1 Work out a way to measure twisting forces; for example, the force needed to turn a doorknob.

2 Design and build your own force meter to measure pushes and pulls. You will need something that returns to its original shape after bending or stretching.

3 The buoyancy force in salt water is much greater than in fresh water. (This means that the upwards force is greater in salt water than in fresh water.) Design a test using a newton spring balance and a mass to show that the buoyancy force is greater in salt water.

CHALLENGE

1 Forces are measured in newtons. How would you explain to someone how big a force of 20 N is?

2 'A car travelling in a straight line at constant speed has no forces acting on it.' Explain why this statement is false.

3 A cricketer hits a ball into the air. Is there a force on the ball while it is in the air? Explain.

4 In which direction will this boat move? Explain your answer.

5 Sam is whirling a ball on a string in a circle above his head. What forces are acting on the ball?

ISBN 978 1 4202 3822 8

3.2 Frictional forces

Friction is an example of a contact force. It occurs whenever two surfaces in contact move past each other.

Friction always opposes motion. Suppose you try to push a bookcase full of books, and it doesn't move. This is because of the friction between the bookcase and the floor. This frictional force is just as large as your push, but in the opposite direction. If you get someone to help you, and your combined push is greater than the maximum frictional force, then the bookcase will move.

Even when you do move an object, friction still opposes the motion. Stop pedalling your bike and the frictional forces soon bring you to a stop.

Friction occurs because objects are never completely smooth. The roughness of the two surfaces means there are many points that catch and stick together.

Figure 3.7 Snowboarders need to reduce friction to go downhill as fast as possible. However, some friction between the board and the snow is needed to allow them to turn quickly.

INVESTIGATION 3.3 Measuring friction

Aim

To measure the force of friction between different surfaces.

Materials

- large block of hardwood with a hook
- a 5 N spring balance
- various surfaces (see step 4)

Surface tested	Force needed to pull block (N)				
	predict	1	2	3	average
desk top					
carpet					

Risk assessment and planning

Read through the experiment and decide who in your group will do what.

Draw up a data table like the one shown.

PART A Measuring frictional forces

Method

1 Hook the spring balance onto the block of wood. Use the spring balance to pull the block slowly over the desk top.

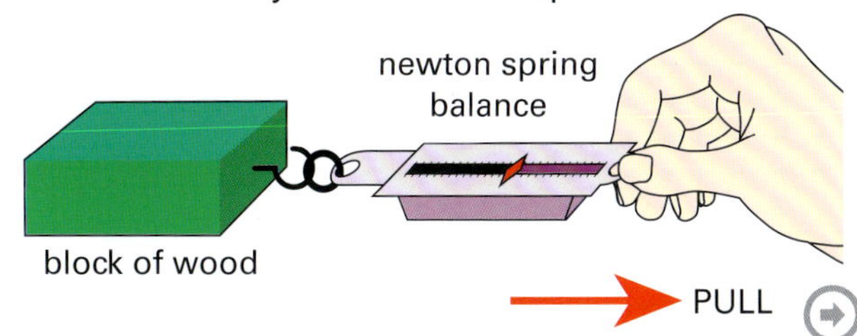

2 Measure the force in newtons needed to just keep the block moving at a constant slow speed. This force is equal to the opposing frictional force.

Record this reading in your data table.

3 Repeat the measurement and check to see whether it agrees with the first result.

Repeat it a third time, and take an average of the three results. (If you have forgotten how to find an average, see page 29.)

4 Predict the force needed to move the block over other surfaces, such as:

- carpet
- lino
- a concrete path
- a bitumen path
- sandpaper
- the smooth and rough sides of masonite.

Record your predictions in the data table.

5 Put the block of wood on each of the surfaces, and record the force needed to move it. (Remember that you need to measure the force three times for each surface, and then find the average.)

Record all your results. How accurate were your predictions?

Discussion

1 Why was it necessary to take three measurements each time, instead of just one?

2 Draw a bar graph of your data.

3 Which surface produced the most friction? Which produced the least?

4 Why does a concrete path produce greater friction than a lino floor?

PART B Inquiry

Investigate one or more of the following. Work in a small group to design the investigation. Write a report using the usual headings.

1 How can the frictional force be reduced? You could put pens, pencils, marbles or wooden dowels under the block. Or you could try lubricants; for example, talcum powder, water, liquid detergent or glycerine.

2 What happens to the frictional force if you increase the mass of the block? You can do this by placing blocks on top of each other.

3 How does the area of contact between the block and the surface affect the frictional force?

As you found in Investigation 3.3, friction depends on the type of surfaces that are rubbing together. Rough surfaces generally produce more friction than smooth surfaces. The friction also depends on the mass of the object. For example, it is much harder to push a bookcase full of books than it is to push an empty one.

Small mass—small friction

Large mass—large friction

Figure 3.8 The larger the weight, the more the friction

ISBN 978 1 4202 3822 8

Reducing friction

A rolling object meets with less friction than a sliding one. This explains how the ancient Egyptians were able to move the huge blocks needed to make the pyramids by putting logs under them. It is also how ball bearings reduce friction between a wheel and an axle.

Figure 3.9 Ball bearings reduce friction.

Lubricants (LOO-bri-kints) such as oil and grease are used to reduce friction between the moving parts of machines. For example, the bearings of bicycles and roller blades are oiled or greased to reduce wear and make the wheels turn more easily.

Our bodies, like machines, also have moving parts. Where bones slide over each other at joints, there is a lubricating fluid between them to make them slide more easily. If this lubricating system doesn't work properly, you get swelling and pain in your joints. This is called arthritis.

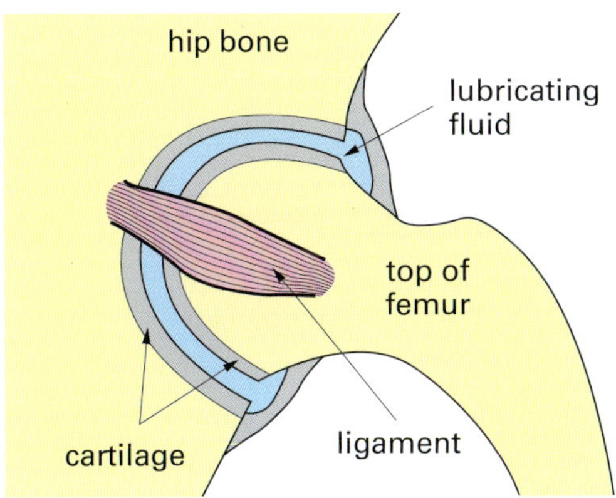

Figure 3.10 The human hip shows how friction is reduced at joints.

Surfaces in contact can be polished so that they slide over one another more easily. For example, the hulls of racing yachts are polished so they slide through the water more easily.

If air is blown between two surfaces, the friction becomes very small. This is how a hovercraft works.

Air resistance is the friction between a moving object and the air it is moving through. Without air resistance, parachutes would not work and kites would not fly. On the other hand, air resistance can be a nuisance, and modern cars have a streamlined shape to reduce this air resistance. Streamlining is also important for objects moving through water; for example, surfboards, speedboats and fast-swimming fish.

Figure 3.11 Surfboards have a streamlined shape and a very smooth surface to reduce the resistance in water.

ISBN 978 1 4202 3822 8

Friction in everyday life

We use friction every day. Sometimes we need friction, and at other times we try to reduce it. When we walk we use the friction between our shoes and the ground. Imagine trying to walk if there was no friction. This would be like walking on ice or a highly polished floor. You could not stop a car without friction. You could not start it moving either—the wheels would just spin without the car moving. And everybody knows what happens if you go too fast around a corner or if the road is slippery. Friction also prevents knots from coming undone, and holds nails and nuts and bolts in place.

Figure 3.12 Friction lights the match.

Figure 3.13 Ice has reduced the friction between the tyres and the road surface, causing the bus to slide off the road.

Friction produces heat. This can be useful when you rub a match on the side of a matchbox, but it can cause a car engine to overheat. Friction also causes wear and tear.

1 Look at the diagrams on the right. Copy and complete the sentences below, selecting the correct word for each.

  a Diagram A shows _____ (sliding/rolling) friction.
  b The friction in B is _____ (greater/less) than in A.
  c Rolling friction is _____ (greater/less) than sliding friction.
  d When an object slides, there is _____ (more/less) resistance to movement than when it rolls.
  e With lubrication (diagram C) you need _____ (more/less) force to move an object.
  f Lubrication _____ (increases/decreases) friction.

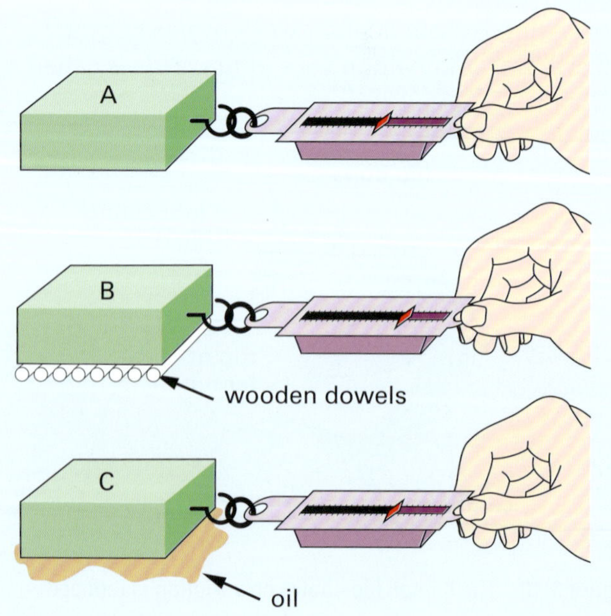

ISBN 978 1 4202 3822 8

2 Which two factors influence the size of a frictional force?

3 Copy and complete the paragraph below. Choose from these words:

carpet	force	rough	more
glass	rubs	smooth	less

Things move more easily across a ________ surface than a _____ surface. This is because of friction. It happens when one surface _____ on another. Rough surfaces like _____ produce _____ friction than smooth surfaces like _____.

4 Use your knowledge of friction to explain the following.
- a Gymnasts put resin on their hands before competing.
- b Cars that travel in snow have to carry chains that fit around the tyres.
- c Surfers wax their surfboards.
- d A car uses more petrol when it is fitted with a roof rack.
- e When you drive a car in city traffic for some time the brakes become quite hot.
- f The underneath of the Orion crew capsule (below) is covered with special heat-resistant material called Avcoat ablator.

Figure 3.14 The new Orion crew capsule has a heat shield on the underside.

5 The cartoons below show friction in action. For each example:
- a name the two surfaces between which the friction acts
- b say what the force of friction is doing
- c say what would happen if the frictional force suddenly disappeared.

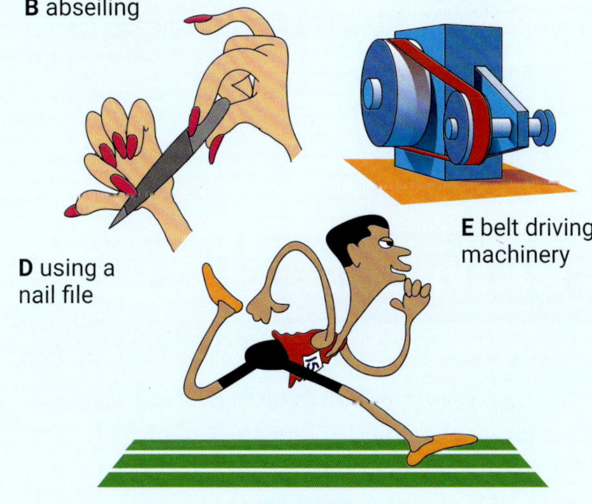

6 For each of the following, describe how friction is reduced.
- a roller blades
- b a water slide
- c a jet flying at high speed
- d a door hinge
- e a bobsled

ISBN 978 1 4202 3822 8

1 What is the purpose of the tread on a tyre? How does it work on a wet road?

2 a Racing cars use 'slicks'—wide tyres with no tread. Discuss the advantages and disadvantages of these tyres.

b How does the rear 'spoiler' improve the car's performance?

3 Explain why frictional forces depend on the:

a surfaces in contact

b mass of the object.

4 The cartoon below shows a perpetual motion machine—a machine that will keep going forever, after an initial force is applied to it. Use what you have learnt in this chapter to explain why such a machine will not work.

EXPLORE ONLINE

Many people over hundreds of years have tried to invent perpetual motion machines. Use the internet to find out more about perpetual motion machines.

Check out the links to **Museum of hoaxes** (search under *perpetual motion*).

EXPLORE

1 Write a short story about a world without friction.

2 Design an experiment to test an oil company's claim that its brand of engine oil reduces friction more than other brands do.

3 Make a hovercraft from a metal lid with a very small hole in it. Glue a 2 cm one-holed stopper over the hole, as shown on the right. Blow up a balloon (not too much), and hold its neck while you fit it over the stopper.

Now hold the hovercraft on a smooth, level desk, and slowly release your fingers.

Use what you have learnt about forces to explain how the hovercraft works.

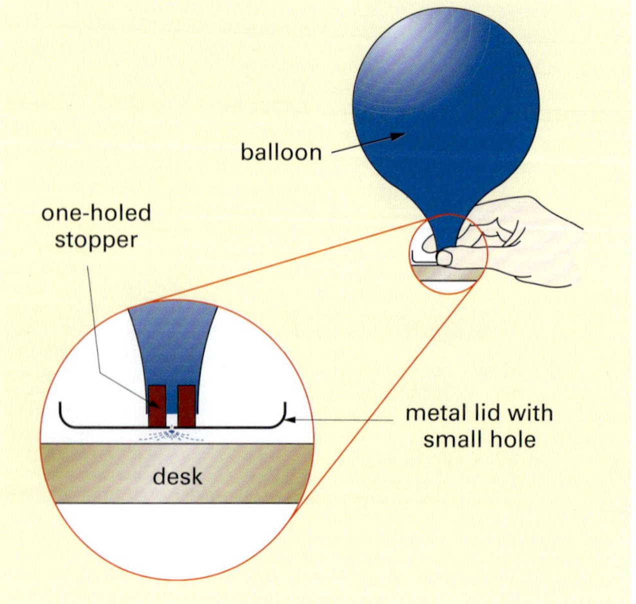

ISBN 978 1 4202 3822 8

3.3 Gravitational forces

Gravity on Earth

You are sitting at the top of the Tower of Death, a 100-metre tower at the amusement park. The catch is released and you plummet towards the ground. You and everyone else in the car are being pulled towards the centre of the Earth by the force of **gravity**.

Figure 3.15 Gravity pulls objects towards the centre of the Earth.

Over the centuries people have suggested many inferences to explain gravity. In the 17th century, Sir Isaac Newton came to the conclusion that gravity is the force of attraction between objects, and that the size of this force depends on the **mass** of the objects. The mass of an object is the amount of matter in it. Mass is measured in grams or kilograms.

All bits of matter attract each other. The greater the masses of the objects, the greater the force between them. You are attracted by the Earth and the Earth is attracted by you, but the Earth has a large mass and attracts you very strongly. This is why you don't fall off the Earth. This gravitational force acts towards the centre of the Earth.

Spring balances and scales actually measure the force of attraction between an object and the Earth. This is what **weight** is. Weight is dependent on the mass of an object and the strength of gravity. Weight is a force measured in newtons (N).

Gravity—a non contact force

Gravitational force is a non-contact force because it exists between objects even when they are not touching. The gravitational force between the Earth and the moon keeps the moon in orbit. Similarly, a gravitational force keeps satellites in orbit around the Earth, and all the planets in orbit around the sun. Gravitational forces can act over the huge distances of space; for example, between stars and between galaxies.

The moon has less mass than the Earth. This is why gravity is less on the moon than it is on Earth. If you jumped on the moon you would not come down as quickly as you do on Earth because your weight is less on the moon, even though your mass is the same. Similarly, larger heavier planets such as Jupiter have more gravity than smaller lighter planets such as Mars.

Figure 3.16 Gravity on the moon is only about one-sixth of Earth's gravity. As a result, walking and jumping on the moon are very different from what they are on Earth.

ISBN 978 1 4202 3822 8

ACTIVITY

We use a range of devices to make use of gravitational force. Look at each of the photos and say how we are using gravity in each case.

ISBN 978 1 4202 3822 8

Drawing lines of best fit

In the following experiment you are going to hang different masses on a spring. After you take measurements, you are going to plot a line graph of the results.

Graphs are very useful when you are looking for a *pattern* in your results. However, sometimes you have to draw a line of best fit so that you can see this pattern clearly.

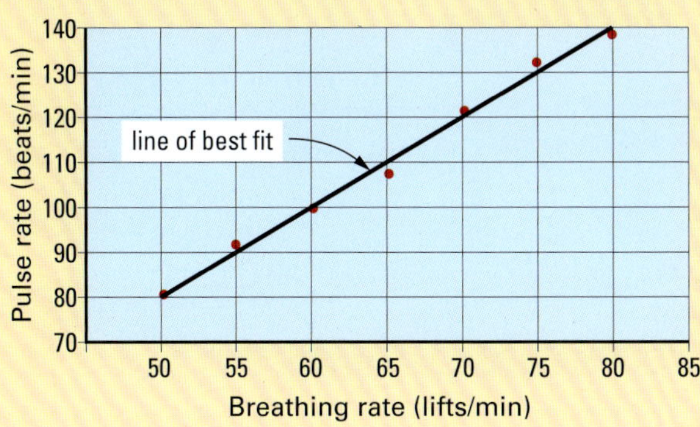

For example, in an experiment on fitness, Matthew recorded Sarah's pulse rate and breathing rate. He then plotted the graph below.

Notice that the points lie roughly on the straight line. This is called a straight line of best fit.

Drawing lines of best fit takes practice! The line doesn't need to go through all the points. It has to pass close to as many points as possible.

Exercise

Draw a straight line of best fit for the following data.

Time (days)	Growth of seedlings (cm)
0	0
1	1.0
2	2.1
3	2.6
4	3.8
5	5.0
6	5.8

EXPERIMENT 3.1 Stretching a spring

In Chapter 2 you designed an experiment to solve the problem: how quickly can you catch a falling ruler?

Here is another experiment in which you have to design tests to solve a problem.

1 The problem to be solved

What happens to the length of a spring when the mass attached to it increases?

2 Designing your experiment

You will be given a spring and some masses. You can use other materials, but you will have to write a list of your requirements to give your teacher.

Work in a small group and decide how you are going to solve the problem. Write a draft of your method, list the materials needed and discuss any safety issues.

3 Results

Design a data table for your results and plot a line graph of the results.

4 Writing your report

Write a full report of the experiment including diagrams where appropriate. In the discussion, use the graph to write a generalisation about the mass on the spring and the stretch of the spring.

You could also take a digital photo of your equipment to include in your report.

ISBN 978 1 4202 3822 8

SCIENCE AS A HUMAN ENDEAVOUR

Sir Isaac Newton

Isaac Newton was born in 1642. His parents lived on a farm in England, and his father died the day before Isaac was born.

Isaac was quite a sickly young child, and very shy. He didn't do very well at school, and he was often bullied by the bright boy in the class. So Isaac worked extra hard until he was the best in the class.

During his boyhood years, Isaac liked making things, and he was very good at solving everyday problems. For example, he made a model windmill which was driven by mice running in a treadmill. He also made a water clock and a sundial, and flew kites with lanterns attached to them.

By the time he was 18, Isaac was very interested in mathematics. His uncle said he would make a poor farmer, and talked his mother into sending him to Cambridge University. He did very well there, and when he was 27 he became a professor of mathematics, and made many important discoveries. Perhaps the greatest of these was his theory of gravitation (see page 65). He also passed sunlight through a prism, and found that white light is a mixture of the colours of the rainbow.

Newton had a good imagination. For example, he imagined a very tall mountain from which he could fire a bullet. His idea was that if you could fire the bullet fast enough, it would continue to circle the Earth, just as the moon does. (Look at the cartoon below.)

One of Newton's faults was that he couldn't take criticism, and he spent a lot of time quarrelling with other scientists. He never married, and throughout his life he avoided publicity. When he was 50 he had a nervous breakdown, and he died at the age of 84. He was the first scientist to be buried in Westminster Abbey in London.

Questions

1. In which century was Isaac Newton born?
2. Did he do well at school?
3. What did he make when he was a boy?
4. What was Newton's greatest discovery?
5. List Newton's 'good' and 'bad' points in a table.
6. Using library resources or the internet, find two other things that Newton discovered. Describe how they are used in today's society.

ISBN 978 1 4202 3822 8

1 Look at the diagram below. Choose the correct word to complete each sentence in your notebook.

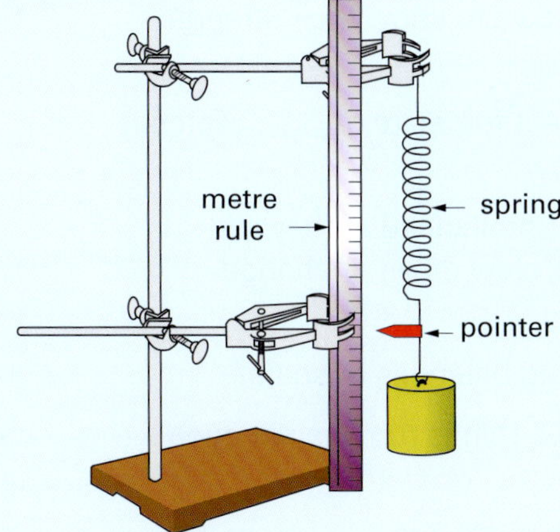

a The force of gravity is a _____ (push/pull).
b A force pulls the spring _____ (up/down).
c The heavier the object, the _____ (less/greater) the force.
d The stronger the force, the _____ (more/less) the spring stretches.
e A 10 N weight is a _____ (larger/smaller) force than a 20 N weight.
f A 20 N force stretches the spring _____ (twice as much/half as much) as a 10 N force.
g Use the data in the table to draw a straight line of best fit. Remember to label the axes and give the graph a title.

Weight (N)	Stretch (cm)
100	8
200	17
300	23
400	32

h Which measurement is the independent one? Give a reason for your answer.
i Using the data in the table or your line graph, predict how much a 50 N weight would stretch the spring.

2 The shop assistant said the bag of apples had a weight of 1 kg.
a Is this correct? What is the weight in newtons?
b If the bag of apples was taken to the moon, would its mass and weight be the same? Explain your answer.

3 Imagine you are going on a space trip. Your travel agent has given you this table showing the force of gravity on the planets, the moon and outer space.

Place	Gravity (compared with Earth)
Mercury	4
Venus	9
Earth	10
outer space	0
moon	1.6
Mars	4
Jupiter	25
Saturn	11
Uranus	9
Neptune	11

a Where would you weigh the most? What difference would this make to getting around on the planet?
b Where would you be weightless?
c Suggest why the gravity is similar on Mercury and Mars.
d On which planet would you need the least fuel for blast-off?
e How high can you jump on Earth? Predict how high you would be able to jump on the moon.

1 The moon has no atmosphere. Suggest a reason for this. (Hint: gases such as oxygen have mass, even though the mass is very small compared to a solid.)

2 If a rocket was launched from Earth and travelled close to the planet Mars, it could go into orbit without the rocket engines being fired. How could this happen?

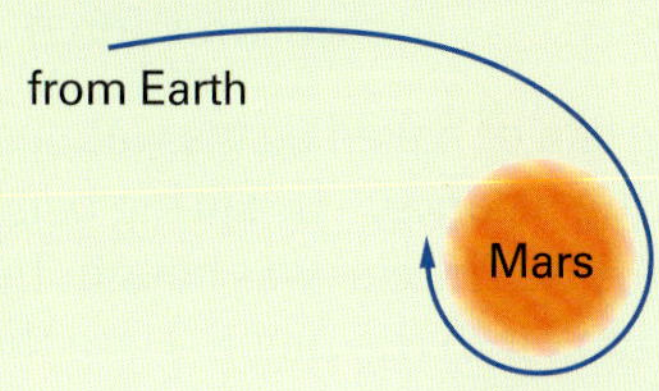

ISBN 978 1 4202 3822 8

MAIN IDEAS

Copy and complete these statements to make a summary of this chapter. The missing words are on the right.

direction
lubricants
friction
pulls
magnetic
unbalanced
mass
newtons
weight
surfaces
gravity

1 Forces are pushes or _____. You can use arrows to show their strength and _____.

2 Contact forces act on contact, and non-contact forces (e.g. _____ forces) act at a distance.

3 The forces on an object may be balanced or unbalanced. If they are _____, the object will start moving, speed up, slow down or change direction.

4 Forces are measured in _____ (N) using spring balances or scales.

5 _____ is a contact force that occurs when two things rub against each other. It slows down or prevents motion.

6 Friction depends on the roughness or smoothness of the _____, and the mass of the object.

7 There are several ways of reducing friction; for example, ball bearings, _____ and polishing.

8 _____ is the force of attraction between any two objects; for example, between the sun and a planet. It depends on the masses of the objects.

9 The _____ of an object is the downwards pull of gravity on it. The _____ of an object is the amount of matter in it.

CH•3 REVIEW

1 Match the words with their correct meanings.

force	unit of force
carpet	reduces friction
friction	a pull from a large body
newton	push or pull
gravity	surface with large friction
spring balance	force that exists when one surface rubs on another
lubricant	produced when there is friction
heat	measures force

2 A 1 kg can of baked beans was suspended from a spring balance. The spring balance reading was 9.8 N. The can was bought by an astronaut, who took it to the moon. Here the can was again suspended from the spring balance, and the reading was only 1.6 N.

Explain the different readings on the spring balance.

3 Give two examples of contact forces and two examples of non-contact forces.

4 A truck has become bogged in mud, and the back wheels are spinning because of a lack of friction. Which of the following actions could help move the bogged vehicle? (There may be more than one answer.)

A Put more mass into the back of the truck.
B Place rubber mats under the wheels.
C Pour water around the wheels.
D Let most of the air out of the back tyres.

For each action you select, say *why* this would increase the friction.

ISBN 978 1 4202 3822 8

5 When Kieran stands on some bathroom scales in a stationary lift, the reading is 700 N. As the lift descends rapidly, what is the reading on the scales likely to be?

A 0 N
B 600 N
C 700 N
D 800 N

6 In which of the following situations is friction an advantage, and in which is it a disadvantage? Explain your answers.

a stopping in a hurry
b pushing a fridge across the floor
c running a car engine
d parachuting from a plane

7 A boat floating on still water is said to be under the action of forces that are balanced because the downwards force on the boat is:

A greater than the upwards force of the water.
B the same as the upwards force of the water.
C less than the upwards force of the water.

8 Sometimes you want to reduce friction, and sometimes you want to increase it. Explain, using the cartoon as an example.

9 Christina was using a spring balance to measure the force needed to move a block of wood over a number of different surfaces. She measured the force four times for each surface. (See the table below.)

Surface	Force (N)			
	Trial 1	Trial 2	Trial 3	Trial 4
concrete	20	23	20	18
newspaper	9	8	8	8
vinyl tiles	4	4	3	5
sandpaper	34	32	36	35
grass	16	18	17	19

a What was the average force needed to move the block for each of the five surfaces?
b Draw a bar graph of the results. (Use the average force for each surface.)

10 The van is travelling to the left at a constant speed.

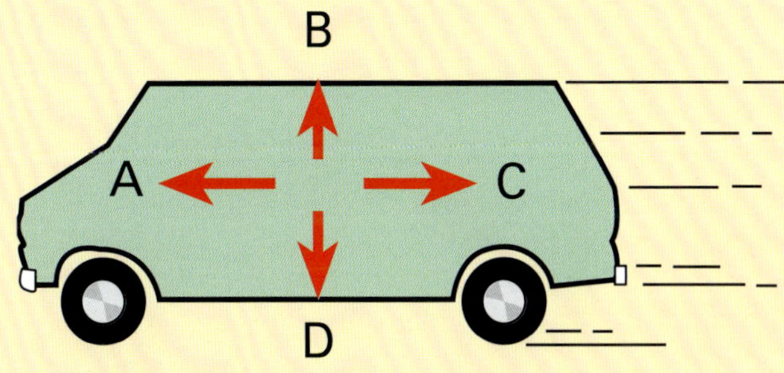

a In which direction do the frictional forces act to slow down the van?
b In which direction does the weight of the van act?
c Are the forces on the van balanced? Explain your answer.

11 Lorna hits a tennis ball. What forces are acting on it as it moves across the net? (There may be more than one answer.)

A the force of gravity
B the force of friction
C the force of the hit
D no forces

Check your answers on page 238.

ISBN 978 1 4202 3822 8

IN THIS CHAPTER

Science Understanding

> develop an understanding of how simple machines can magnify a force, change a force's direction, or make things go faster
> investigate levers, pulleys, ramps, inclined planes and other simple technologies

Science Inquiry Skills

> use measurements to calculate the mechanical advantage of different types of levers and pulleys
> investigate how the gears of a bicycle work
> invent a useful device containing simple machines
> use a model to show how air moving over an aircraft wing creates lift

CH•4 Simple machine technology

ISBN 978 1 4202 3822 8

GET STARTED: *EXPLORE*

Work in a small group to solve one or more of the following problems.

- A 200 kg meteorite lands in your driveway, and you and a couple of friends have to move it. In the garage, you find a crowbar, some pulleys, rope and a few round posts. Using these things, work out a way to move the meteorite off your driveway.
- You notice that the more you blow up a balloon, the faster it goes when you let it go. Try to explain your observations.
- You are riding your bike when you come to a steep hill. To ride up the hill, you change gears. Do you use the largest gear wheel on the back wheel or the smallest? Do you use the largest gear wheel on the pedals or the smallest? Try to explain your answers.

ISBN 978 1 4202 3822 8

4.1 Simple machines

If you have used a door handle or a bottle opener today, or you have ridden a bicycle, you have been using **simple machines**.

The most common simple machines are levers, pulleys, inclined planes (ramps), screws and gears. Complex machines such as cranes, winches, clocks and bicycles are a combination of many simple machines. Machines help you do things more easily. They do this in three different ways.

1 *Machines magnify the force you use*

When you have to remove a nail from a piece of wood, it is easier to use a claw hammer than to use your fingers. The larger the hammer, the smaller the force needed to remove the nail. Machines that magnify the force you use are said to have a *force advantage*.

Figure 4.1 A hammer gives you a force advantage when removing a nail and when hammering.

2 *Machines change the direction of the force*

When you have to pull up the sails on your sailing boat, the pulley at the top of the mast makes it easier. You pull the rope *down* and the sail goes *up*.

Figure 4.2 A pulley changes the direction of a force.

3 *Machines make things go faster*

When you push on the pedals of a bicycle, the back wheel turns much faster than the pedals. Machines that make things go faster are said to have a *speed advantage* or a *distance advantage*.

You can't get more energy out than you put in

Some machines give you a bigger force than you apply and others make things go faster. But no machine can give you a bigger force and go faster at the same time. In other words, a machine cannot give you a force advantage as well as a speed (or distance) advantage.

If this were to happen, it would mean that you were getting more energy out of the machine than you were putting in. Machines transfer energy from one point to another; for example, from the pedals to the back wheel of a bicycle. And because of friction, machines always lose some energy as heat and sound.

ISBN 978 1 4202 3822 8

Levers

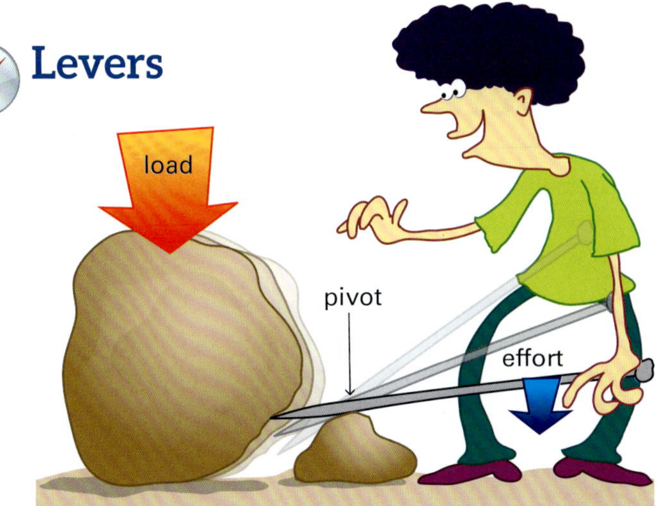

Figure 4.3 You want to move a large rock with a crowbar.

A crowbar is a type of simple machine called a **lever**. A lever moves around a fixed point called the **pivot** (or fulcrum), in this case a small rock. The large rock is called the **load**, and the force you apply is called the **effort**.

The crowbar has a *force advantage*. This is because the force (effort) you apply to the crowbar is less than the force used to move the rock.

Sometimes you might want a lever to have a *distance advantage* rather than a force advantage. You can use a lever to catapult a small rock into the air by placing the pivot close to the effort. Here the effort is much larger than the load, but the distance the lever moves at the load end is much greater than the distance the lever moves at the effort end.

Figure 4.4 Using a lever in this way provides a distance advantage.

Everyday levers

A wheelbarrow is another lever. It is an example of a lever where the pivot (the wheel) is at the end of the lever and the effort is at the other end. The load is between the pivot and the effort.

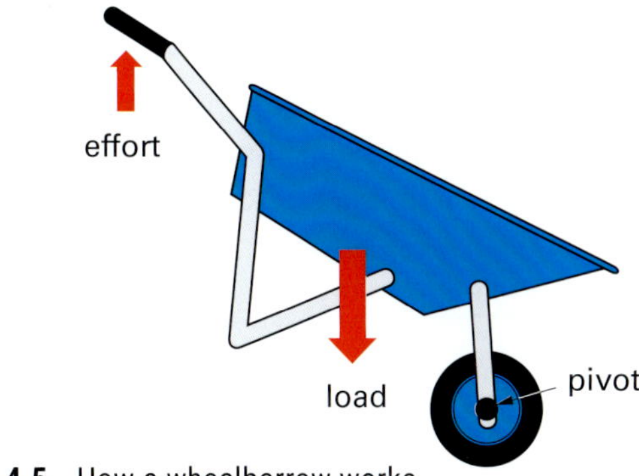

Figure 4.5 How a wheelbarrow works

Pliers are made of two levers held together at the pivot. They work in the same way as a crowbar. The pivot is close to the load, which means you need only a small effort to exert a large force between the jaws of the pliers.

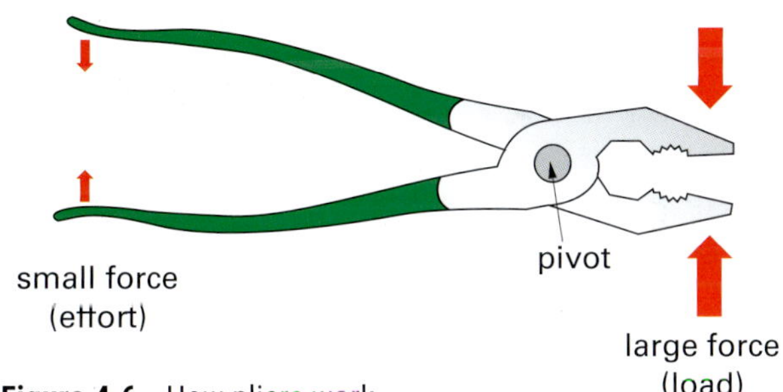

Figure 4.6 How pliers work

Mechanical advantage

You can get a measure of how useful a simple machine is by dividing the load you move by the effort you use. This measure is called the machine's **mechanical advantage**.

For example, suppose you used a crowbar and applied a 100 N force to move a 500 N (50 kg) rock.

$$\text{mechanical advantage} = \frac{\text{load}}{\text{effort}} = \frac{500\text{ N}}{100\text{ N}} = 5$$

ISBN 978 1 4202 3822 8

Using levers

Aim

To investigate different types of levers.

Materials

- metre ruler (preferably with holes every cm)
- brass masses and two hangers
- large paperclips (bent to form an S)
- spring balance (5 N)
- stand and clamp

PART A

1 Use the diagram as a guide to find out what effort you need to balance a 100 g load.
Draw up a data table like the one shown below and record your results.

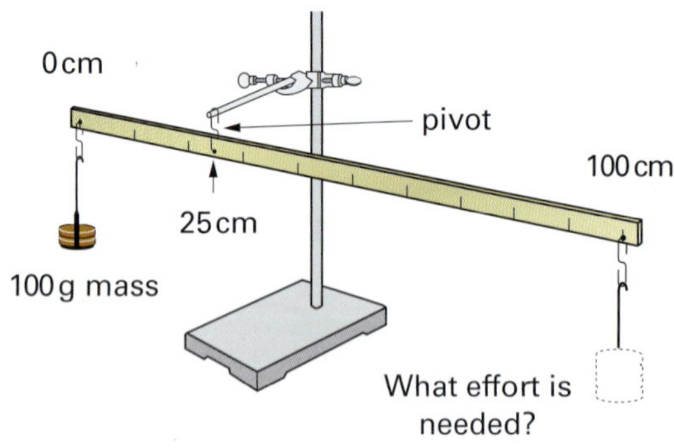

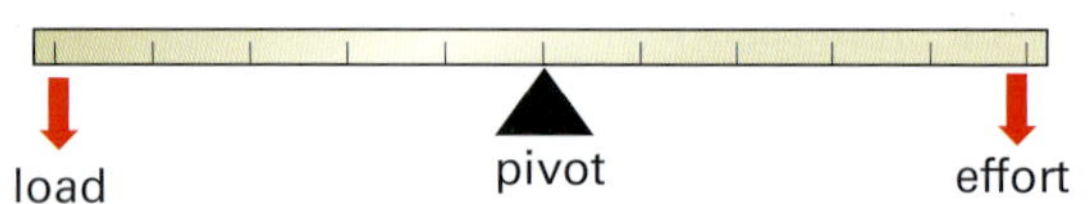

Figure 4.7 The arrangement of pivot, load and effort for this type of lever

2 Find the effort needed when you increase the load. Try masses of 150 g, 200 g and 250 g.
Record your results.

3 Calculate the mechanical advantage of the lever for each load.
Record your results.

4 Investigate the effort needed for each of the loads in the data table when you change the position of the pivot. Calculate the mechanical advantage of the lever for each load.

PART B

1 Set up the ruler, stand and clamp, spring balance and masses to model a wheelbarrow, as shown in the diagram here. The spring balance will measure your upwards effort to balance the load.

2 Experiment with this arrangement. Which position of the load makes the best wheelbarrow?
Write a report of your findings. Make a drawing of this type of lever. (Use Figure 4.7 as a guide.)

Discussion

1 Write a generalisation about effort, load and position of the pivot for the type of lever used in Part A.

2 There is a third way to arrange the effort, load and pivot. Work out this arrangement and calculate the mechanical advantage for this type of lever.

Load (mass in g) at 25 cm	Downward force produced by load (N) (Divide the mass by 100)	Effort (N)	Mechanical advantage MA = load/ effort
100 g	100 g/100 = 1 N		
150 g			
200 g			
250 g			

ISBN 978 1 4202 3822 8

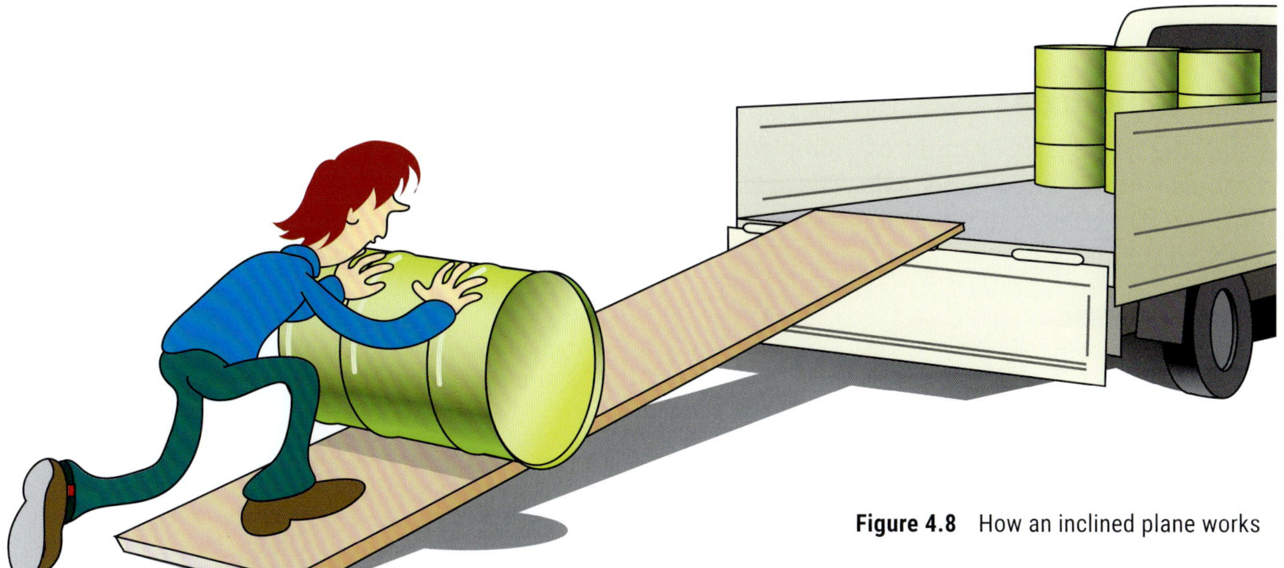

Figure 4.8 How an inclined plane works

Other simple machines

Inclined planes

You want to lift some drums onto the back of a truck. It is far easier to use a ramp and roll them on than to lift them straight up.

The ramp is an example of an **inclined plane**—another simple machine. Inclined planes have a force advantage—you apply a small effort to move a heavy load, but you have to move the load over a large distance. In other words, you use less effort this way, but you have to roll the drums further.

Axes

To split a block of wood you need an axe, which has a wedge-shaped blade. A wedge is a pair of inclined planes that changes the direction of a force. You push the axe downwards, and the wedge-shaped blade pushes the wood apart.

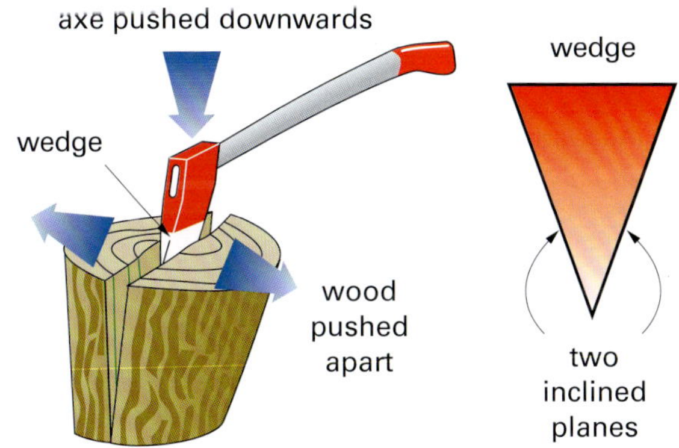

Figure 4.9 An axe is made up of two inclined planes.

Screws

A screw is a special type of inclined plane that spirals round and round to form a thread. To drive the screw into a piece of wood, you use a small force and turn the screw many times. This is changed into a larger force where the thread touches the wood, pushing it in a small distance.

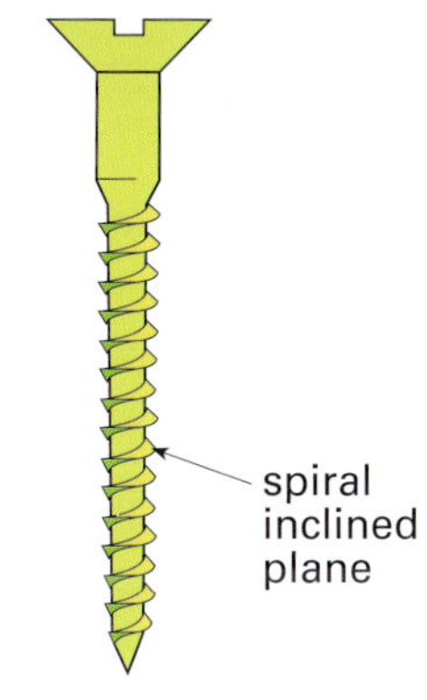

Figure 4.10 Screws have a spiral inclined plane.

Jacks

A jack is an example of a screw. You use a small force to lift the heavy load of the car. However, you have to turn the jack handle many times to raise the car by a small amount.

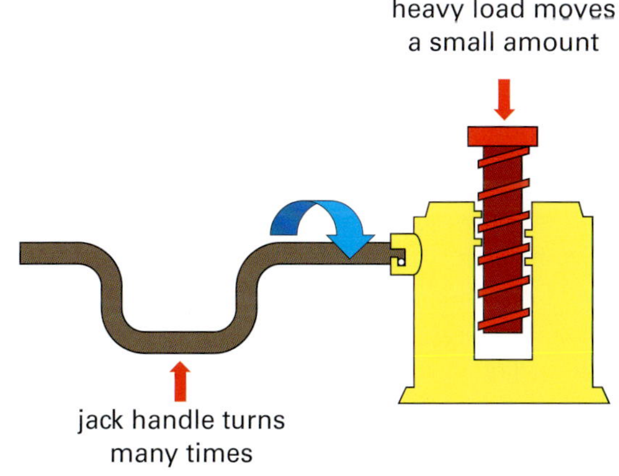

Figure 4.11 A jack is basically a screw.

ISBN 978 1 4202 3822 8

Wheel-and-axle

A steering wheel is an example of a simple machine called a **wheel-and-axle.** The axle is the central rod or column and the steering wheel is attached to this axle.

Without a steering wheel it would be very difficult to turn the axle. The force needed would be too great. With a steering wheel, applying a small force will turn the axle. The bigger the steering wheel, the easier it is to steer.

Door knobs and screwdrivers

A door knob is an example of a wheel-and-axle. The handle is the wheel. When you turn the handle, there is a large force applied to the axle. This opens the door. A screwdriver is another example of a wheel-and-axle.

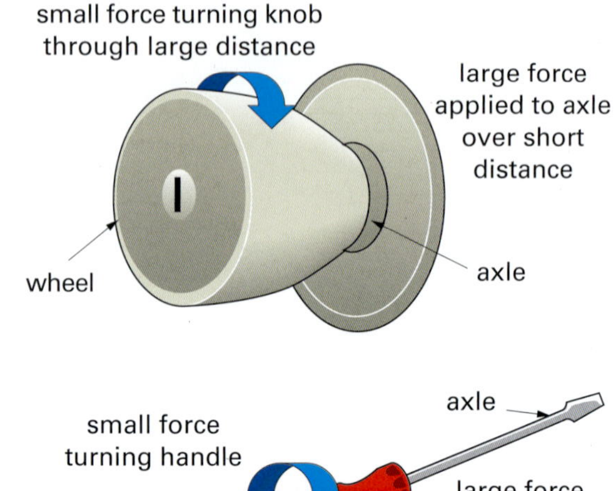

Figure 4.12 Two examples of a wheel-and-axle

CHECK

1 Copy and complete the following sentences. Choose from these words:

magnify effort screw
pivot lever direction

- a A crowbar is a type of _____ .
- b The _____ is the force you put into a simple machine.
- c Some machines help you by changing the _____ of the force.
- d The _____ is the fixed point around which a lever moves.
- e Some machines _____ the force you use.
- f A _____ is a type of an inclined plane.

2 The following statements are false. Rewrite them to make them correct.

- a A simple machine can give you a bigger force and make things go faster as well.
- b Pliers are an example of two levers with the pivot at the end.
- c If you apply a 50 N force to a lever to lift a 250 N load, it has a mechanical advantage of 25.
- d It is much easier to remove a tight screw with a screwdriver with a small handle than one with a large handle.

3 In which one of the three ways listed on page 74 does an inclined plane help you? What is the disadvantage of an inclined plane?

4 When you lift an object in your hand, your arm is acting as a lever. Where is the pivot? Draw a diagram like the one in Figure 4.7 on page 76 showing the load, effort and pivot.

1 kg

5 A lever can be set up to have a force advantage. It can also be set up to have a distance advantage. Draw labelled diagrams to show these two different types of levers.

ISBN 978 1 4202 3822 8

6 Each of the devices below works as a lever. For each one, draw a simple diagram of how it works, labelling it with effort, load and pivot. The first one is done for you.

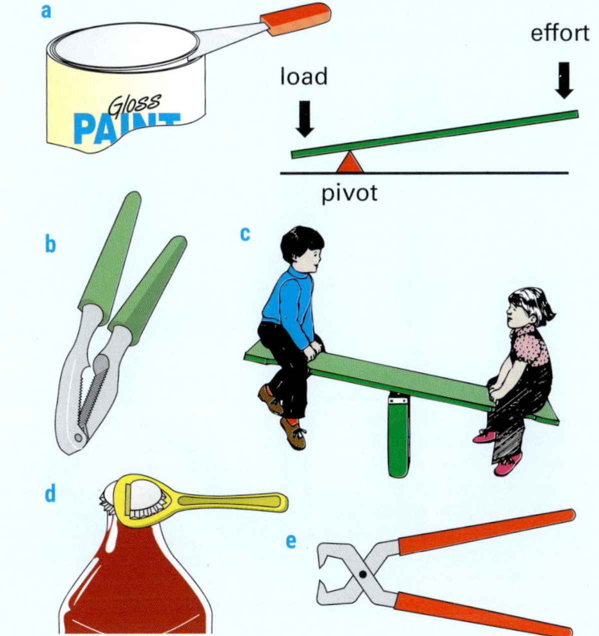

7 You want to remove a nail from a piece of wood with a claw hammer. To do this the hammer is acting as a bent lever. Draw and label a diagram to show how this happens.

8 Label the wheel-and-axle in each of these devices.

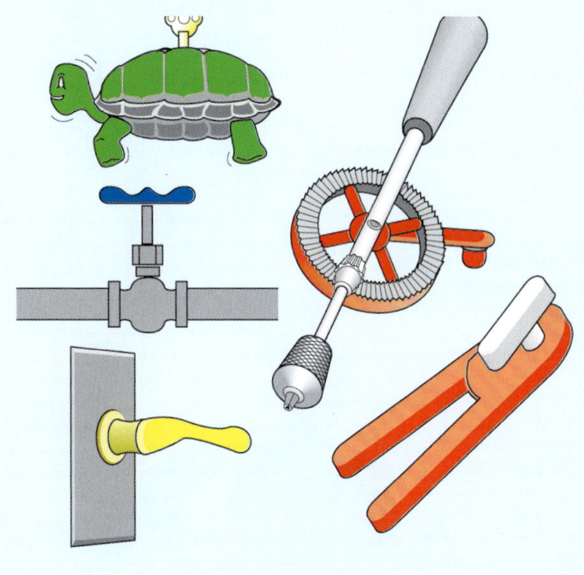

CHALLENGE

1 The diagram shows an old-fashioned well.

a What type of simple machine is this?

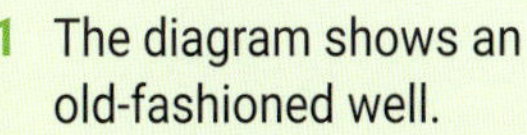

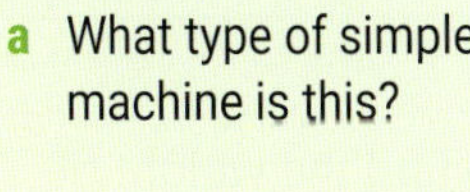

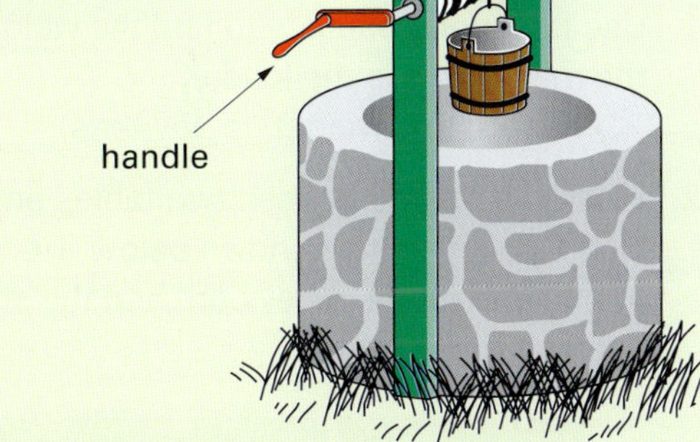

b Suppose the bucket was full of water at the bottom of the well. You tried as hard as you could to turn the handle but the bucket was too heavy. Suggest ways to lift the bucket.

2 You try to remove a screw from a piece of timber with a small screwdriver but it won't budge. When you use a larger screwdriver the screw turns. Explain in terms of simple machines why this happens.

3 Jodie uses a jack to change a flat tyre on the rear wheel of her mother's car. The jack has a mechanical advantage of 10 and it has to lift a load of 3000 N. How much force does Jodie have to apply to lift the rear wheel of the car?

4 Explain how you can set up a lever to give a force advantage, then rearrange it to give a distance advantage.

5 Design an experiment to show how you could measure the mechanical advantage for a wheel-and-axle.

6 Nathan uses a crowbar to lift a box of tools. The box weighs 600 N. He uses a triangular metal block as a pivot and places it 15 cm from the load end of the crowbar. The distance from the pivot to the effort end of the crowbar is 75 cm. Use this information to calculate the force Nathan has to use to lift the box.

7 A trebuchet (TREB-oo-shay) was a destructive ancient weapon of war used in Europe between 800 CE and 1350 CE. Use the internet to find out how this device used simple machines.

ISBN 978 1 4202 3822 8

4.2 Pulleys and gears

Pulleys

Lifting a bucket of sand is hard work. However, by using a pulley, you can pull *downwards* to move the bucket *upwards*. This is easier than lifting the bucket upwards directly.

A **pulley** is a grooved wheel with a rope over it. A single pulley does not magnify your force. It simply changes the direction of the force.

Figure 4.13 How a single pulley works

To magnify your force you need to use more than one pulley. This lets you lift very heavy loads by using only a small effort. However, you need a lot of rope to lift the object only a little way.

Figure 4.14 Many pulleys together make up a block and tackle.

INVESTIGATION 4.2 Using pulleys

Aim

To find out how pulleys make it easier to lift loads.

Materials

- 2 single pulleys and 2 double pulleys
- 2 short pieces of cord
- 1 piece of cord several metres long
- mass (up to 1 kg)
- newton spring balance
- piece of timber to act as a beam
- metre rule

Risk assessment and planning

- Carefully read through the Method and draw up the data table on the next page before starting.
- What safety issues do you need to discuss with your group before starting this investigation?

Method

1 Use the spring balance to find the weight of the mass. This is the load.
Record this weight (in newtons).

2 Place the beam between two tables and tie a single pulley to it as shown below. Tie the load to one end of the pulley cord and the spring balance to the other.

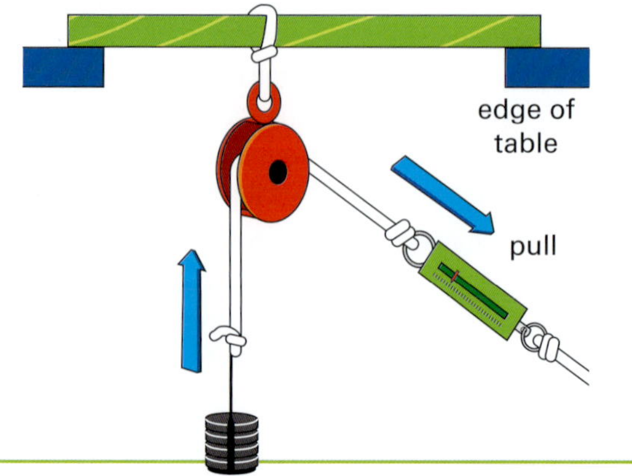

ISBN 978 1 4202 3822 8

3 Slowly pull the spring balance and find what force is required to lift the load.

4 Measure how far the spring balance end of the cord moves (the distance the effort moves) to lift the load 50 cm (the distance the load moves).
Record the results in the data table.

5 Now set up two single pulleys as shown below. Repeat steps 3 and 4.
Record these results.

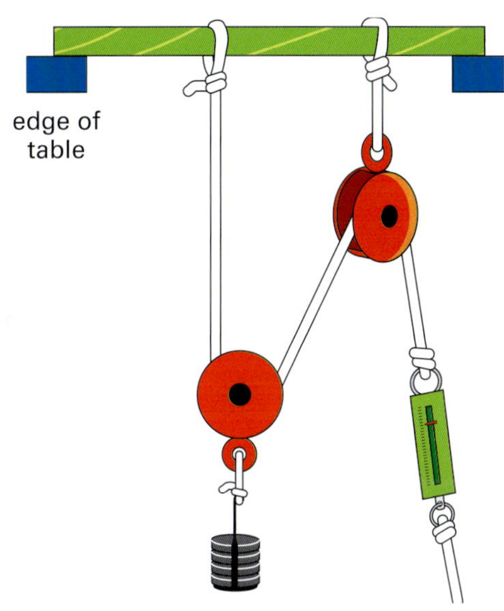

Figure 4.15 A two pulley system (two single pulleys)

6 Set up two double pulleys. This is equal to four single pulleys.
What force is needed to lift the mass this time? How far does the end of the cord move to lift the mass 50 cm?

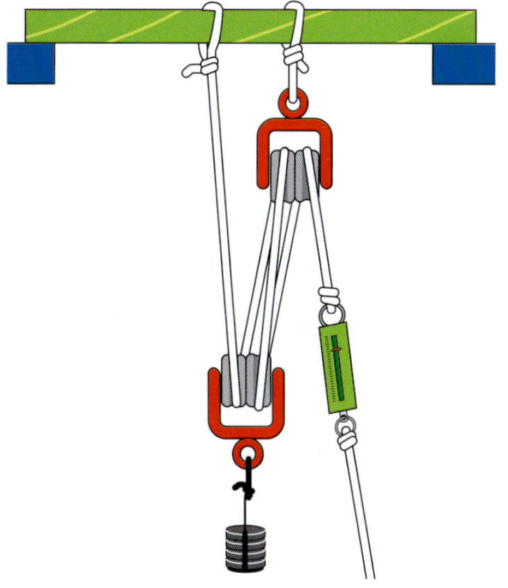

Figure 4.16 A four-pulley system (two double pulleys)

Discussion

1 Write a generalisation about the way the effort required to lift a load changes as the number of pulleys changes.

2 Calculate the mechanical advantage of the set-ups using one pulley, two pulleys and four pulleys. Put your answers in the data table.

3 Write a generalisation about how the effort distance changes with the number of pulleys used.

4 Predict the effort needed to lift the mass if you used eight pulleys.

5 Design and test a set-up to lift a load using one pulley only in which the mechanical advantage is greater than one.

No of pulleys	Load lifted (N)	Effort needed (N)	Mechanical advantage	Distance load moves (cm)	Distance effort moves (cm)
1				50	
2				50	
4				50	

ISBN 978 1 4202 3822 8

Figure 4.17 A block and tackle contains a number of pulleys and can be used to lift very heavy loads. Here it is used to control the boom and mainsail on a large yacht.

Gears

Gear wheels are wheels with teeth on them. The teeth of one gear fit into the teeth of another. Gears are used to transfer the force from one wheel to another.

In a bicycle, the large gear wheel attached to the pedals is connected by a chain to the smaller gear wheel on the back wheel. The gear wheel attached to the pedals is called the **driving gear** because it supplies the force. The gear wheel on the back wheel is called the **driven gear**.

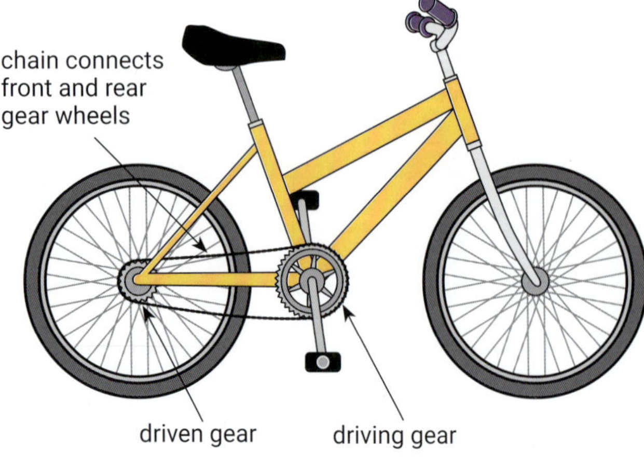

Figure 4.18 How a bicycle works

Gears can speed things up

In the diagram below, the large gear wheel has more teeth on it than the smaller gear wheel. This arrangement of gears makes the propeller spin faster than the motor spins. Gears speed things up or slow things down.

Machines that use a larger driving gear include kitchen hand beaters and hand drills. In these machines the beaters and drill bits have to rotate very fast to do their job. Therefore, the driving gear often has four or five times as many teeth as the driven gear.

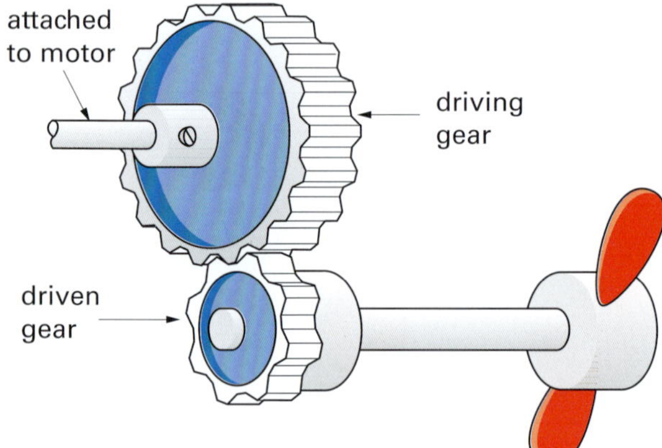

Figure 4.19 Gears speed up this propeller.

Gears can slow things down

In the diagram below, the driving gear has fewer teeth than the driven gear and the propeller spins more slowly than the motor.

Machines that have a driving gear smaller than the driven gear include winches and rotating displays found in shops, where the displays rotate slowly so that you can see the goods on sale.

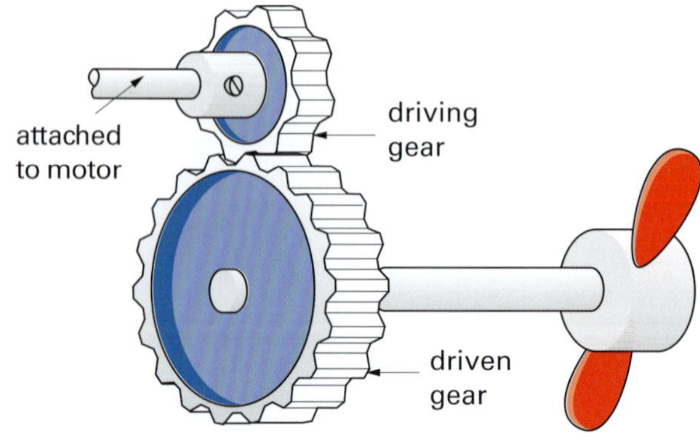

Figure 4.20 These gears slow the propeller down.

ISBN 978 1 4202 3822 8

For this activity, you will need a technical model set that includes gear wheels (e.g. Lego).

1 Use the model set to make the gear arrangements in Figure 4.19 and 4.20 on the previous page.
Which arrangement could be used to make a paint stirrer?

2 Slowly rotate the driving gear.
Does the driven gear turn in the same direction as the driving gear?
Use your model set to make an arrangement in which the driven gear turns the same way as the driving gear.

3 You may have seen a winch on a boat trailer being used to pull the boat up onto the trailer. The winch uses a combination of gear wheels to do this. Use the model set to make a winch that can pull heavy objects with the smallest effort.

4 Your teacher will give you a hand beater or hand drill. (Or you could make a model gear arrangement.)
How many times does the small wheel turn for each turn of the large wheel? How does this number compare to the ratio of the teeth on the wheels?
What is the relationship between the number of teeth on the gear wheels and how fast they turn?

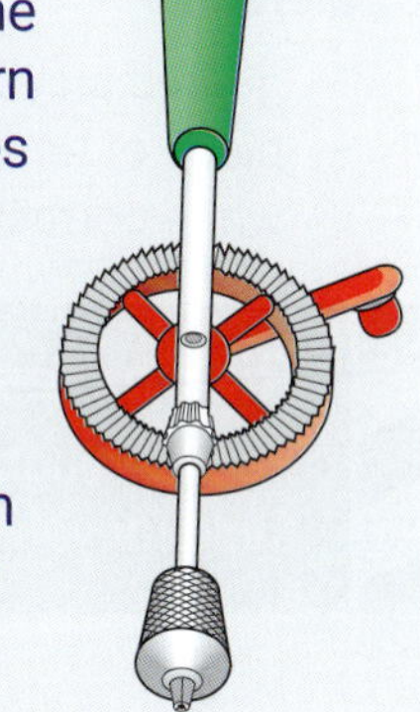

The bicycle

A bicycle is a **system** of many parts that work together as a whole. For example, the pedals and crank are attached to the front gear wheel called the *chain wheel*. The pedal and crank act as a wheel-and-axle and make it easier to turn the back wheel.

The gear wheels on a bicycle are called *sprockets*. The front sprocket is always larger than the back one. On a 27-speed bicycle shown below, there are three sprockets on the chain wheel, and nine on the back wheel.

Figure 4.21 A bicycle is made up of many simple machines that work together.

ISBN 978 1 4202 3822 8

INVESTIGATION 4.3 Bicycle gears

Aim

To investigate how the gears on a bicycle work.

Materials

- geared bicycle (per group)
- piece of timber to act as a beam
- white correcting fluid

Risk assessment and planning

Read through the Method and design a data table for your results.

PART A Investigating gears

Method

1 Lift the back wheel of the bicycle off the floor by sliding the piece of timber under the metal bracket that connects the seat and the back forks. Support the timber on two desks or stools.

2 Turn the pedals and use the gear levers to place the chain on the largest front sprocket and the largest rear sprocket.

3 Put a spot of correcting fluid on the back tyre. Now slowly turn the pedals so that the front sprocket travels through one complete turn. Count how many revolutions the back wheel makes during this time.
Record this result.

4 Count the number of teeth on the front sprocket and also on the back sprocket.
Record this result.

5 Suppose you now use the gear lever to put the chain on the smallest back sprocket. Will the number of revolutions of the back wheel be greater or less than before? Discuss this with your group.
Record your prediction and why you made that prediction.

6 Now test your prediction.
Record your results.

Discussion

1 Using which rear sprocket would you travel furthest with one revolution of the front sprocket?

2 Which rear sprocket would you use to ride up a steep hill?

3 Which combination of front and rear sprockets would you use to ride downhill?

PART B Inquiry

1 The number of teeth on the front sprocket divided by the number on the rear sprocket is called the *gear ratio*.
Calculate the gear ratio for the largest front sprocket and the largest rear sprocket. What is the ratio for the largest front sprocket and smallest rear sprocket? Are high gear ratios best for going uphill or downhill?

2 You can find out exactly how far you travel for one revolution of the pedals for a particular gear by calculating the circumference of the back wheel and multiplying this by the number of revolutions it turns.
Try this for a particular gear. Then change gears and calculate the distance again.

ISBN 978 1 4202 3822 8

Bicycle gears

When you want to ride up a hill, it is best to use a low gear. The lowest gear on a 12-speed bicycle uses the smaller front (driving) sprocket and the largest rear (driven) sprocket. With this combination of gears you have a *force advantage*—the effort will be less than the load, but the chain wheel turns faster than the back wheel.

If you are pedalling downhill you should use a high gear. The highest gear on a 12-speed bicycle uses the larger front sprocket and the smallest rear sprocket. In this gear, you have a *distance advantage*—the back wheel turns faster than the chain wheel. You can also reach the highest speed using this gear.

Figure 4.22 The largest rear sprocket is the lowest gear.

Figure 4.23 The smallest rear sprocket is the highest gear.

ACTIVITY

Gear ratios

Low gear and high gear in bicycles (and other machines that use gears) refer to the gear ratio that you may have calculated in the previous investigation.

The gear ratio is found by dividing the number of teeth on the driving sprocket by the number of teeth on the driven sprocket.

For example, if the driving sprocket has 52 teeth and the smallest driven sprocket has 14 teeth, the ratio is 3.7. When the largest driven sprocket is used (28 teeth), the ratio is 1.9.

The smallest driven sprocket is called high gear because it has the highest gear ratio of 3.7.

The lowest gear on most 12-speed bicycles is obtained by using the smaller front sprocket (e.g. 40 teeth) and the largest rear sprocket (e.g. 28 teeth). This gives a gear ratio of 1.4.

Questions

1 A track bike has a rear sprocket with 14 teeth on it. If the gear ratio is 5.0, how many teeth does the front chain wheel have?

2 The two front sprockets on a bike have 42 and 56 teeth. The six back sprockets have the following number of teeth—14, 15, 17, 20, 24 and 28. Which combination of sprockets will give you gear ratios of 2.0 and 3.0?

ISBN 978 1 4202 3822 8

EXPERIMENT 4.1 Your invention

In this experiment you have to invent useful devices containing simple machines, such as levers, gears and pulleys.

Planning

Work in a group and decide which device you would like to design and build. You can do all three if you like! Show your designs to your teacher before you start. Then write a report for each invention using the guidelines on page 15.

Invention 1

1 Study the device below. Explain to another student how it works.

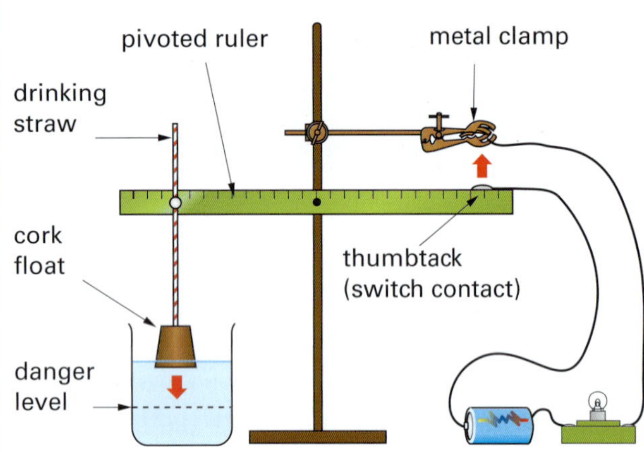

2 You want to use this device to build a water supply for animals. Suppose there is a large tank of water nearby and you want a constant supply of water in the water trough.
Design a device that includes levers, pulleys or gears that will do this.

Invention 2

1 Use your imagination to design a machine using levers, pulleys and/or gears that will raise a weight, move it 50 cm and then lower it.

2 Check your design with your teacher, then go ahead and make a model for it. (You may be able to work on your invention at home.)

Invention 3

1 Your task is to build a device that uses simple machines (e.g. gear wheels, wheel-and-axles, and/or pulleys) to pull a block of wood over the surface of a table. The best device will do this with the least amount of effort.

2 The diagram below will help you design your device. Remember, the winning device will be the one that starts the block of wood moving over the table with the least mass. This device has the greatest mechanical advantage.

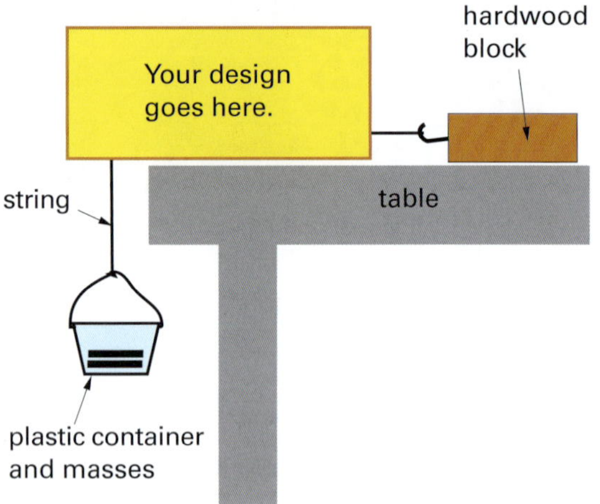

How does your design compare with those of other groups in the class? Whose design is the most efficient? What is your device's mechanical advantage?

Follow the links to **Simple machines**. This site is lots of fun with plenty of animations.

ISBN 978 1 4202 3822 8

1 Copy and complete the following sentences. Choose from these words:

less	greater	driving
direction	high	driven

a The bicycle sprocket on the front chain wheel is the _____ sprocket.
b A single fixed pulley only changes the _____ of the force.
c With a two-pulley system, the effort is _____ than the load.
d To pedal a bicycle downhill, you would usually use _____ gear.
e The mechanical advantage of a two-pulley system is _____ than one.

2 Give two examples of where you would find:
a pulley systems
b gears.

3 What does the term *mechanical advantage* mean? Use the words *load*, *effort* and *force advantage* in your explanation.

4 Hand winches on boat trailers are used for pulling boats from the water. Which arrangement of gears below would be suitable for a winch? Give a reason for your answer.

A B

C D

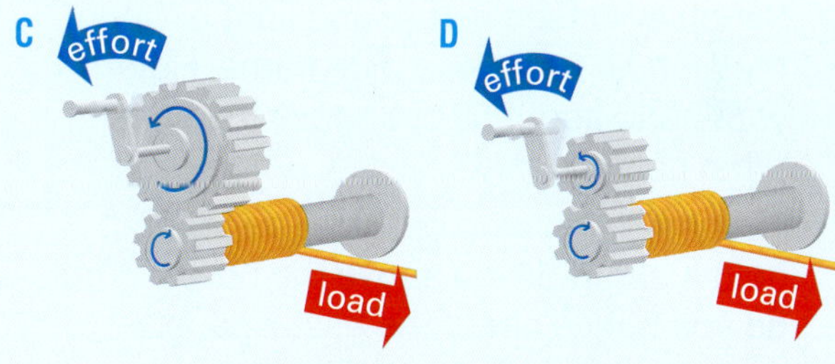

5 Katia wanted to lift a weight of 50 N using pulleys. Which arrangement in the diagrams above right would need the least effort?

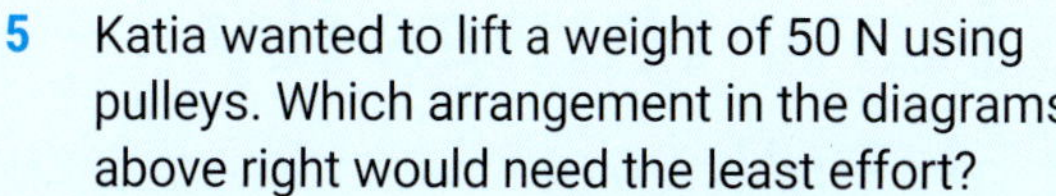

a b

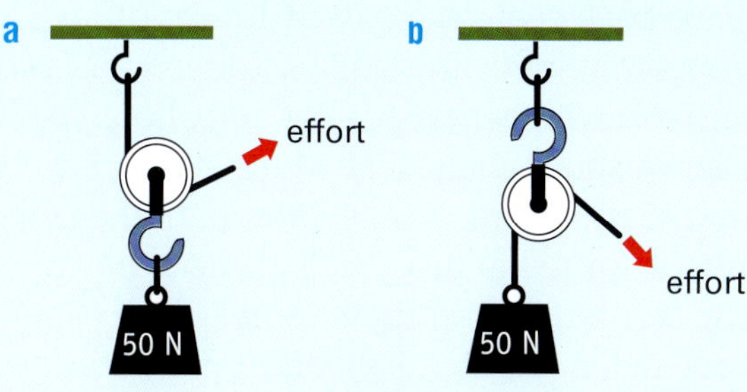

c

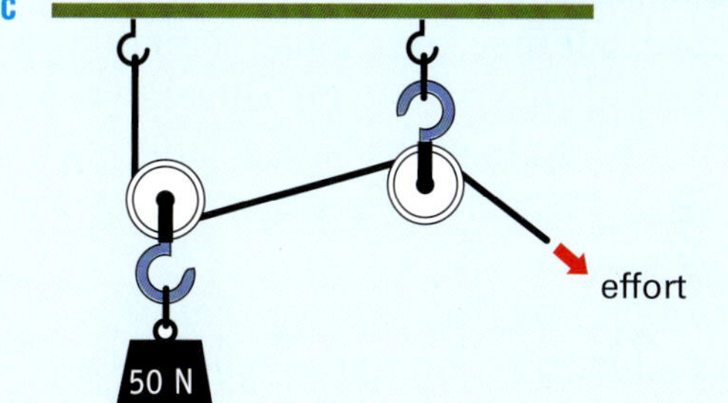

1 Three gears are connected together as shown below. Gear A is the driving gear and turns in a clockwise direction. In which direction does gear C turn?

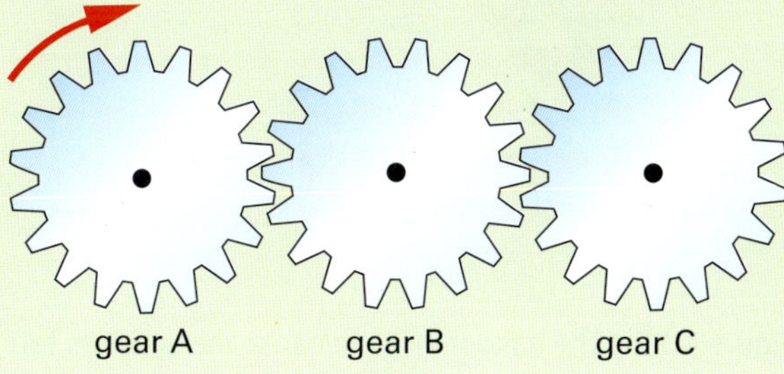

2 The front sprocket on a bike has 45 teeth. The back wheel sprocket has 15 teeth on it. What is the gear ratio for this bike? How many times does the back wheel turn around for one turn of the pedals?

3 You have an electric motor that spins at 200 revolutions per minute and a gear wheel with 50 teeth on it. You want a large advertising sign to rotate five times a minute.
What other gear wheel would you need?
Draw a sketch of the set-up.

ISBN 978 1 4202 3822 8

4.3 How things fly

Hot air balloons

Humans have been trying to fly for centuries. Early attempts to mimic birds and bats failed miserably. But on 21 November 1783, a hot air balloon took brothers Jacques and Joseph Montgolfier, into the air for the first time. The 25-minute flight sailed over Paris for 12 km before returning to Earth.

The Montgolfier brothers were French paper makers and made their balloon from layers and layers of paper. They were not sure why the balloon rose, but they knew it had something to do with the fire underneath the balloon. They also knew that the weight of the balloon affected its flight and tried to make their paper balloon as light as possible.

A hot air balloon rises because there are unbalanced forces acting on the balloon. Hot air is less dense and is pushed up by the denser cooler air.

ACTIVITY

Teacher demonstration

1. Set up a Bunsen burner and light it. Open the air hole to make a blue flame. Hold your hand about 50 cm above the flame. What can you feel?
2. Tear off a thin strip of tissue paper and hold it about 50 cm above the flame. What do you notice about the tissue paper? Write an inference to explain this.
3. You can follow the movement of the air above a burner by using smoke from a wax taper.
4. Cut off the string and top of a tea bag. Empty out the tea and open the bag to form a tube. Stand it on its end on a heatproof mat. Light the top of the tea bag. Observe the remains of the tea bag rising like Montgolfier's balloon.

See the tea bag fly on **Planet Science**.

Figure 4.24 Hot air balloons were one of the first inventions to fly.

The rising air creates an upwards force on the balloon. This upwards force is often called *lift*. Gravity acts on the balloon's mass, causing a downwards force. If the upwards force is greater than the downwards force, the balloon rises.

When balloonists want to come back to Earth, they turn down the burner and this reduces the amount of hot air produced. This in turn reduces the lift. Gravity is now the greater force and the balloon moves downwards.

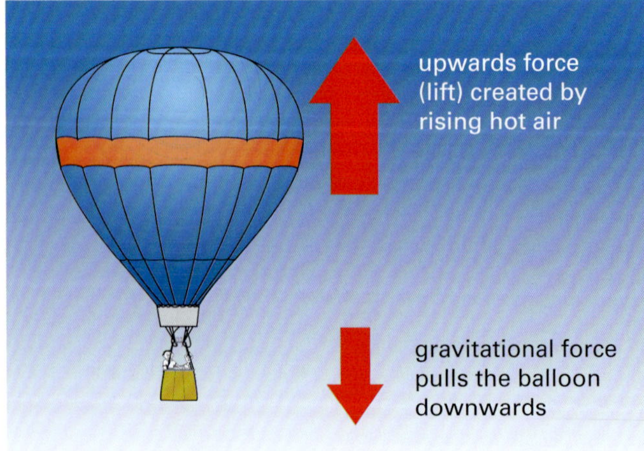

Figure 4.25 The balloon rises because the upwards force (the lift) is greater than the downwards force.

ISBN 978 1 4202 3822 8

Rockets

If you blow up a balloon and let it go, it moves off quickly. What makes it move?

When you let the balloon go, the air inside the balloon rushes out. You can feel the force of this air when you hold your hand over the hole. The force of the air rushing out of the balloon creates an equal and opposite force that pushes the balloon forwards. These twin forces are called **action and reaction**.

The action force in a rocket is supplied by its engine. This force is also called *thrust*.

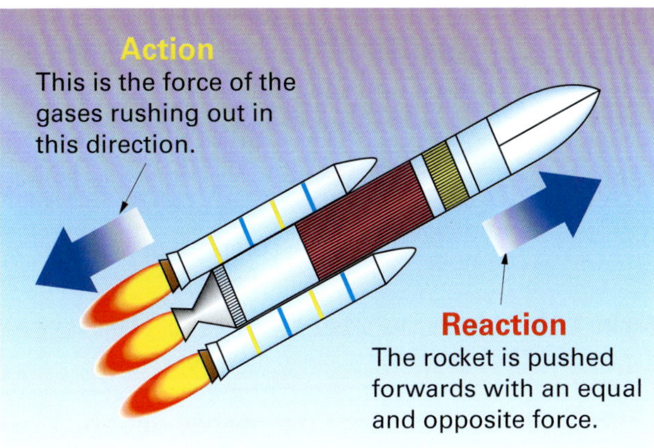

Figure 4.26 The force of the hot exhaust gases rushing out creates an equal and opposite force that pushes the rocket forwards.

Aeroplanes

An aeroplane has four basic parts—an engine that makes it move, a body, wings and a tail. Each of these parts helps in making the aeroplane fly.

A flying aeroplane has four forces acting on it—gravity, lift, thrust and drag.

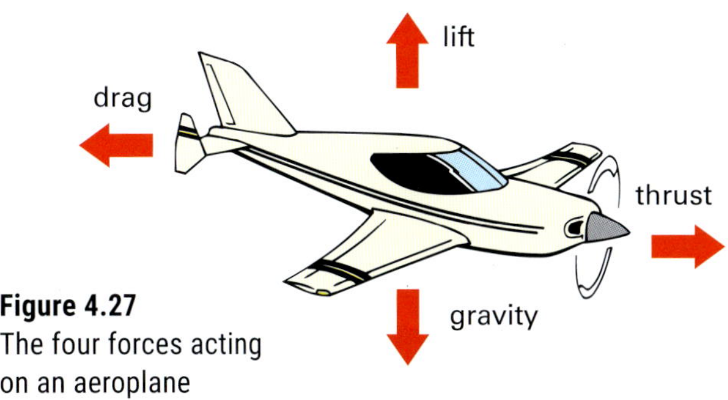

Figure 4.27 The four forces acting on an aeroplane

Gravity pulls the aircraft downwards, while *lift* is the force that pushes it upwards. *Thrust* is the force that pushes the aircraft forwards and is caused by the engine. *Drag* is the frictional force of the air on the aircraft, which slows the aircraft down. Drag acts in the opposite direction to the thrust.

How does the aeroplane create a lift force?

Explaining how a wing works is actually complicated. There are different things that can contribute to the lift. If you look at a cross-section

ACTIVITY

Part A

You will need a sausage-shaped balloon, a drinking straw, about 5 metres of fishing line and some adhesive tape for this activity.

1 Use the diagram below as a guide to make a model rocket from the balloon and straw.

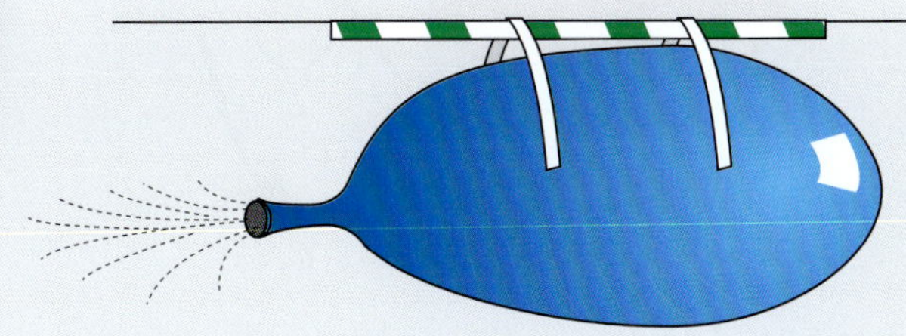

2 Experiment with your rocket so that it shoots down the fishing line in the fastest possible time.
Compare your rocket with those made by other students.

Part B—Inquiry

Design and build a balloon rocket that will travel in a straight line without the aid of the fishing line.

ISBN 978 1 4202 3822 8

of a wing, you will see that it is more curved on the top than it is on the bottom. As the aeroplane speeds up along the runway, the air rushes over the wings. Because the air passing over the top wing surface has to travel further, it moves faster than the air passing under the wing. This fast-moving air causes a lower air pressure on the top of the wing. The difference in air pressure produces the upwards force called lift.

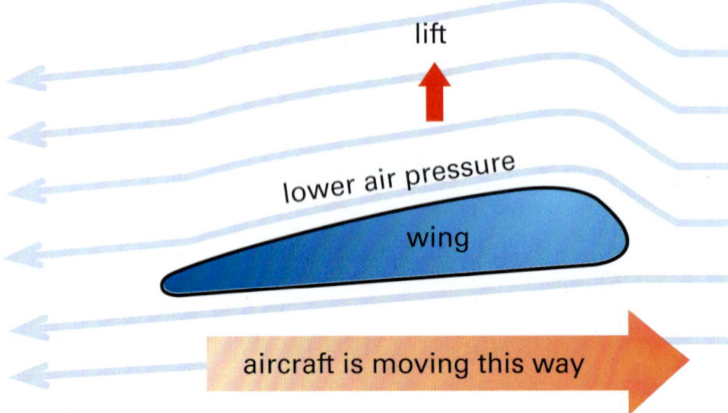

Figure 4.28 Air moves faster over the top of the wing, creating a pressure difference.

But the pressure difference alone is not enough to make an aeroplane fly. A wing must also deflect air downwards. This can be done by using a wing shape that pushes air down and by adjusting the angle of the wing, known as the angle of attack. Air deflects off the underside of the wing and is pushed down, as in Figure 4.29. This creates an opposite force in the upwards (and backwards) direction. The engines must push hard enough to overcome the backwards force produced, and the lift must be larger than gravity to allow the aeroplane to rise.

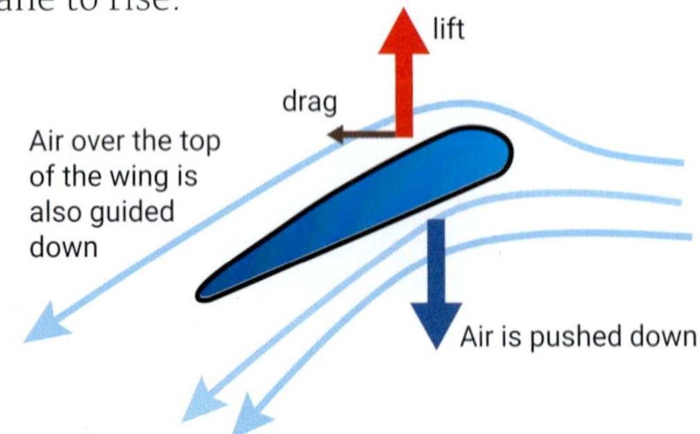

Figure 4.29 Air is pushed down by the wing, producing an opposite force of lift.

Thrust

The engines of an aeroplane create the thrust. The engine can be propeller-driven or a jet. Both work on the same basic principle—action and reaction. The jet engine drives the aeroplane forwards by pushing hot exhaust air backwards. Air is sucked in at the front of the engine. Fuel is ignited in the combustion chamber and the hot exhaust gases are blasted out of the rear of the engine. The force of the gases (action) pushes the aircraft forwards (reaction).

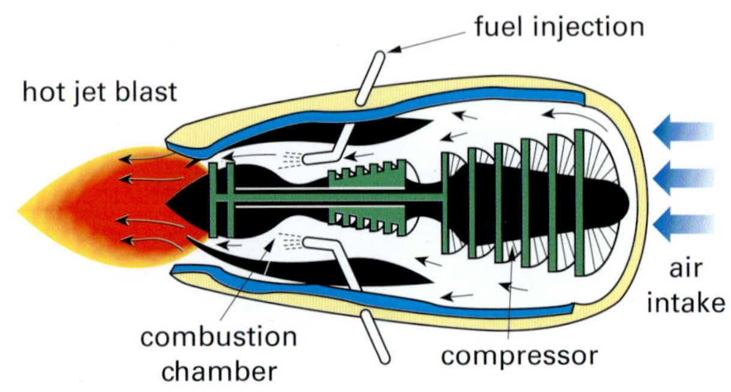

Figure 4.30 A jet engine blasts hot gases out of the back of the engine. This causes the plane to move forwards.

Propellers are shaped so that they push air backwards. The force of the air moving backwards (action) pushes the aircraft forwards (reaction). The faster the propellers turn, the greater the force of the air being pushed backwards, and the greater the force pushing the aircraft forwards.

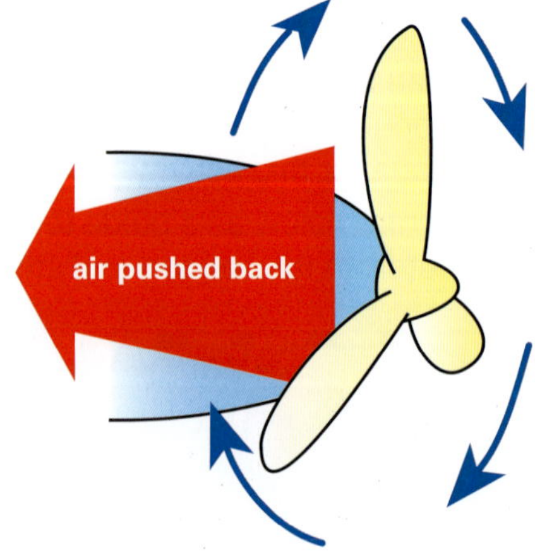

Figure 4.31 A propeller pushes air backwards.

ISBN 978 1 4202 3822 8

How an aeroplane controls flight

Figure 4.32 Controlling flight

INVESTIGATION 4.4 How an aeroplane flies

Aim

To use models to show how an aeroplane flies.

PART A A model wing

Materials

- piece of plain A4 paper
- wax taper and matches

Method

1 Hold the two corners on the short side of a piece of paper as shown in the diagram. The paper will curve like the top surface of a wing. Now blow on the paper and observe what happens.

2 Blow harder and observe any differences in the movement of the paper.

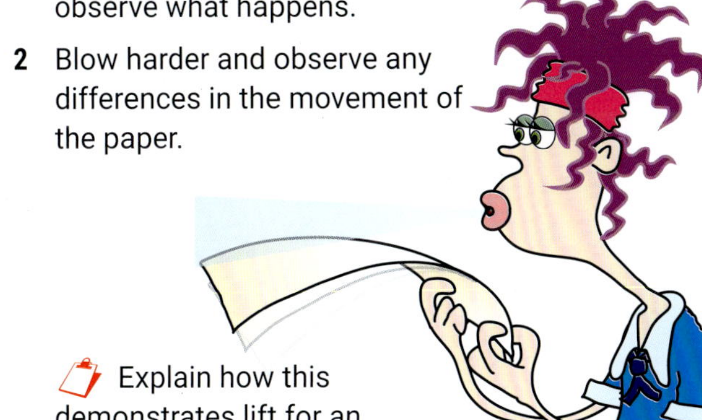

Explain how this demonstrates lift for an aircraft.

PART B The world's best paper glider

Challenge: Can you make a paper glider that is judged by your class as the world's best paper glider?

Use the websites below or search the internet to find designs for paper gliders, or design your own glider.

Work in a group to research, design and build your paper glider. As a class, you will need to agree on the rules used to judge the 'world's best paper glider'.

Write a report of your glider design and test flights.

EXPLORE ONLINE

Paper aeroplane designs
This site has many designs for paper gliders.

Making paper aeroplanes
This site has a number of free instructional sites where you can download designs and building instructions for many types of paper gliders.

ISBN 978 1 4202 3822 8

CHECK

1 The following sentences are false. Rewrite them to make them correct.
 a A hot-air balloon comes back to Earth when the force due to the hot air rising is greater than the gravitational force.
 b The force exerted by the gases coming out of a rocket is called the reaction force.
 c In an aeroplane, the lift force opposes the drag force.
 d To land a plane, a pilot would have to increase the thrust and decrease the lift.

2 A propeller and a jet use the same basic principle to make an aeroplane fly. Explain the similarities in terms that a younger child would understand. Draw diagrams to help your explanation.

3 Various parts of an aeroplane help to control its movements. Which parts help to:
 a make it airborne from a runway?
 b make it turn sideways?
 c raise or lower the nose of the plane?

4 Use action and reaction to explain the following.
 a A rotary sprinkler turns when the tap is turned on.

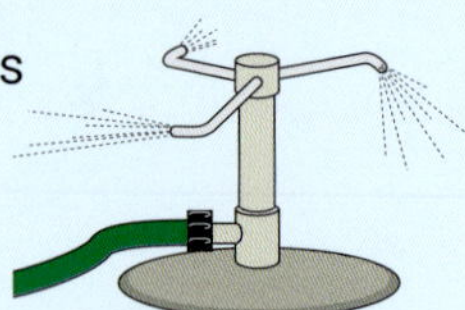

 b A space capsule moving from left to right turns in the direction shown when rocket A is fired.

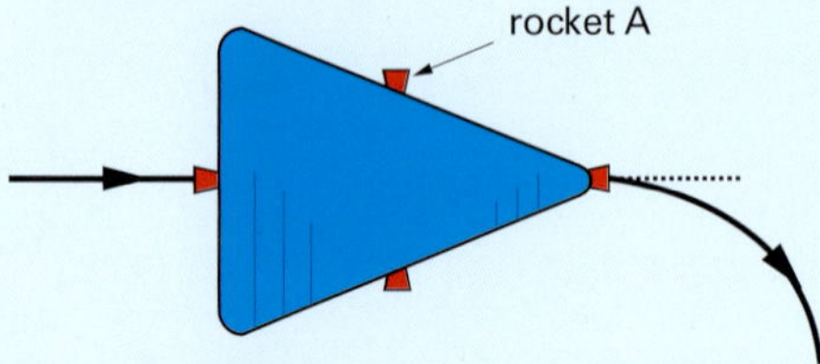

 c The cardboard and rollers shoot backwards when a wound-up toy car is placed on them and let go.

CHALLENGE

1 You are an aircraft designer. What factors—drag, lift, thrust and gravity—do you take into account when you want a plane to:
 a fly as fast as possible?
 b be able to take off quickly on a short runway?

 Use diagrams to help your answers.

2 Tony made a model wing out of foam and tested it with air blown from a hair dryer. Tony expected the wing to rise up the glass tubing, but it didn't. Use the diagram to the right to explain why the wing did not rise.
You could make a wing yourself and try to get it to rise.

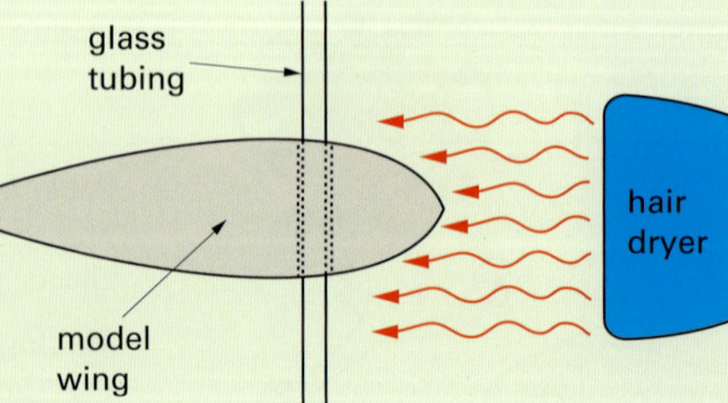

3 Jet engines have replaced propeller-driven engines on almost all commercial passenger-carrying aircraft. What do you think are the advantages of jet engines? Why do small aircraft still use propellers?

4 Rocket engines are usually used to carry spacecraft out of the Earth's atmosphere. Why can't jet engines be used? (You will need to know how rocket engines work.)

ISBN 978 1 4202 3822 8

SCIENCE AS A HUMAN ENDEAVOUR

STEM and robotics

STEM is a common term for **S**cience, **T**echnology, **E**ngineering and **M**athematics. STEM is about how scientists, technology experts e.g. engineers and mathematicians work together to solve problems and invent things.

STEM is widely used in robotics as it requires many scientists to work together to produce a robot. Robotics requires knowledge of scientific principles from many fields, including physics, biology and chemistry. It also requires engineers to be able to design the components of a robot, and technology experts to write computer code, all of which requires a deep knowledge of mathematics.

Robotics

Robots are complex machines. They are a system made up of individual machines, a computer and various other parts that work together. A robot may be a simple one that spot-welds parts of a car on a production line. Such a robot just does the same thing over and over and reduces errors that humans may make, while relieving humans of a repetitive, boring or unsafe task.

Other robots can be more complex, such as those used to perform surgery. Even more sophisticated robots may use artificial intelligence (AI) to solve problems. AI is progressing at a very fast rate, and robots are beginning to be able to

Figure 4.33 Most schools have robotics kits that will allow you to explore coding and to build your own robots.

Figure 4.34 Robots can be used on production lines such as for building cars.

learn new things on their own and respond to their environment.

InMoov

The cover of this book shows a robot called InMoov. Originally developed as an artistic project, InMoov is made from open-source technology and components, and many of its parts are able to be produced using a 3D printer. This means anyone is allowed to access the programming code and plans to build their own robot. Many schools and universities around the world now make this robot and use it to explore robotics further.

InMoov is able to speak, see, hear and move independently. It can detect its environment through micro-cameras in its eyes and recognise voice commands from its controllers. Its fingers have touch sensors, so it can detect when it touches something in order to pick it up. It also has a motion-sensing device that can allow the robot to sense its surroundings in 3D. The programming is done through MyRobotLab, which is available to anyone on any computer.

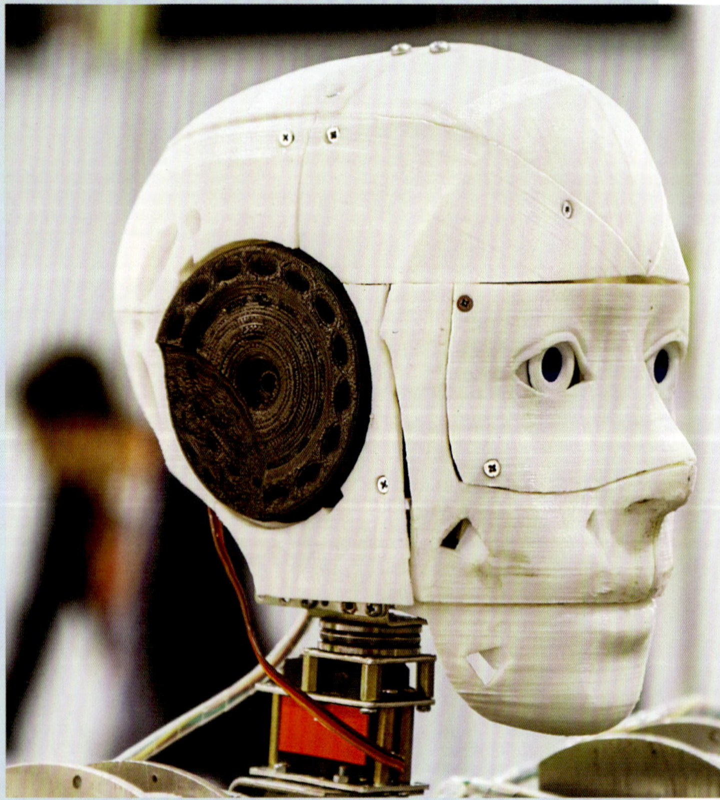

Figure 4.35 InMoov can be made very cheaply using parts that are easily available.

ISBN 978 1 4202 3822 8

3D printing

A 3D printer is another complex machine that was produced through STEM. These printers can be used to print objects using plastic or other materials, building up the object layer by layer until complete.

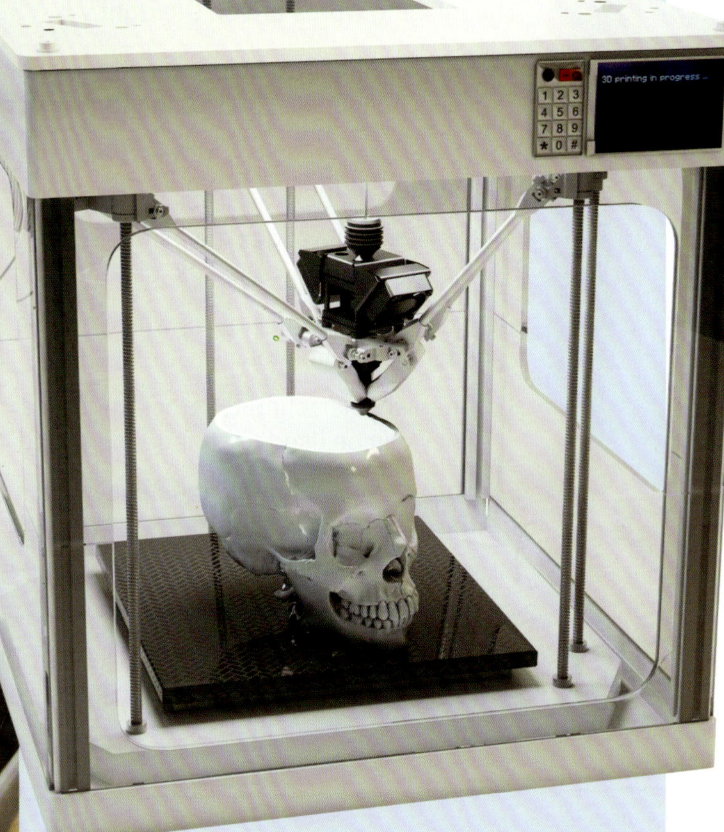

Figure 4.37 3D printers can create very complicated objects that may contain moving parts.

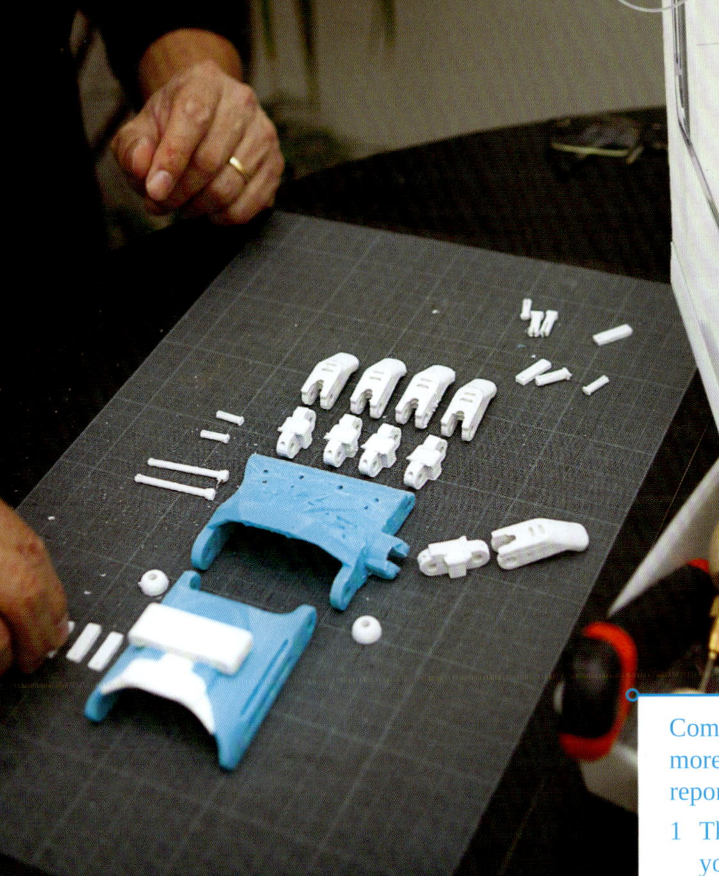

Figure 4.36 The e-NABLE project puts volunteers with a 3D printer in touch with handicapped children who need a hand prosthesis. The volunteers produce and assemble the parts and send the hand to the child.

EXPLORE ONLINE

Complete one of the following activities by researching more information about it, and produce an interactive report on your findings. Some links are provided for you:

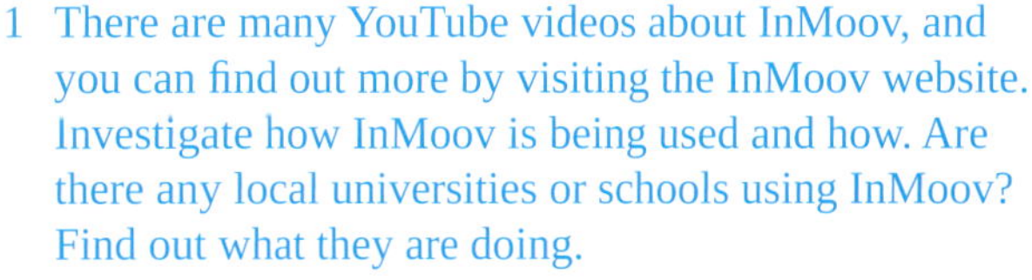

1 There are many YouTube videos about InMoov, and you can find out more by visiting the InMoov website. Investigate how InMoov is being used and how. Are there any local universities or schools using InMoov? Find out what they are doing.

2 Investigate a robot you are interested in. There are many different types that can do many different things. If you are unsure where to start, investigate who Elon Musk is and what he is trying to achieve with robots.

3 Investigate advances in AI, and find out what it is being used for and how it may be used in future.

3 Find out more about 3D printers, how they work, how they are already being used and how they may be used in the future.

ISBN 978 1 4202 3822 8

Copy and complete these statements to make a summary of this chapter. The missing words are on the right.

1 Simple machines help you in three ways. They can _____ the force you use, change the _____ of the force, or make things go _____.

2 The effort you use to move a load with a lever changes when you shift the position of the _____.

3 Screws and wedges are examples of _____. These simple machines reduce the effort but increase the distance the effort must move.

4 _____ are special wheels which can be arranged to magnify a force and to change its direction.

5 Gear wheels acting together transfer a _____ from one place to another. They can be used to change the speed of things.

6 The _____ of a simple machine is calculated by dividing the load moved by the effort used.

7 The up and down movement of hot air balloons is due to the _____ forces acting on the balloon.

8 The movement of rockets and jet aeroplanes is due to twin forces called _____ and _____. When gases are forced backwards, the aircraft moves forwards.

action
pivot
faster
inclined planes
unbalanced
force
magnify
reaction
pulleys
direction
mechanical advantage

CH•4 REVIEW

1 Name the simple machines marked A, B and C on the diagram.

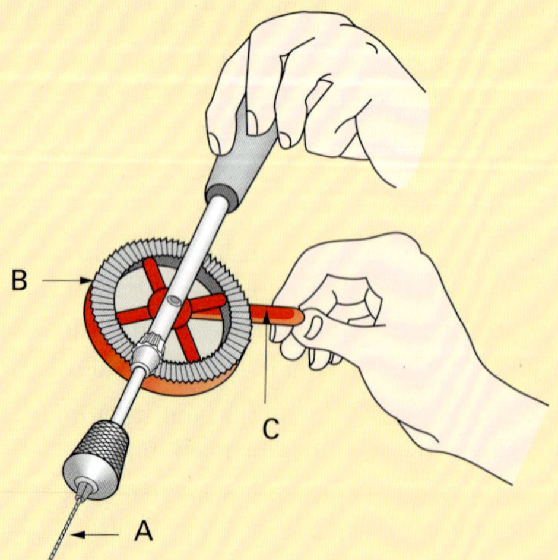

2 Which of these machines is an example of a lever? (There may be more than one answer.)
- **A** steering wheel
- **B** bicycle sprockets
- **C** scissors
- **D** ramp
- **E** bottle opener

3 Gina wants to get the lid off a paint tin. Which would make the best lever?
- **A** 10-cent coin
- **B** screwdriver
- **C** fingernail file
- **D** ice-cream stick

4 Describe the three ways in which simple machines help you. For each way give two examples.

ISBN 978 1 4202 3822 8

5 The photo shows a brake lever on a bicycle.

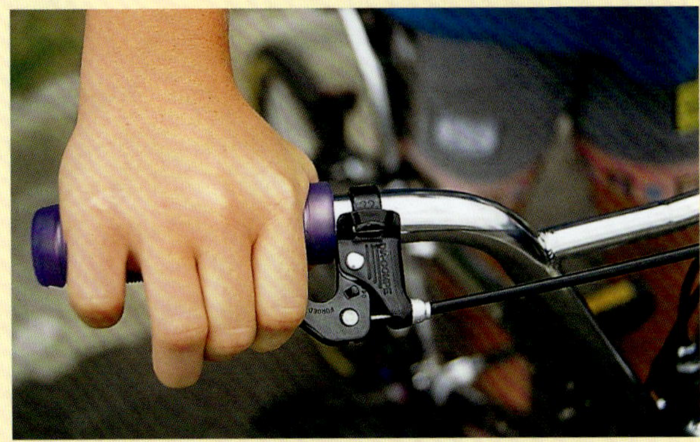

Which of the following statements is *false*?

A The effort is at the end of the lever.
B The pivot is at the end of the lever.
C This is the same type of lever as a wheelbarrow.
D This is the same type of lever as a crowbar.

6 Two gear wheels are connected as shown below.

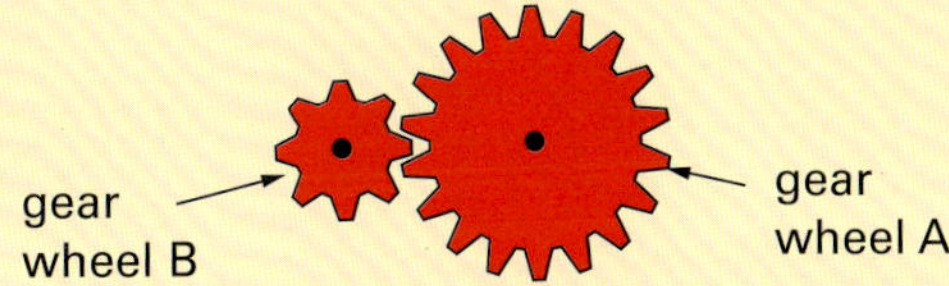

a What will you observe happen to gear wheel B if gear wheel A turns one complete revolution?
b Name a machine that uses this type of gear arrangement.

7 John steps off a boat to reach the jetty. What will probably happen to the boat? Why? How can John reach the jetty without falling in the water?

8 An aeroplane is travelling down a runway. What design features of the aeroplane help it to take off? Why is speed necessary on take-off?

9 Draw a diagram to show how you would use two pulleys to pull downwards on a rope to lift a 10 N object with an effort less than 10 N.

10 Liana wanted to measure the effort required to lift a 20 N load placed in various positions along a metre rule pivoted at one end.

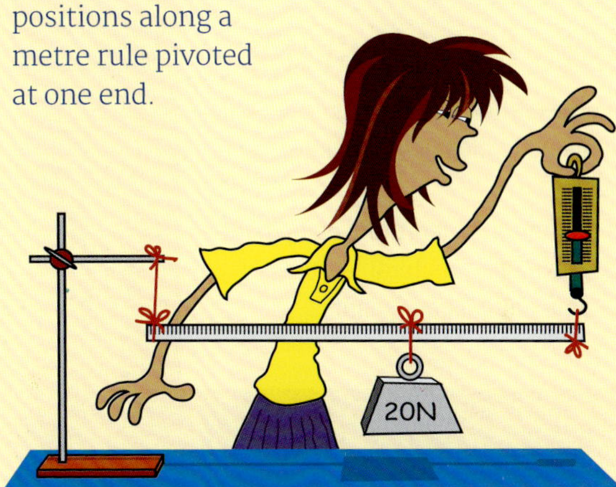

For each position, she measured the effort and the distance of the load from the pivot. Her results are shown in the table.

Position of load (cm)	Effort (N)
20	4
40	8
60	
80	16
100	20

a One of Liana's results is missing. What should it be?
b What type of simple machine uses this lever arrangement?
c At which position does the machine have a mechanical advantage of 2.5?
d Suppose the ruler was 150 cm long. What effort would be required if the load was placed at the 40 cm mark?

Check your answers on page 239.

ISBN 978 1 4202 3822 8

IN THIS CHAPTER

Science Understanding

- group a variety of organisms on the basis of similarities and differences in particular features
- use a key to classify organisms
- consider how biological classifications have changed over time
- recognise that some organisms consist of a single cell
- use scientific conventions to name species

Science Inquiry Skills

- use observations of various organisms to classify them into groups
- use appropriate laboratory equipment to observe the structural features of the same animals and plants

CH•5 Classifying living things

ISBN 978 1 4202 3822 8

GET STARTED: *EXPLORE*

Look at the pictures below. Create a card for each living thing by drawing a picture of it, or writing its name on a card. You should have 22 cards.

frog, snake, pine tree, magpie, spider, fern, worm, mosquito, dolphin, bat, horse, rose, starfish, moss, gum tree, fish, lizard, chicken, seaweed, sugar glider, platypus, shark

- Sort the organisms into three or four groups so that the organisms in each group have similar features.
- Which features did you use to group the organisms? Compare the way you grouped them with the way other people did.

ISBN 978 1 4202 3822 8

5.1 Classifying things

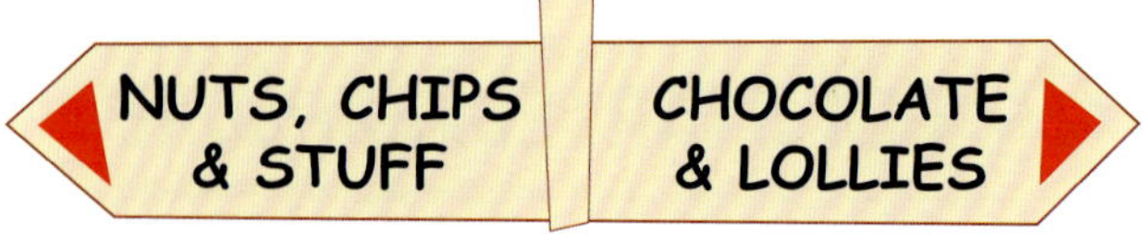

Suppose you wanted to buy some scorched almonds, corn chips and peanuts for a party. Fortunately, your local shopkeeper, Mr Smith, has organised these party foods into groups in his shop to make finding and selecting the goods a lot easier.

One group contains sweets that have chocolate in them—bars of plain chocolate, nut chocolate, caramel chocolate and many others. This group also contains chocolate-covered nuts and sultanas, as well as sweets that have chocolate centres, like Jaffas and Smarties.

Mr Smith uses certain *characteristics* or features of the confectionery to sort them into groups. This process is called **classification**. Each group contains items with similar features. The diagram below shows how Mr Smith classifies his party foods.

Classifying foods makes it easier to find goods in your local store or supermarket because you know that each group contains things with similar features.

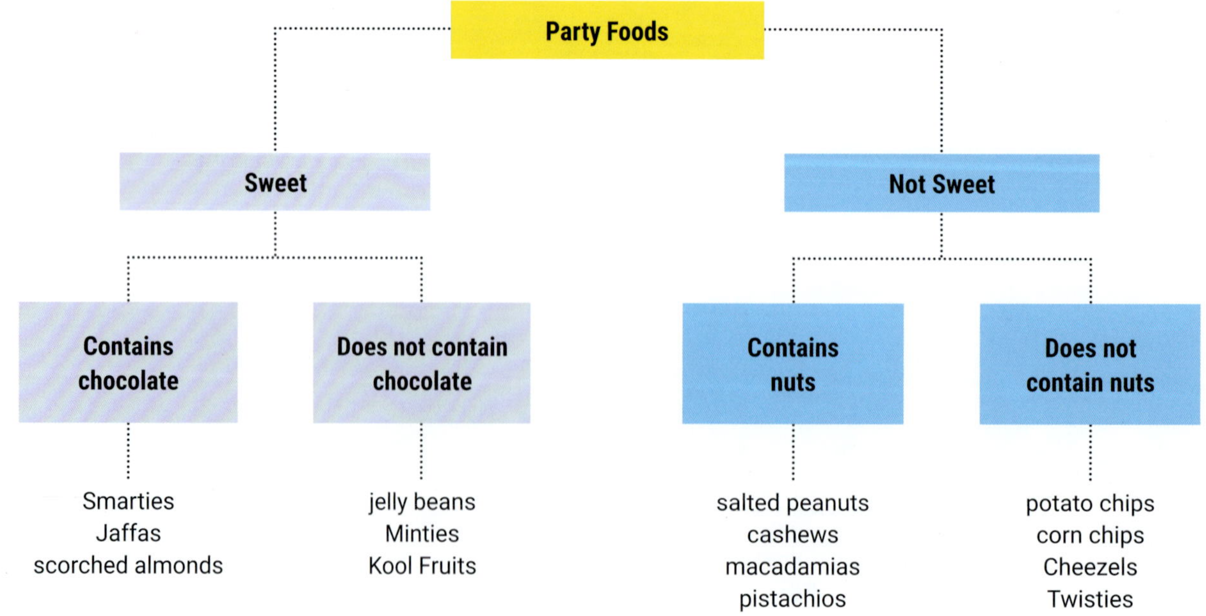

Figure 5.1 A key to classify snacks

ISBN 978 1 4202 3822 8

Like the items in Mr Smith's shop, the living things on this planet can be classified into groups. For example, the ancient Greeks used their observations to classify living things into two large groups—animals and plants. They further classified the animal group into three smaller groups, as shown below.

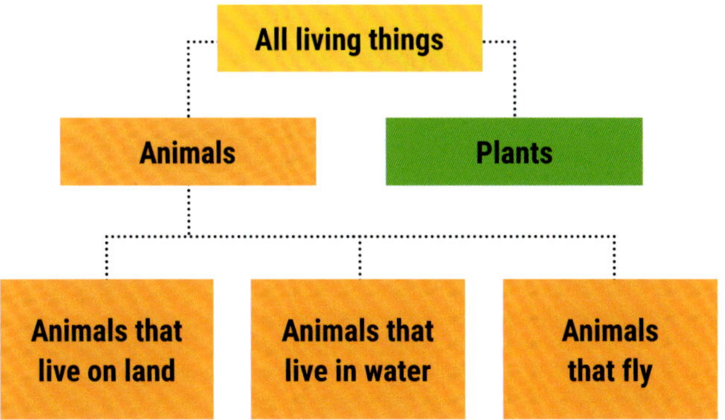

Figure 5.2 Ancient Greek key to classify animals and plants

Using keys

Objects can be classified using a **key**. Mr Smith used a key to classify his party foods. The Greeks used a key to classify living things. By using a key, you can easily classify objects or identify an unknown object, like the buttons in the key below. The best way to make a key is to have *two* alternatives for each characteristic. For example, in the key below, buttons are first classified into two groups—plastic and non-plastic. Each of those groups is then classified into two smaller groups, and so on.

ACTIVITY

1. Into which groups would you place the following food items using Mr Smith's method of classification?

   Freddo frogs, jelly babies, Kool Mints, caramel popcorn, Burger Rings, M&Ms, butterscotch, Maltesers, jelly snakes, Mars bar, rice crackers, Crunchy bar, beer nuts, Cherry Ripe, nougat, licorice

   Draw a key and add the foods to the appropriate group.
2. Use the ancient Greek method of classifying animals to place the animals in the list in **Get started** on page 99 into their appropriate group.

   Can you see any problems with this classification method? Explain with examples.
3. Work in a small group for this activity.

   Your teacher will give you 10 or 12 assorted buttons.

   Use the button key below or make up your own to classify the buttons so that each button is in a separate group.

   If you have made up your own key, draw it on a large piece of paper and present it to the class.

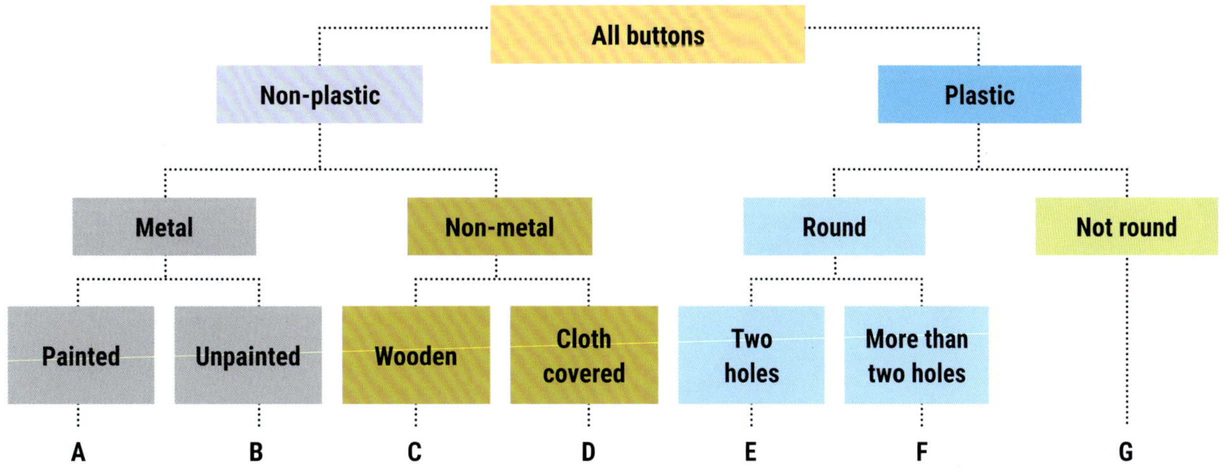

Figure 5.3 Button key

ISBN 978 1 4202 3822 8

Work in a group of three or four for this activity. Your task is to make a key which you can use to classify the people in your class into a number of different groups.

1 Look for characteristics for which the differences are clear-cut, permanent and likely to be agreed upon by others.

2 Make sure there are two alternatives for each characteristic. For example, male and female, or can roll tongue and cannot roll tongue.

Here are some other characteristics that you may find useful.

- earlobe attached/unattached
- folds arms left over right/right over left
- freckles on nose/no freckles
- second toe longer/shorter than big toe
- light-coloured hair/dark hair
- blue eyes/not blue eyes

3 Make a draft copy of the key.

Compare your key with those made by other groups.

4 Test the key by classifying the people in your class. Modify your key if necessary and test it again.

What makes things living?

In **Get started**, you devised a way to classify 22 living things. How do you know something is living?

Look at the rock-like things in the photo below. These things are actually alive. They belong in the same plant group as cactuses, and live in very dry areas of Australia and other countries. They are called *rock plants*.

There are seven characteristics used to tell whether a thing is living or not living.

Living things:

- are able to move
- need oxygen
- need food or nutrients
- produce and eliminate wastes
- grow
- respond to changes
- reproduce.

Biologists (scientists who study living things) know that rock plants are alive. They have all of the seven characteristics in the list above, even though some of the characteristics, such as their movement and their response to changes, are very hard to see!

Figure 5.4 Rock plants are actually living.

Figure 5.5 Euglena is a single-celled organism.

ISBN 978 1 4202 3822 8

Any living thing is called an **organism**, regardless of whether it is the size of a massive blue whale that weighs 170 tonnes, or a microscopic euglena (you-GLEEN-a) that weighs only a millionth of a gram.

The euglena in Figure 5.5 consists of one cell. A **cell** is the basic unit of organisms, and all organisms are made of one or more cells. The euglena is called a *unicellular* organism, and organisms made of many cells are called *multicellular* organisms.

A blue whale is made of billions of cells that have different shapes and functions. There are skin cells, liver cells, muscle cells, and bone cells.

Classifying living things

You may have found in the activities on page 101 that there are problems in classifying animals using the ancient Greek method. For example, animals that live in water include dolphins, starfish, fish, platypuses and frogs, but these five animals have little else in common. You could also use colour or size to classify these animals. However, there is so much variation in colour and size that this method would also prove unsatisfactory.

Over the last 200 years or so, biologists throughout the world have developed a better method of classifying living things. What characteristics do a dolphin, platypus, fish and frog have in common? One of these is the presence of bones, including a backbone. Animals that have backbones are called **vertebrates** (VER-te-brates). This is similar to the word *vertebra*, which is one of the bones in the backbone. Animals without backbones are commonly called **invertebrates**.

The presence of a backbone is part of an animal's *structure*. The use of structural characteristics is one way in which biologists classify organisms. The number of legs, the presence or absence of lungs or gills, feathers and a scaly skin are all structural characteristics.

The way an organism *functions* is also used to classify living things. For example, mammals and birds have a fairly constant body temperature, while all other animals have a body temperature that changes with the outside temperature. Body temperature is a functional characteristic.

The key on the next page can be used to classify animals.

Figure 5.6 Vertebrates have a backbone.

Figure 5.7 A dolphin is a mammal and has a constant body temperature.

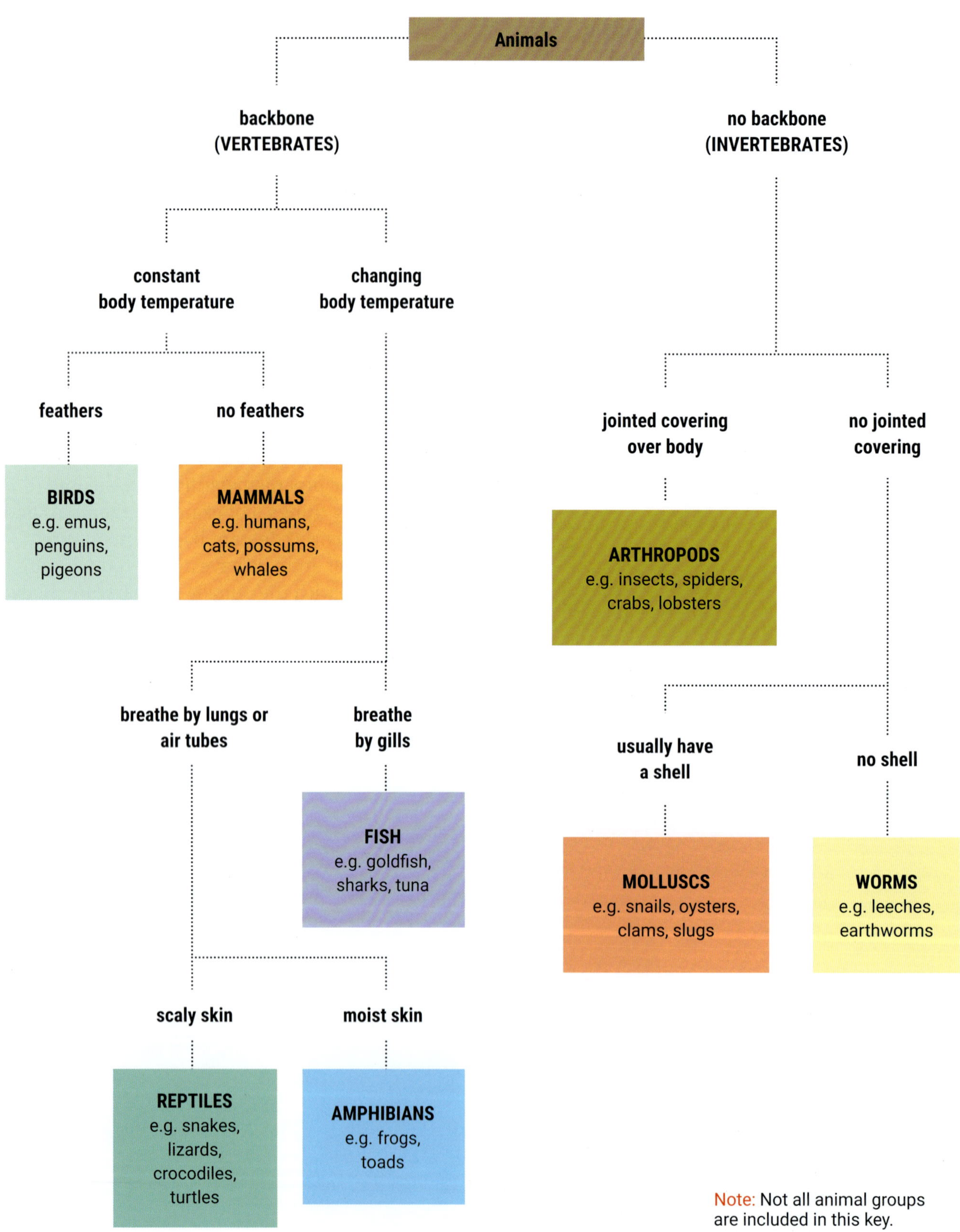

Figure 5.8 Classifying animals

ISBN 978 1 4202 3822 8

INVESTIGATION 5.1 Animal keys

Aim

To classify animals using a key.

Materials (per class)

- at least 20 different live or preserved animals, each with a number
- hand lens (optional)

Risk assessment and planning

- Work in pairs and read through the Method. Then design a data table for at least eight animals that you have to classify.
- Many of your animals will not be alive, so you will have to research some of the functional characteristics of these animals or rely on your general knowledge of them before you can classify them fully.

Method

1 You have to classify at least eight animals. Choose one animal and work through the animal key on the previous page. Discuss the animal's characteristics with your partner and then classify it. Record the name or number of the animal and the group in which you have classified it.

2 Use the key to describe the animal. For example, SPECIMEN 2—REPTILE (lizard) has a backbone, changing body temperature, breathes by lungs and has a scaly skin.

3 Repeat steps 1 and 2 for each of the other animals. Record all your observations and descriptions. Be prepared to discuss your results with other members of the class.

CHECK

1 Each group below contains one item which has different characteristics from the other three. Choose the odd one out and give a reason for your choice.
   a iron, steel, copper, plastic
   b shirt, tablecloth, socks, skirt
   c pencil, felt pen, rubber, crayon
   d surfboard, skateboard, bicycle, rollerblades

2 Copy and complete the following sentences.
   a The process of sorting things into groups with similar characteristics is called _____.
   b Animals with backbones are called _____.
   c Animals are classified using _____ and _____ characteristics.
   d Living things are called _____ .
   e Living things can be classified using a diagram called a _____ .
   f There are _____ characteristics used to tell whether a thing is living or non-living.

3 Use the button key on page 101 to answer the following.
   a Describe all the buttons in group E.
   b Into which group would you place a painted metal button?
   c Into which group(s) would you place these two buttons? How would you change the key to classify them?

   wood

   d Describe the differences between the buttons in groups A and D. In which ways are they similar?

4 Classify the objects in each of the lists below into two groups, and write down the characteristics you used to classify them.
   a apple, pear, capsicum, banana, tomato
   b brown snake, sea snake, turtle, tree snake, lizard, python, goanna
   c surfboard, sailboard, canoe, skateboard, dinghy, surf ski, catamaran

ISBN 978 1 4202 3822 8

5 The list of characteristics below could be used to classify organisms.
- feeds its young with milk
- has two large eyes on the front of its head
- changes body colour and pattern with different backgrounds
- has two large canine teeth in each jaw
- squirts out black ink when disturbed
- hibernates during very cold weather

For each characteristic, decide whether it is structural or functional and make two lists. For those characteristics that you are uncertain about, list them under the heading 'uncertain'. Discuss your decisions with your partners.

6 Why do biologists use body structure and function instead of size, colour or behaviour when classifying animals?

7 List the seven characteristics used to decide whether something is living or non-living. (The order is not important.)

8 Use the key on page 104 to describe the characteristics of each of the animals in the diagram.

9 Use the animal key to name the group to which each of the following animals belong.
- a This animal has no backbone and has a soft body with a shell.
- b This animal has a backbone, a changing body temperature and gills.
- c This animal has a hard, jointed covering over its body and no backbone.
- d This animal is a vertebrate with a constant body temperature and feathers.

10 Write a sentence using the word 'multicellular' so that a reader will know what the word means. Give examples of multicellular organisms.

11 In which ways are birds and mammals similar? In which ways are they different?

12 How can you tell a reptile from an amphibian, and a fish from an amphibian?

1 Look at the arthropods below.
- a Use the animal key on page 104 to describe the features of arthropods.
- b Design a key that could be used to classify the arthropods below. Did you put more than one in the same group? Why?

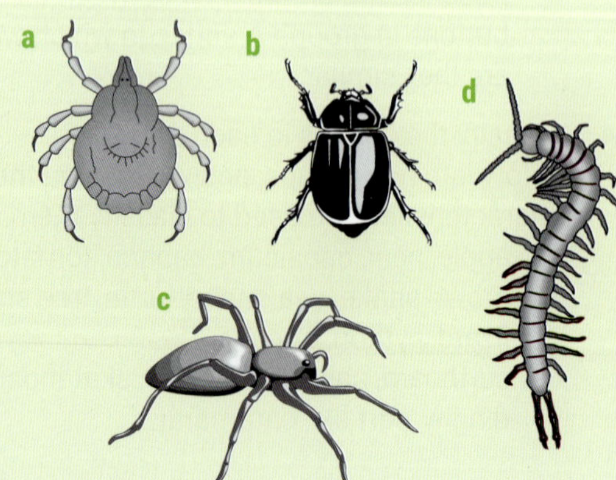

2 A fruit bat and a parrot are about the same size, they both have wings and fly, and both eat the same sorts of foods. Suggest why biologists classify them in different groups.

3 Not all animal groups are shown in the key on page 104. For example, the groups to which starfish and jellyfish belong are not shown. Use the library or the internet to find out the names of these two groups and the characteristics of the animals in these groups.

ISBN 978 1 4202 3822 8

5.2 The five kingdoms

Until the beginning of last century, biologists classified all living things into two groups—animals and plants. These large groups are called **kingdoms**.

When bacteria (microscopic organisms) were first observed and identified, biologists did not know which kingdom to put them in because they had features that were quite different from those of microscopic plants and animals. Some biologists began using a three-kingdom system of classification. Bacteria were grouped with microscopic plants and animals, and these were placed in the third kingdom.

However, with the invention of very powerful microscopes and new scientific techniques, other important differences between organisms in these three kingdoms were identified. It became obvious that the three-kingdom system was not a satisfactory method of classification. Most biologists throughout the world now recognise *five* kingdoms.

Fungi were originally placed in the plant kingdom. But fungi cannot make their own food like plants. Because of this important difference fungi were placed in a kingdom of their own.

The five kingdoms

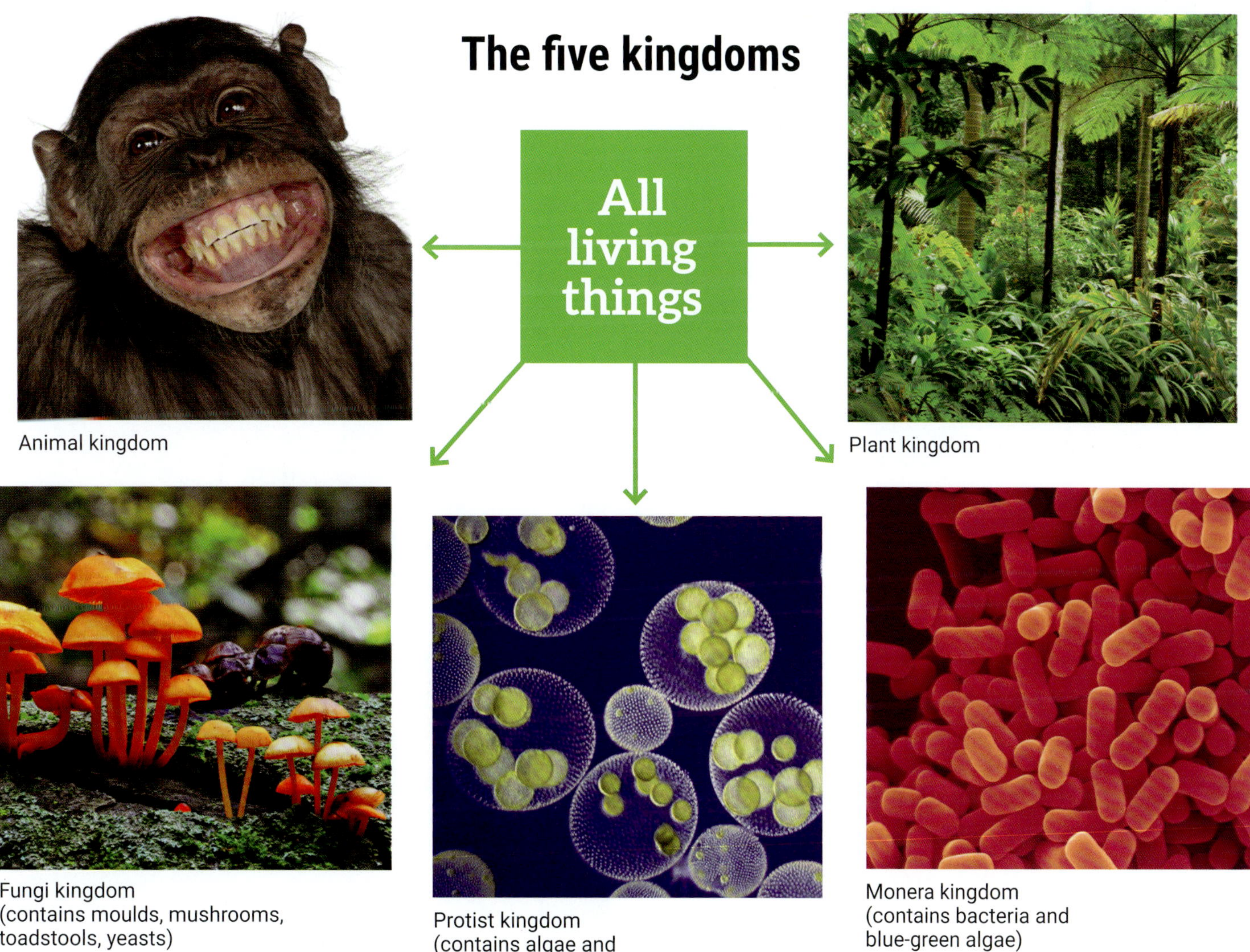

Animal kingdom

Plant kingdom

Fungi kingdom (contains moulds, mushrooms, toadstools, yeasts)

Protist kingdom (contains algae and microscopic organisms)

Monera kingdom (contains bacteria and blue-green algae)

ISBN 978 1 4202 3822 8

Animals

The organisms in the animal kingdom eat other organisms to obtain energy and materials for growth and movement. There are many different types of animals, but they are all multicellular organisms. Some live on land, others live in the sea or in fresh water, and others can fly.

All large land animals are vertebrates. The system of bones in these animals gives support and allows them to live on land successfully. The largest vertebrate that has ever lived on Earth is thought to be the blue whale. It can measure up to 35 metres in length and weigh 170 tonnes! The water of the ocean helps support its huge weight.

Figure 5.9 A blue whale

Plants

These multicellular organisms contain the green pigment **chlorophyll** (KLOR-a-fill). This substance is able to absorb the energy from sunlight. The plants use this energy to make food, in the form of sugars, from carbon dioxide and water, and give off oxygen. This process is called **photosynthesis** (foe-toe-SIN-thu-sis). The word is made up from the words *photo*, meaning 'light', and *synthesis*, meaning to 'make'.

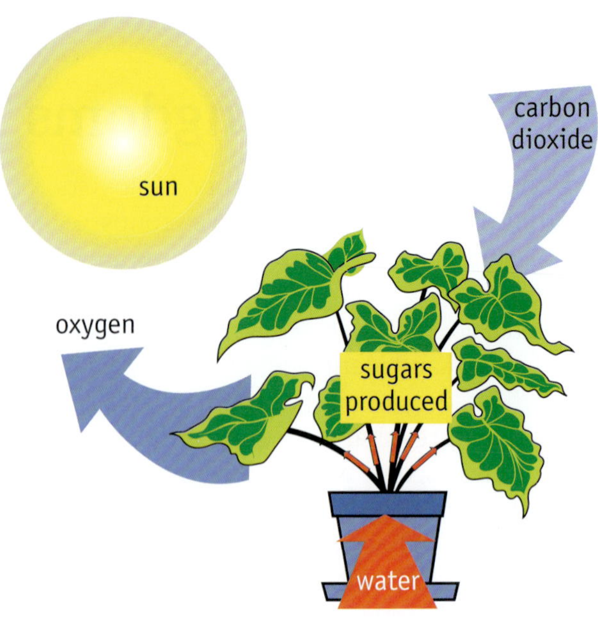

Figure 5.10 Photosynthesis occurs when sunlight is absorbed by the green chlorophyll in plants.

Plants cover much of the surface of the Earth. They vary in size from very small mosses a few millimetres wide to the largest living thing—the mountain ash of southern Australia, which grows to over 100 metres in height. The plant kingdom also contains the *oldest* living organism—King's Lomatia, which is found in the rainforests of Tasmania and is thought to be 43 000 years old.

The plant kingdom is divided into four groups: mosses, ferns, conifers and flowering plants.

Figure 5.11 King's Lomatia

ISBN 978 1 4202 3822 8

Fungi

The organisms in this kingdom include mushrooms, toadstools, bread mould and yeasts. They are similar to plants in that they are generally fixed to the ground and do not move around.

Fungi do not contain chlorophyll, so they cannot make their own food. Therefore, they have to obtain nutrients from other sources. They do this by growing on things they can use as a source of nutrients, such as dead plants or animals. Chemicals released from fungi break down the remains of the plant or animal into simpler substances that can easily be absorbed by the fungi.

Fungi reproduce by **spores**. These are made in caps or bulbs that stick up from the rest of the fungus. For example, in a mushroom, the dark-coloured gills under the cap are the organs that make spores. The rest of the mushroom grows on or under the ground (see Figure 5.12 below).

Spores are tiny cells with a hard coat around them to stop them from drying out. They are very light and are easily carried on the wind. A single mushroom can produce up to 2000 million spores!

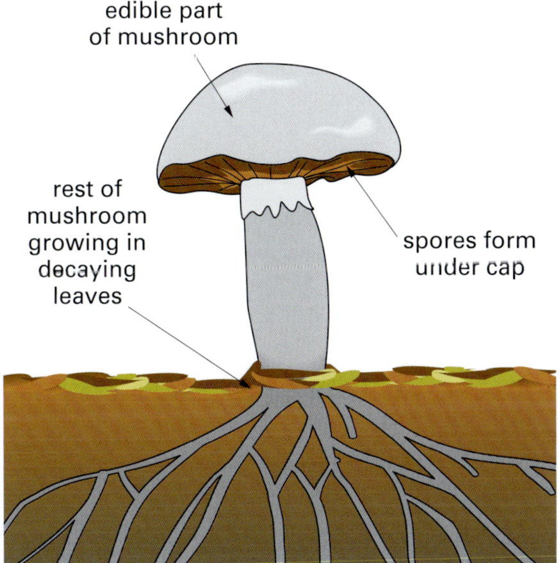

Figure 5.12 The edible part of a mushroom is where the spores are made. The rest of the fungus grows in the material it breaks down for food.

Helpful and harmful fungi

Fungi are very important organisms because many of them break down or decompose dead organisms. These fungi are called *decomposers*.

Fungi such as mushrooms can be eaten, and yeasts are used for making bread, beer and wines. Other fungi are used to make medicines such as antibiotics.

Some fungi grow on living things and are called *parasites*. They obtain all their nutrients for growth from the organism they grow on. For example, ringworm is a fungus that grows on human skin. It takes its food from the cells in the skin and makes the skin itchy, inflamed and sore. Powdery mildew is a fungus that grows on leaves, and it may eventually kill the plant.

Figure 5.13 The fungi growing on this oranqe will decompose it until very little remains.

Figure 5.14 Powdery mildew is a parasitic fungus that grows on the leaves of some plants.

ISBN 978 1 4202 3822 8

For these activities you will need a large, flat field mushroom, some bread mould and a hand lens or microscope.

1 To grow bread mould, moisten some stale bread and leave it in an open container for a day. Then cover the container and leave it in a warm place for a few days.

Place a small piece of bread mould on a slide. Use a hand lens or microscope to observe the thread-like filaments of the mould and the round spore cases.

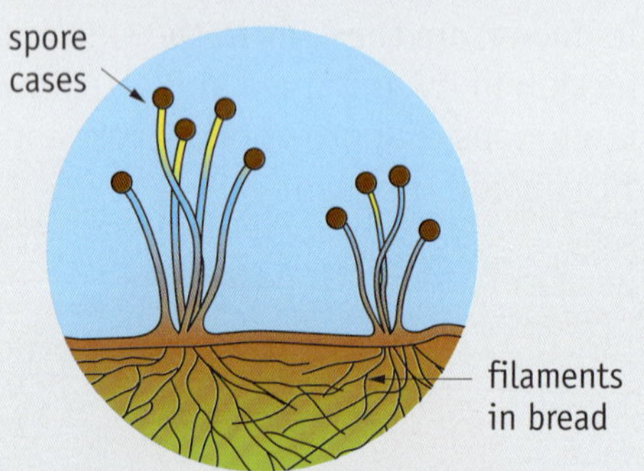

2 Observe the dark gills on the underside of the mushroom cap. To collect the spores, tap the cap over a piece of white paper. You may need a hand lens or microscope to observe the spores.

Fun facts about fungi

Interesting and easy-to-read site with information on types of fungi, examples and photos.

Australian fungi

A very informative site that covers types of Australian fungi; uses of fungi, including Aboriginal uses, and information about interesting examples of fungi.

Protists

The **Protist** kingdom includes organisms that have a very simple structure. Most of them are unicellular and most live in water—either fresh water or sea water. Algae (singular: *alga*) are included in this group. Many types of algae are unicellular, but some, like the seaweeds you see at the beach, are multicellular.

Like plants, algae contain chlorophyll and can photosynthesise. However, algae are classified as protists because they have a much simpler structure than plants—they have no roots, stems or leaves.

Figure 5.15 Algae are different from plants because they do not have stems, roots or leaves.

ISBN 978 1 4202 3822 8

Monerans

The organisms in the kingdom Monera (MON-er-a) have the simplest cell structure of all living things. They are all microscopic, unicellular and have a very simple cell structure. They include bacteria and blue-green algae. Organisms in this kingdom are called **monerans**. (Note: Blue-green algae are different from green algae, which belong to the Protist kingdom.)

Bacteria are very important because many of them break down dead animals and plants. Some bacteria cause diseases in animals and plants; for example, tetanus and tuberculosis. Other bacteria are used to make cheese, yoghurt and antibiotics. Many of the differences between the organisms in the five kingdoms can be seen in the structure of their cells.

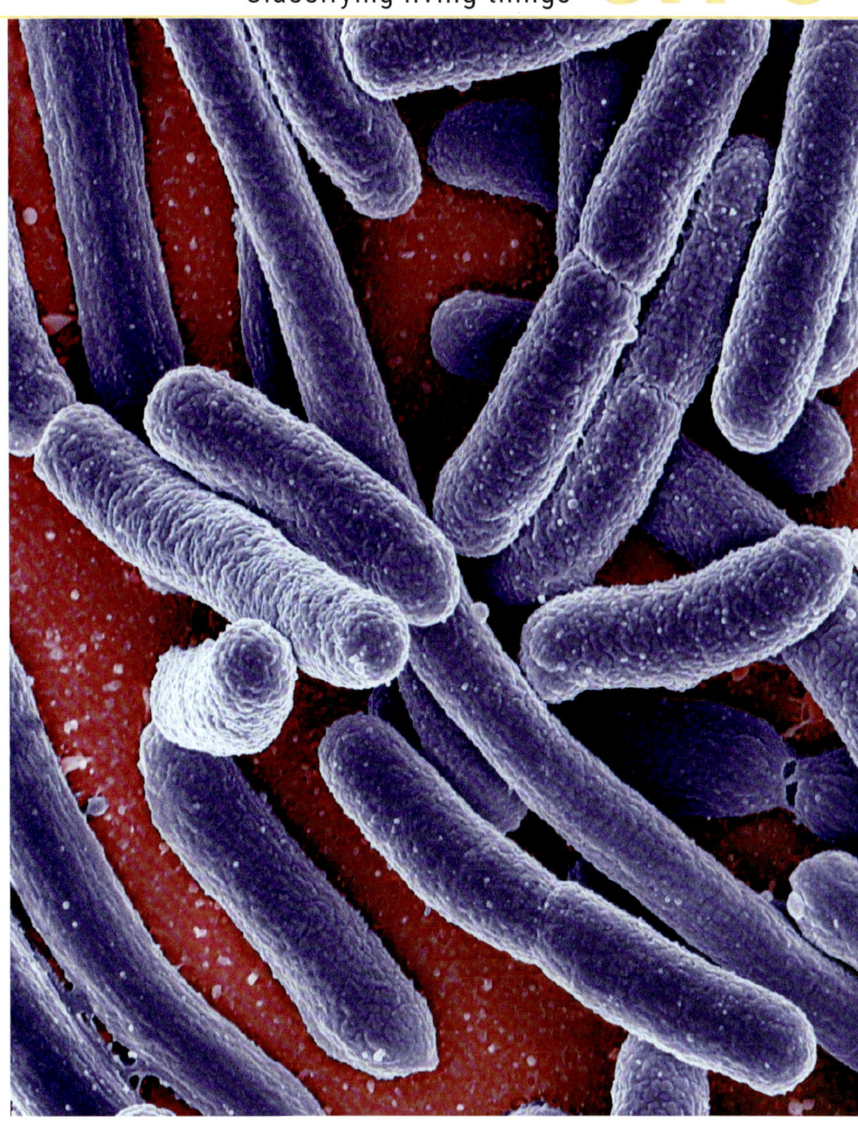

Figure 5.16 These rod-shaped bacteria, magnified 20 000 times, are found in the wastes of animals including humans.

ACTIVITY

Types of cells

All cells are held together by a structure called a cell membrane. Plant cells have a firm cell wall made from *cellulose* around the cell membrane, whereas animal cells have no cell wall. Fungi also have cellulose cell walls, but the cells do not contain chlorophyll as plant cells do.

The table below shows the characteristics of the cells in each of the five kingdoms.

Use the information in the table to design a key that can be used to classify the cells of organisms from the five kingdoms.

You observe a unicellular organism that has chlorophyll and no cellulose in its cell wall. What problems would you have in classifying it?

Animals	Plants	Fungi	Protists	Monerans
Multicellular	Multicellular	Multicellular	Mostly unicellular	Unicellular
No cell wall	Cellulose cell wall	Cellulose cell wall	Some have a cellulose cell wall	No cellulose in cell wall
No chlorophyll	Have chlorophyll	No chlorophyll	Some have chlorophyll	Some have chlorophyll

ISBN 978 1 4202 3822 8

Viruses—are they alive?

Viruses are extremely small (much smaller than bacteria) and are not made of cells. Viruses have features of both living and non-living things. For example, they can form crystals like non-living matter, but they reproduce like other living things. Viruses are completely parasitic because they rely on another organism (called the host) for all their requirements. They can reproduce only inside another organism, where they invade the organism's cells and use the cell's materials to make new viruses. In this process, some of the cells are destroyed, making the organism sick or causing its death. Human diseases caused by viruses include influenza, mumps and AIDS (HIV).

INVESTIGATION 5.2 Classifying organisms

Aim

To classify various organisms into kingdoms.

Materials (per class)

- about 20 stations around the laboratory, each containing a numbered specimen or photo
- hand lenses or stereomicroscopes

Risk assessment and planning

- Do not remove from its container any specimen that has been preserved in formalin. Formalin is harmful to the skin and has harmful vapours.
- Read through the Method and draw up a data table for your results.
- To observe some of the specimens, you will need a hand lens or stereomicroscope.

Method

1 There will be a living or preserved organism, or a photo of one at each station. For each organism, record its number and observations about its structure, size, colour and any other features that may help you classify it.

2 Observe at least 10 organisms. Then work in a group to classify the organisms into kingdoms, using the information on pages 104–111.

Discussion

1 Your group may be asked to present your results for two or three selected specimens to the class. For each specimen, give its number, its kingdom and the reasons why you placed it in this group.

2 Make a list of the kingdoms represented in this investigation and the special characteristics of the organisms in each.

ISBN 978 1 4202 3822 8

SKILLBUILDER

Becoming a trained observer

1 Look at the sketch on the right.
- What does it look like to you?
- What do other people see in it?
- Why do you think different people see different things?

Did you see two different animals—one facing the right and one facing the left? Different people will make different observations, depending on their past experiences.

2 A medical student watches two radiologists studying a chest X-ray and discussing, in technical terms, what is wrong with the patient's lungs. The student is puzzled because all he can see in the X-ray are the shadows of the heart and ribs and a few spidery blotches between them. However, as the weeks go by and he looks at many different chest X-rays he realises that he has to forget about the ribs and try to see the lungs behind them. Once he does this, he sees so much more than he could before.
- Why can the student now make better observations of chest X-rays than he could before?

3 Look at the photo on the right. The red material on the rocks is a living organism called a lichen.
- Into which of the five kingdoms described in this chapter would you classify it? Why?

Without a good knowledge of biology your answer is probably not much more than a guess. For hundreds of years, biologists thought lichens were single organisms. However, they now know that a lichen is two different organisms living together. Most of the lichen is a fungus, but it also contains unicellular algae.
- With this knowledge, how would you now classify the lichen? Explain your answer.

ISBN 978 1 4202 3822 8

CHECK

1 Explain what each of the following words means by writing a sentence to show its meaning. Then check your explanation with the one in the text or in the glossary.

photosynthesis	vertebrate
decomposer	parasite
kingdom	spores

2 Copy and complete the following sentences.
   a The green substance _____ absorbs the energy of sunlight and uses it in the process of _____.
   b Bacteria are classified as _____ because they have a very simple _____ structure.
   c Fungi do not contain _____ therefore they rely on other organisms for _____.
   d The kingdom Monera contains _____ and _____.
   e Seaweeds are a type of _____ that belong in the _____ kingdom.
   f Spores are very tiny _____.

3 Use the list below to match each organism to its description.

animals	fungi	algae
bacteria	protists	plants

   a The organisms which belong to this kingdom are mostly unicellular.
   b These organisms are multicellular and contain chlorophyll.
   c These organisms are plant-like, but do not contain chlorophyll.
   d These organisms are very small and have a very simple structure.
   e The organisms in this kingdom are multicellular and eat other organisms for food.
   f These organisms contain chlorophyll, but do not have the structures common to plants.

4 Fungi are often called decomposers.
   a Why is this? What other organisms could also be decomposers?
   b Is there a difference between a decomposer and a parasite? Explain your answer.

5 The photo below shows an organism growing on a dead tree. To which kingdom do you think this organism belongs? Give reasons for your answer.

CHALLENGE

1 Below are two organisms; one is classified as a protist and the other as an animal.
   a Both organisms are microscopic, but one is five times larger than the other. Which one is smaller? Suggest why.

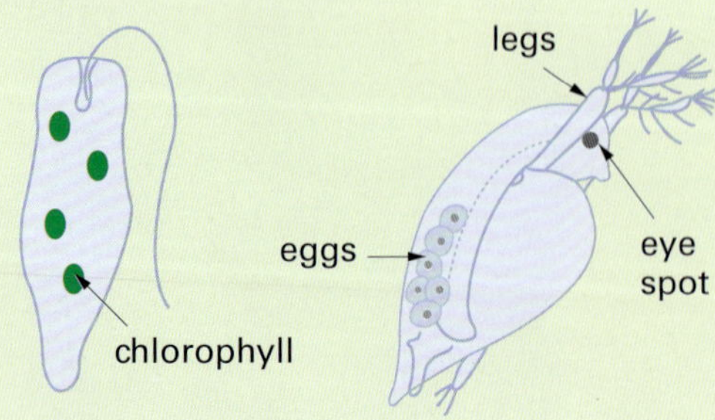

   b Suggest why they are placed in two separate kingdoms.
   c Suggest how the protist is able to move.

2 In a science fiction story, organisms called blobs have the characteristics of protist organisms but are as large as a car or a house. Suggest why protist-like organisms could not be this size in real life.

 ISBN 978 1 4202 3822 8

5.3 Animals and plants

Classification system

The kingdoms are the top level of classification, after which organisms are then grouped into sub-categories based on how similar they are. The levels of classification are as follows:

- kingdoms are divided into **phyla** (singular = **phylum**)
- phyla are divided into **classes**
- classes are divided into **orders**
- orders are divided into **families**
- families are divided into **genera** (singular = **genus**)
- genera are made up of different **species** that are similar to each other; the genus level is as similar as organisms can get without being exactly the same species
- species; a species has unique features and is different from all other species.

As you move down through the groups, the organisms in them become more similar to each other. Also, there are fewer organisms in each group as they get divided up into different groups.

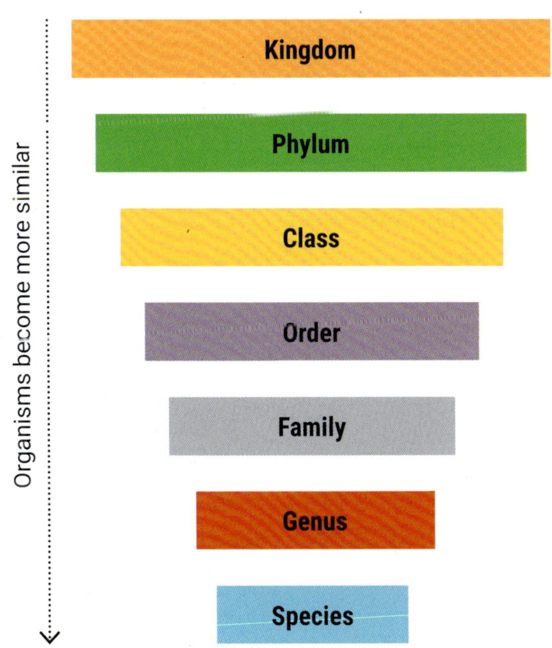

Figure 5.17 Classification from kingdom to species

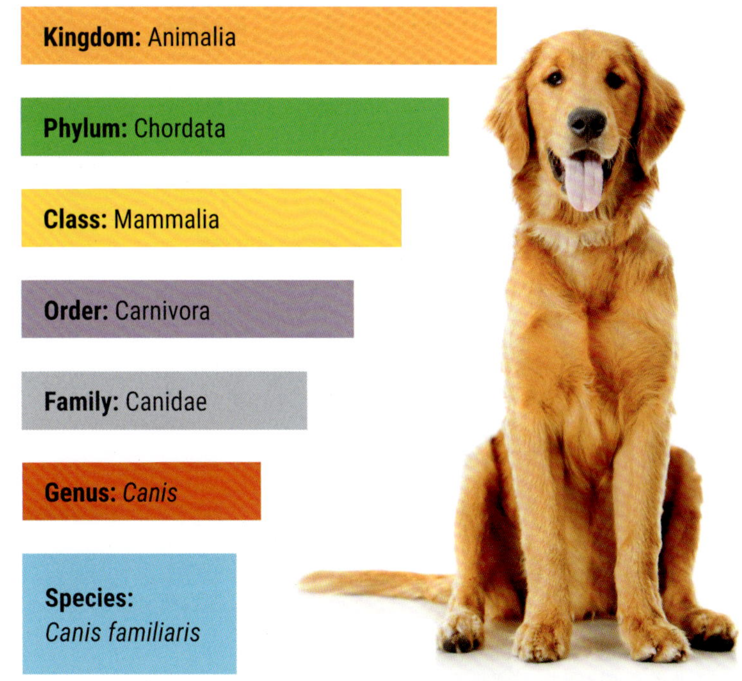

Figure 5.18 Classification of domestic dogs

Naming species

Many plants and animals have a common name, but in order for scientists to communicate about organisms without confusion, special scientific naming is used for each species. Species are named using a **binomial** (two-part) naming system. For example, the domestic dog is named *Canis familiaris*.

The first part of the name tells you which genus the organism belongs to and starts with a capital letter. The second part starts with a lower case and is a name that is unique to each species.

If you look at Figure 5.19 you can see that *Canis latrans* and *Canis lupus* are very similar in appearance as they belong to the same genus, *Canis*. You will also notice that those in the genus *Canis* are very different from those in the other genus *Vulpes*.

ISBN 978 1 4202 3822 8

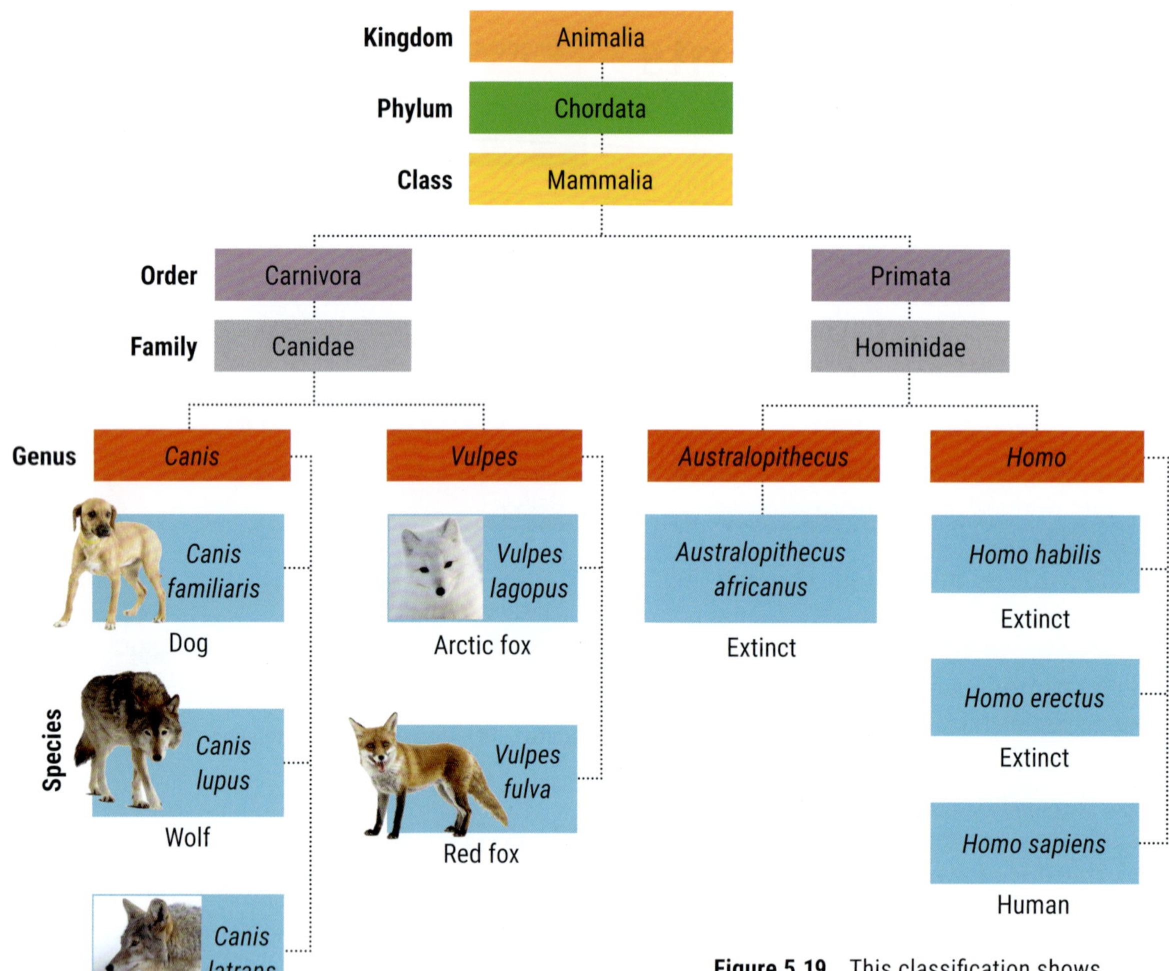

Figure 5.19 This classification shows examples of some species.

The animal kingdom

The key on page 104 shows you how animals can be divided into two large groups—the vertebrates and the invertebrates. Of these, the invertebrates contain many more types of animals than the vertebrates. There are about 950 000 different types of animals on Earth. Of these, about 800 000 are arthropods!

Invertebrate classification

There are seven phyla of invertebrates:

- kingdom: Animalia (animals)
- phyla (the following phyla all have no backbone)
  - poriferans
  - cnidarians
  - echinoderms
  - nemotodes
  - platyhelminthes
  - molluscs
  - arthropods

In this section we will look at arthropods and molluscs in detail.

Arthropods

Arthropods are invertebrate animals with a jointed body-covering that supports and protects their bodies. This covering is called an **exoskeleton** (*exo* means *out*) because it is on the outside of the arthropod's body.

Most arthropods are *insects*. Members of this group of arthropods have six legs and three distinct body segments—a head, thorax and

ISBN 978 1 4202 3822 8

abdomen. *Arachnids* belong to another group of arthropods that includes spiders and ticks. These animals have eight legs and only two distinct body segments. Crabs, prawns and lobsters are called *crustaceans* and breathe through gills. Most crustaceans live in water.

Figure 5.20 A lobster is an arthropod. Its exoskeleton supports and protects its body.

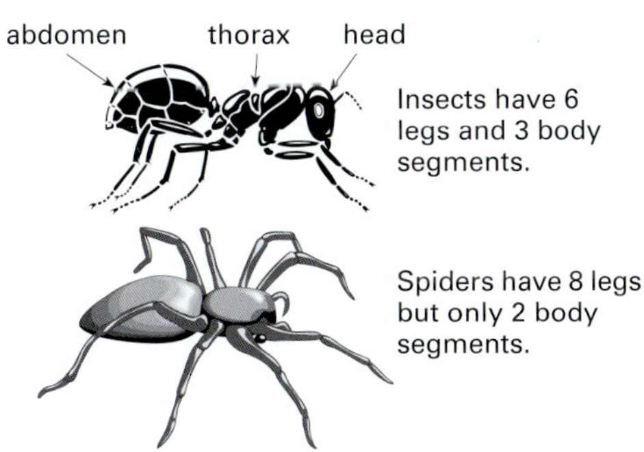

Figure 5.21 The differences between insects and spiders

Molluscs

The second largest group of animals is the **molluscs**. Most of the animals in this group have shells; for example, snails and oysters. Molluscs live in water (both sea water and fresh water) or in moist surroundings. This is because they take in oxygen through a delicate membrane underneath their shells, which has to be kept moist. When conditions are dry, many molluscs can withdraw their bodies into their shells. They seal the opening and can stay like this for long periods of time until water is again available.

Some molluscs have no shell (e.g. slugs and octopuses) or a small internal shell (e.g. squid and cuttlefish).

Figure 5.22 Molluscs, such as these periwinkles, move by sliding over the ground on a film of mucus.

Complete the following table about the other invertebrate phyla, by doing your own research.

Phyla	Description of main features	Examples	Photo
Poriferans			
Cnidarians			
Echinoderms			
Nematodes			
Platyhelminthes			

ISBN 978 1 4202 3822 8

Vertebrate classification

Vertebrates have a backbone and are animals with an internal skeleton (or endoskeleton). There are five classes we will explore in this section:

- kingdom: Animalia (animals)
- phyla: Chordata (have a backbone)
- classes:
    - fish (includes bony fish and cartilaginous fish)
    - Amphibia (amphibians)
    - Mammalia (mammals)
    - Aves (birds)
    - Reptilia (reptiles)

Fish

The animals in this group live in water and breathe the oxygen dissolved in the water through gills. Fish have a changing body temperature (they are incorrectly called cold-blooded). Most fish lay their eggs in water and the young hatch outside the mother's body. Sharks and rays are fish, but they have a skeleton made from cartilage instead of bone.

Amphibians

These animals have a moist skin and include frogs, toads and salamanders. Their eggs have no protective covering and are laid in water. The larvae of amphibians live in water and breathe through gills, while the adults live on land and breathe through lungs. These animals have a changing body temperature.

Mammals

These animals have a constant body temperature and usually have hair or fur that keeps them warm in cold weather. They breathe air through lungs. Most mammals give birth to live young and feed them on milk. Humans are mammals.

Birds—Aves

All the animals in this group have a constant body temperature (they are warm-blooded). They have feathers and breathe air through lungs. They lay eggs with a hard outer shell.

Reptiles

These animals have a dry, scaly skin and a changing body temperature. Turtles, snakes, lizards and crocodiles belong to this group. They all lay eggs with a tough, flexible covering, and all breathe air through lungs.

Figure 5.23 The five classes of vertebrates

fish

Amphibia

Mammalia

Aves

Reptilia

ISBN 978 1 4202 3822 8

INVESTIGATION 5.3 Observing animals

Aim

To observe the features of animals that belong to different groups.

Materials

- a freshly-killed fish, preferably with gills (from the fish markets or a fish shop)
- part of a cooked crab, e.g. a leg or claw
- dissecting board or dish
- dissecting scissors, probe and forceps
- disposable gloves
- an insect and a spider (either freshly killed, preserved or a good photo) for Part C
- hand lens
- stereomicroscope (optional)
- glass dish, e.g. Petri dish

PART A Observing a fish

1 Observe the outside of the fish.
Sketch the shape of the fish and label the various structures that help it live successfully in water.

2 Look inside the fish's mouth and observe the gills. Then open the gill covers on the outside behind the head.
What do you think is the function of the gill covers? Suggest why fish open and close their gill covers when they swim.

3 Use scissors to cut one gill section from the fish. Use the hand lens (or stereomicroscope) to observe the gills.

4 Place the gills in a shallow dish of water. Observe the gills again with the hand lens.
What differences do you see when the gills are in water? Suggest how this helps the fish survive.

5 Use scissors to cut the flesh away from the backbone. This flesh is the muscles that move the backbone.
Observe the flexibility of the bones and the joints in the backbone.

PART B Observing an exoskeleton

1 Use the crab leg to observe the exoskeleton (shell). Look at the joints to see how the hard pieces of exoskeleton are connected.
Sketch the crab's leg and label the hard and softer parts.

2 Break some of the shell away to expose the white or pinkish flesh. This is a muscle which moves part of the leg.
Keep breaking away the shell and remove the muscle until you find a piece of hard, shiny, white tendon. Try pulling on this tendon to move the leg.

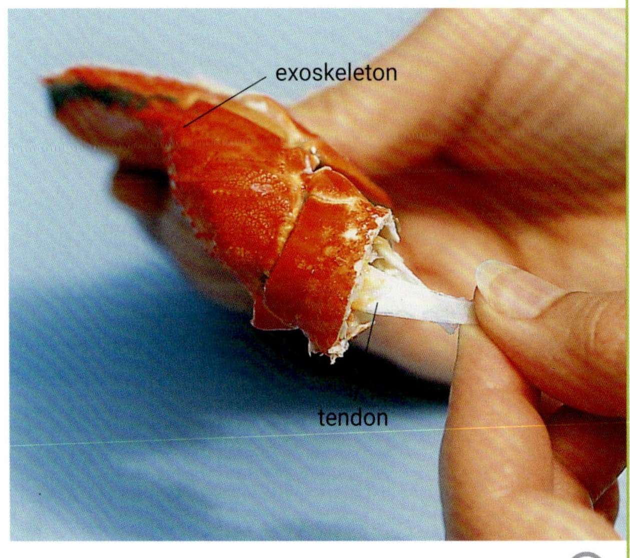

ISBN 978 1 4202 3822 8

PART C Observing other arthropods

1 Look at the insect and the spider. Compare the number of legs and the number of body segments.

2 Compare the thickness of the exoskeleton of the insect and the spider with that of the crab.

3 Use the hand lens to observe the various structures of each arthropod. If you can find a grasshopper, look along the side of its abdomen with the hand lens. You will see tiny holes through which it breathes.

Make a labelled sketch of interesting structures on each arthropod.

Discussion

1 What happens to gills in and out of water? Suggest why fish suffocate and die when they are left out of water.

2 Suggest why fish have such large muscles along their backbones. Do they have a similar bone arrangement to humans? For example, do they have ribs?

3 Suggest why the thicknesses of the exoskeletons of the crab, insect and spider are different. Would it help the insect to have a very thick exoskeleton?

4 Arthropods do not grow as large as most vertebrates. Use your knowledge of exoskeletons to suggest why this is so.

SCIENCE AS A HUMAN ENDEAVOUR

The quokka (genus *Sentonix*) is a small, cat-sized animal native to Western Australia and is classified as a marsupial. Marsupials, like possums, koalas and wallabies, give birth to immature young that then develop further in a pouch.

Quokkas are commonly found on Rottnest Island off the Perth coastline, but also live on the mainland. Like most marsupials, they are herbivores, meaning that they do not eat meat. They need lots of fresh water, but get most of that from the grasses and shrubs they eat.

Originally, Rottnest Island was named 'Rotte nest' which means 'rat's nest'. This is because the Dutch settlers who first visited the island thought that the quokkas were large rats. The name quokka is thought to come from a Nyungar word, which is the language spoken by the Noongar people of the southwest corner of Western Australia.

Quokkas are currently considered to be a vulnerable species, meaning that they are at risk of extinction. Work is being done to protect their habitat so that the quokka does not die out, like other Australian animals have.

Use the website links to answer the questions below.

1 What are some of the factors that are affecting the conservation status of the quokka?

2 The Tasmanian tiger is a species that has become extinct. Research some possible ways that we can protect the quokka from extinction.

3 How are the quokkas on Rottnest Island different from the quokkas on the mainland?

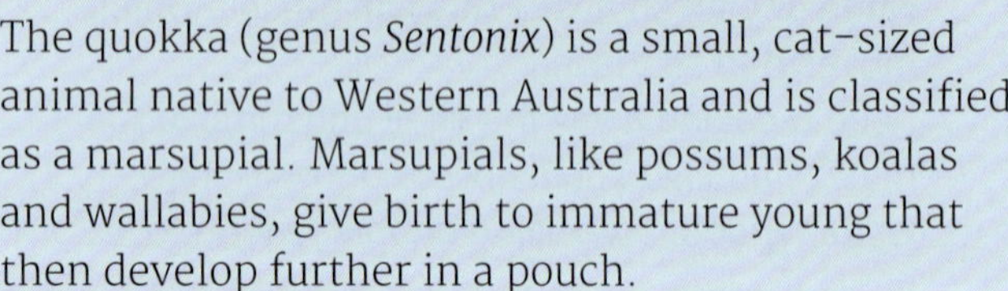
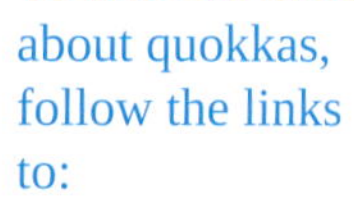

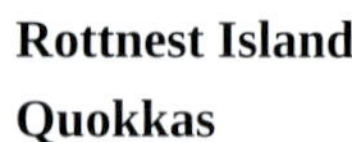

EXPLORE ONLINE

To find out more about quokkas, follow the links to:

Rottnest Island

Quokkas

The plant kingdom

This kingdom includes all the multicellular organisms that can photosynthesise and make food from carbon dioxide and water, using the energy of sunlight.

Look at the plant key below. This is another way to draw a key. Both types of keys are used by biologists when studying living things.

You can use this plant key to identify the four main groups in the plant kingdom.

Plant key

1 No stem . . . . Mosses
Stem . . . . go to 2
2 Makes spores . . . . Ferns
Makes seeds . . . . go to 3
3 Has flowers . . . . Flowering plants
No flowers . . . . Conifers

Plant classification

There are three phyla and four classes we will explore in this section:

- kingdom: Plantae (plants)
- phyla
  - mosses
  - ferns
  - seed-producing plants

There are four classes of seed-producing plants:

  - cycads
  - ginkgo
  - conifers
  - flowering plants

Mosses

Mosses are the simplest plants. They have simple leaves, very simple roots and no stem.

In larger plants, water from the ground is carried up to the leaves in the stem. Because mosses have no stem, their leaves have to be close to the water on the ground. This is why most mosses grow only to a few millimetres high and live in moist places.

Mosses reproduce by spores. These tiny cells are found in spore cases that grow at the top of the plant. When the conditions are right, the spores are released. If they fall onto moist ground they will form new moss plants.

Figure 5.24 Mosses are very simple plants. They are small and have a very simple leaf and root structure.

Ferns

Ferns are much larger than mosses. They have a stem as well as leaves and roots. The stem is called a *rhizome* (RYE-zome) and it grows horizontally under the surface of the soil.

The fern that you see is the leaf or *frond*. These grow up from several places along the rhizome. The new fronds are curled up, but as they grow they uncurl (see Figure 5.25).

If you cut a rhizome and observe it under a microscope you will see tiny tubes. These tubes carry water and food to all parts of the fern. Because it has these tubes in the stem, roots and

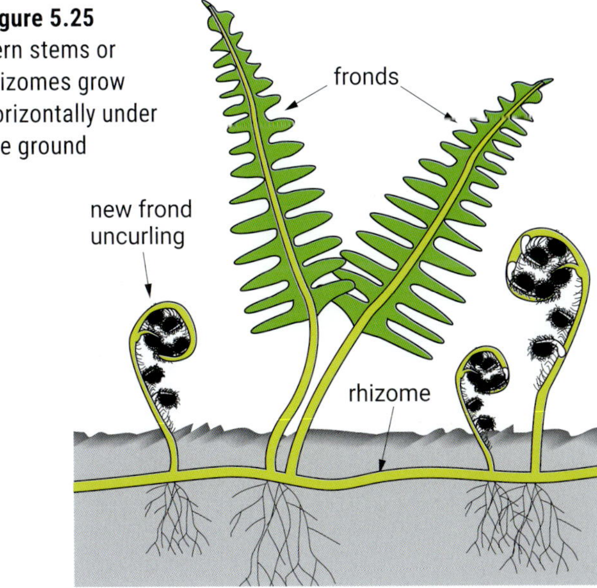

Figure 5.25 Fern stems or rhizomes grow horizontally under the ground

ISBN 978 1 4202 3822 8

fronds, ferns can grow much taller than mosses. Mosses have to be close to the ground so that all parts of the plant are near a supply of water.

Ferns, like mosses, reproduce by spores. At certain times of the year, ferns grow tiny rows of brown spots under their leaves. These brown spots are spore cases and are filled with thousands of spores. When the spores are mature, the spore cases break open, and the spores fall out and are dispersed by the wind.

Conifers

Conifers have stems, roots and leaves and reproduce by seeds instead of spores. Seeds are larger and more complex than spores. Conifers include pines and fir trees. These all have cones that contain the seeds. Male cones are small and produce pollen, while female cones are mostly large and woody and produce eggs.

Flowering plants

Flowering plants have stems, roots and leaves and reproduce by seeds. They include grasses, bushes, shrubs, most trees and even water plants such as waterlilies. The flowers produce pollen and eggs, although some types of flowering plants produce only pollen or only eggs. When the eggs are fertilised by the pollen, they develop into seeds. These seeds are contained in a fruit that may be fleshy and edible—such as in a pumpkin, apple or grape, or hard and woody as in a walnut, wattle or eucalypt.

Figure 5.26 The seeds develop inside fleshy, edible fruit.

ACTIVITY

Complete the following table about the other classes of plants by doing your own research.

Classes	Description of main features	Examples	Photo
Cycads			
Ginkgo			

Figure 5.27 Spore cases on the underside of a fern frond

Figure 5.28 Cones on a conifer

Figure 5.29 Waterlilies are flowering plants.

ISBN 978 1 4202 3822 8

SCIENCE AS A HUMAN ENDEAVOUR

Plant medicines

For thousands of years, plants have been the source of medicines for illnesses and injuries.

Early Europeans scraped the bark of the willow tree and used it to ease headaches and pain. We now know that the bark contained a chemical called salicylate, which chemists use to make aspirin.

Australian Aborigines had a wealth of knowledge about plant medicines. Medicines were prepared by crushing the plant and soaking it in water, often for a long time. The patient would drink it or have it rubbed on a wound. Ointments were made by mixing crushed leaves in animal fat.

Today chemists often use the active substance from a plant to make a synthetic 'copy' that is then used in modern medicines.

EXPLORE ONLINE

Follow the links to **Aboriginal medicines** to find out more about plant medicines.

Figure 5.30 Tamarind fruit pulp is rubbed on the head to treat headache.

CHECK

1. List the levels of classification from phylum to species.
2. a. What does a binomial name mean?
   b. Give an example of a species name.
3. Use Figure 5.19 on page 116 to answer the following questions:
   a. Which of the following pairs of species are more similar—*Canis familiaris* and *Canis lupus* or *Canis latrans* and *Vulpes fulva*? Explain your answer.
   b. *Homo sapiens* belong to which family?
   c. *Homo sapiens* and *Canis familiaris* both belong to the same kingdom, phylum and class. What are some of the characteristics that are similar in both species?
4. Use the list below to match the organism to its description.

mosses	arthropods	ferns
reptiles	conifers	mammals

   a. These plants produce seeds in a cone.
   b. The organisms in this group have a constant body temperature.
   c. These organisms contain chlorophyll but do not have a stem.
   d. These organisms include arachnids and crustaceans.
   e. Female organisms in this group lay eggs with a tough, flexible covering.
   f. These plants produce spores but have a stem and leaves.
5. How would you tell the difference between:
   a. a conifer and a flowering plant?
   b. a fern and a conifer?
   c. a moss and a fern?
6. *Amphibians usually live close to water. In times of drought, they often burrow into moist soil. In the colder months of the year they hide in burrows or under logs and rocks.*
   For each of the three sentences in the description above, suggest why amphibians show this behaviour.
7. Why are snakes and earthworms classified in different groups (see page 104)?

ISBN 978 1 4202 3822 8

8 Use the key on page 121 to write a description for the plants in the photos below.

9 There are two ways of drawing keys. Use the information in the key on page 101 to draw the other type of key.

10 Whales and dolphins spend all their life in water. Why are they classified as mammals?

11 In the two polar areas on Earth where snow and ice are present all year round, some of the following animals might be seen on the icepacks—polar bears, penguins, seals and sea lions. To which groups do these animals belong? How can they survive when other animals such as insects, reptiles and amphibians cannot?

1 The animal key on page 104 contains three invertebrate groups—arthropods, worms and molluscs. However, there are a number of other invertebrate groups. Use the information below to redesign the invertebrate part of the animal key on page 104.

Arthropods (insects, spiders, crabs): jointed covering over body

Molluscs (snails, clams, oysters, mussels, squid, octopuses): soft body, not segmented, usually with a shell

Echinoderms (starfish, sea urchins, sea cucumbers): hard, spiny skin, all live in the sea

Flatworms and roundworms (tapeworms, liver flukes, threadworms, nematodes): long, flat or round soft body with no body segments, poorly developed gut

Segmented worms (earthworms, leeches, beach worms): long, round and soft body divided into segments, well-developed gut

2 Tan and Kif were studying mangroves—flowering plants that grow in salty water along river banks. They used the key in Figure 5.31 to identify the mangroves they observed when they were on a field trip.

a Tan observed a mangrove that had salt crystals on its leaves, and its leaves were growing alternately on the stem. Which mangrove was Tan observing?

b There may be terms in the key that you have not seen before. Draw a sketch of what you think buttress roots and 'knobbly knee' roots are. Discuss your sketches with other people.

c Write a description for the yellow mangrove.

d Kif noticed a mangrove with opposite leaves and aerial prop roots. Which mangrove was he observing? What further observation would be necessary to be sure of the type of mangrove it was?

e In which way is a black mangrove different from an orange mangrove?

ISBN 978 1 4202 3822 8

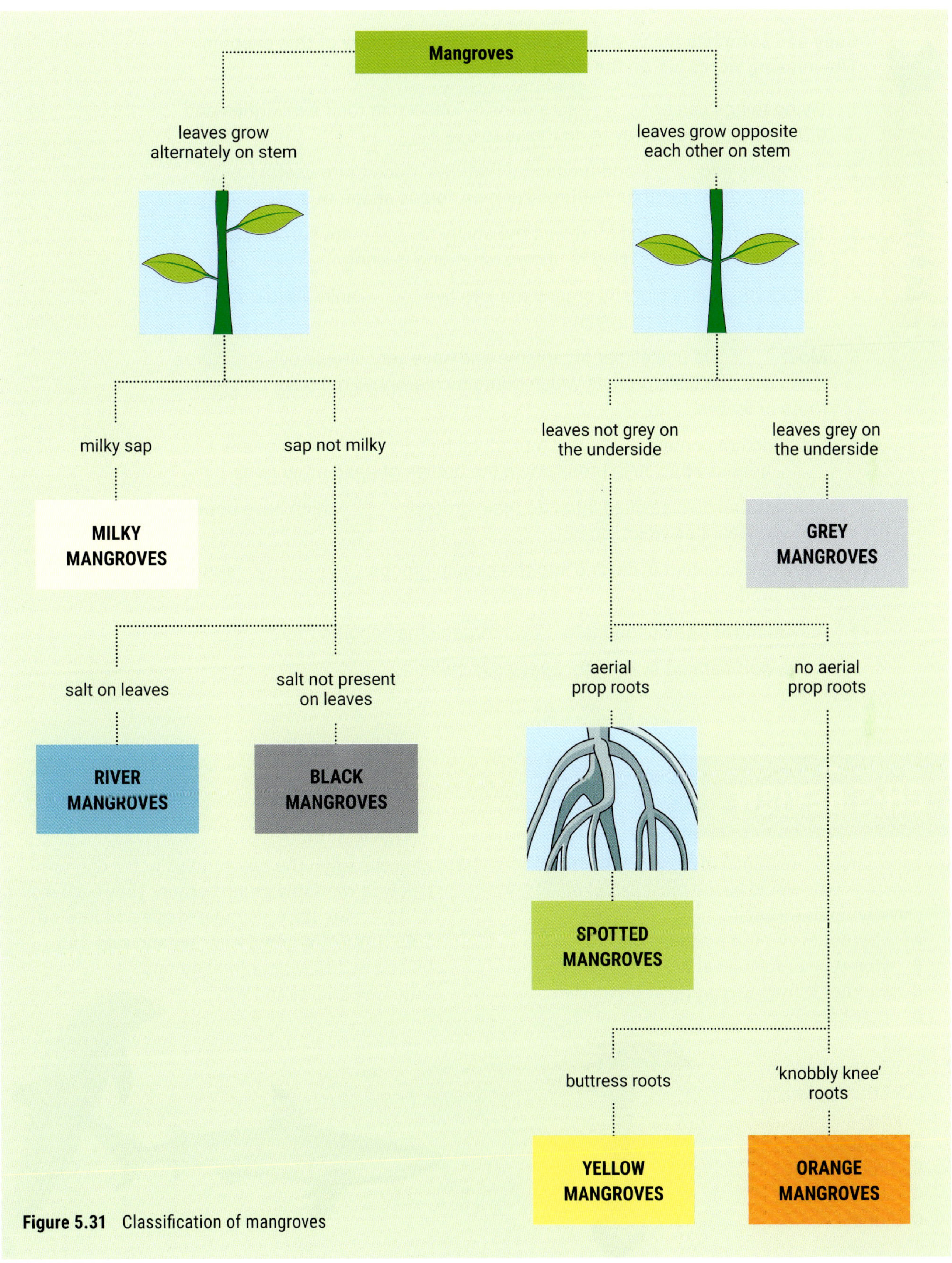

Figure 5.31 Classification of mangroves

ISBN 978 1 4202 3822 8

MAIN IDEAS

Copy and complete these statements to make a summary of this chapter. The missing words are on the right.

1 Living things can be _____ into groups by observing their similarities and differences. A good way to do this is to use a _____ .

2 Biologists find _____ and functional features much more useful to classify organisms than features such as colour, shape or size.

3 The need for _____ and food, and the ability to _____ are some of the seven characteristics used to show something is living.

4 Biologists usually classify organisms into five _____: animals, plants, _____ , protists and monerans.

5 Most _____ are unicellular organisms and have very simple cell structures. _____ are plant-like protists which contain chlorophyll but have no stem, roots or leaves.

6 Fungi do not contain _____ and reproduce by spores. Most fungi are _____ because they help break down the bodies of dead organisms.

7 Animals can be classified into two main groups: _____ which have bones, and invertebrates which do not.

8 The plant kingdom is divided into three main groups: _____ , _____ and seed-producing plants.

9 As you move from kingdom to _____ , organisms become more _____ .

10 A two-part naming system for species is called _____ .

decomposers
chlorophyll
key
structural
fungi
similar
protists
kingdoms
vertebrates
reproduce
mosses
oxygen
algae
classified
species
binomial
ferns

CH•5 REVIEW

1 Leon catches an animal in a pond. Which characteristic would be the most useful in classifying the animal?
A whether or not it has a backbone
B what type of food it eats
C whether it lives in a group or on its own
D its colour

2 In which kingdom does this organism belong?
A Plant
B Protist
C Monera
D Fungi

3 Zian classified the two animals in the diagram below in the same group because they both live in the ocean, have a similar shape and feed on fish. Bruno disagreed with her and said they belong in different animal groups.
Who was correct and why?

ISBN 978 1 4202 3822 8

4 Which characteristic can be used to tell a fern from a conifer?
   A where it grows
   B whether it has a stem or not
   C its colour
   D whether it produces seeds or spores

5 How can you tell the difference between a reptile and an amphibian?

6 Into which group would you put this organism? It is green in colour, has very small root-like structures, is quite small (about 10 mm across) and lives in moist places.

7 Which of the following statements is *incorrect*?
   A Turtles and lizards belong to the reptile group of the animal kingdom.
   B Flowering plants, conifers and ferns all produce seeds.
   C Mammals and birds are two groups of vertebrates which have a constant body temperature.
   D Bacteria and fungi are decomposers because they break down the bodies of dead organisms.

8 Look at the diagrams of the four organisms.
   a In which *two* ways are they similar?
   b How is organism D different from the others?
   c Make up a key that could be used to classify these four organisms.

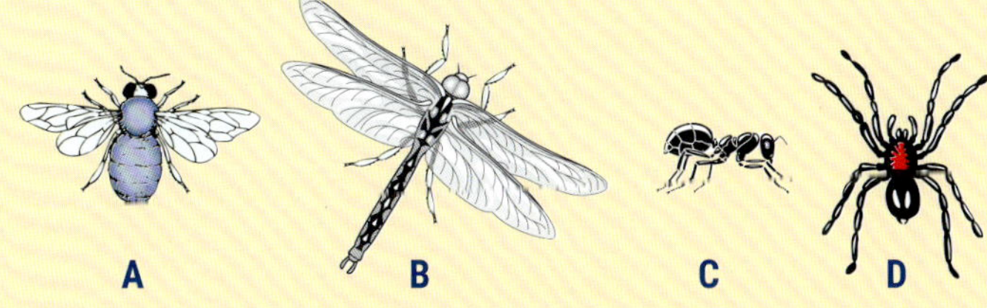

9 The following key was used to classify some organisms in a pond.
   a Describe in one sentence the characteristics of mites and water spiders.
   b How are flatworms and leeches similar? How are they different?
   c You observed a frog and some fish in a pond. Why would this key be unsuitable for classifying these animals?

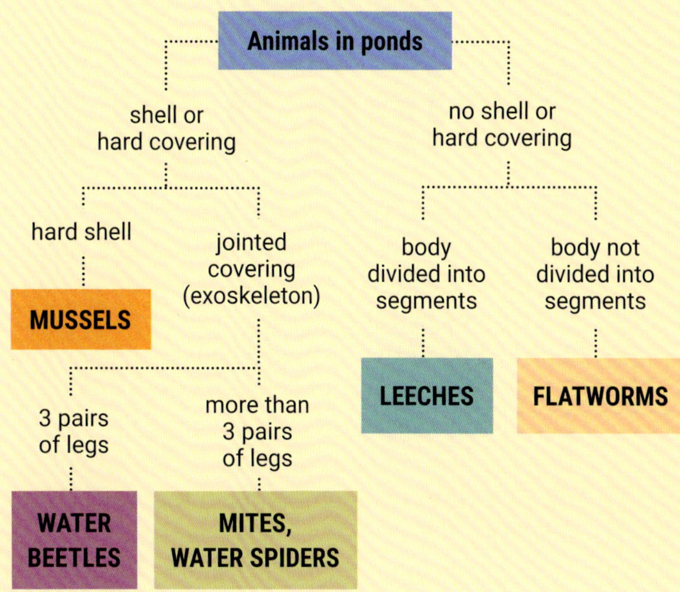

10 Look at the following table and answer the questions that follow:

Category	Lion	Common chimpanzee	Bearded dragon
Kingdom	Animalia		
Phylum	Chordata		
Class	Mammalia		Reptilia
Order	Carvinora (carnivores)	Primata (apes)	Squamata (lizards)
Family	Felidae (cats)	Hominidae (apes, humans)	Agamidae
Genus	Panthera (big cats)	Pan	Pogona
Species	*Panthera leo*	*Pan troglodytes*	*Pogona vitticeps*

   a What is the scientific name for the common chimpanzee?
   b Which two of the three animals are most similar? Explain your answer in relation to the classification system.
   c Name the class that both lions and chimpanzees belong to, and list some of the characteristics of this class.
   d Humans also belong to the family Hominidae. Identify some of the features that humans have in common with apes that may support this classification.
   e Describe what characteristics the animals in the phylum Chordata have in common.

Check your answers on page 240.

ISBN 978 1 4202 3822 8

Doing a project

As you worked through the first chapters of this book you may have found something of special interest to you that you would like to work on by yourself or in a small group. You might like to extend one of the investigations you have done. For example, you could try to make a good wood glue using milk. You could collect and classify fingerprints. Or you could carry out an experiment to solve one of the problems in the Activity on page 43.

If you do a *student research project* like this you will not simply be following instructions in a book. A project involves you deciding what needs to be done, then doing it, and reporting your results to others.

Derrick Roberts, a Year 10 student at Newington College in Sydney, won a BHP Billiton Science Award for investigating whether the diet supplements we buy really work.

Choosing a project

There are many different types of science projects that you can do, such as:

- laboratory experiments
- designing and building things
- field work
- surveys and library research.

The diagram below shows how one student chose her project by starting with the topics in *ScienceWorld 7*.

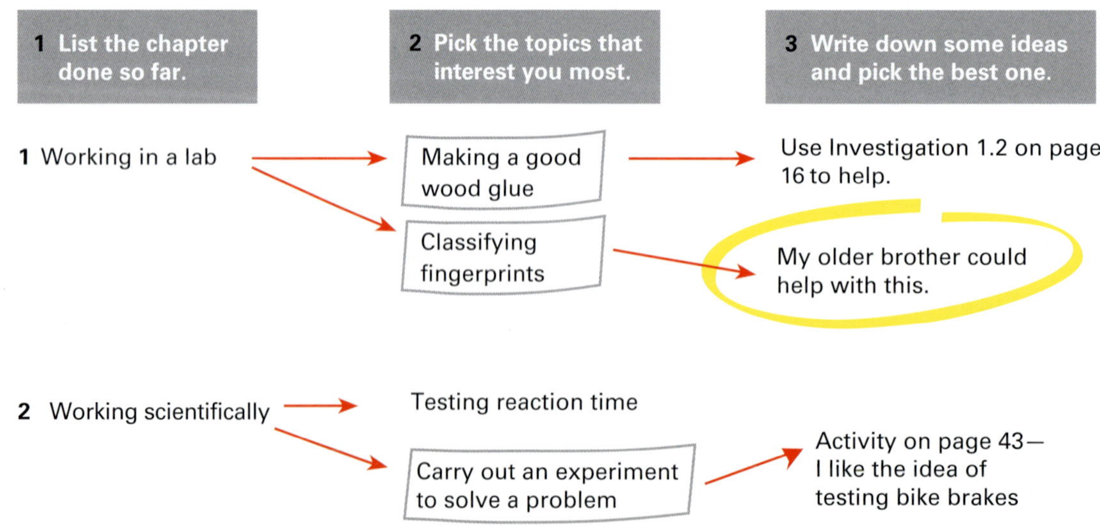

ISBN 978 1 4202 3822 8

Ideas

1 Look through this textbook and note down the topics that interest you the most and that you would like to know more about. Mark the ones you think would be suitable for a student research project.

2 Write down some ideas for projects. There are suggestions in some of the 'Challenges' and 'Try this activities', and many of the investigations and experiments can be extended into projects.

3 Read through the following list of projects that other students have done. You might find something that interests you. Check the websites suggested in the Explore Online box on page 131 for further ideas to get you started.

> *How does temperature affect the germination of seeds?*
> *Which washing powder works best?*
> *Does a wireless mobile phone charger work faster than a plug-in one?*
> *Make bricks out of different materials and test their strength.*
> *Does fruit last longer when stored in a refrigerator?*
> *Make batteries from fruit and vegetables. Which fruit and vegetables work best?*
> *Which type of battery lasts the longest?*
> *How is heart rate affected by exercise?*
> *Which materials decompose fastest in rubbish?*
> *Which material is best for keeping you warm?*
> *Are artificial sweeteners as sweet as real sugars?*

4 Consider each of your ideas in turn. You will probably ask yourself questions like these:

- Am I really interested in this topic? Will anyone else be interested?
- Can I get the equipment and materials I need?
- Can I get extra information on the topic?
- Can I complete the project in the time available?

Talk with teachers and other people about your ideas. Finally, pick the project you think would be best for you.

5 Read through the hints on the next page, then discuss with your teacher when you will do your project.

Your project could investigate seed germination or electronic circuits.

ISBN 978 1 4202 3822 8

Hints on doing your project

1 Write up a brief outline of what you plan to do. Show this to your teacher, who will give you helpful suggestions and then give you approval to go ahead. *Do not attempt the project without your teacher's approval.*

2 Plan what equipment and materials you will need for your project. Check what is available at school, what you can get from home and what you can borrow. You don't always need proper laboratory equipment. For example, you could use glass jars instead of beakers, or measuring jugs instead of measuring cylinders.

3 You will be doing most of the project at home so make sure your parents know all about the project. Plan your time carefully, by dividing the project into stages. An example of a plan for a 7- to 8-week project is shown below.

Selecting a topic	**1 week**
Finalising topic and preparing plans	**1 week**
Doing project	**4–5 weeks**
Evaluating results and preparing report	**1–2 weeks**
Presenting report	**1 day**

4 Keep a project log book. This will be very useful when you prepare your report. It could include notes you have taken from books, photographs, video or drawings of your experiments, tables of data, drawings or plans of a model at various stages of construction etc.

5 If you are going to do an experiment, you should follow these steps:
- Write down the aim of the experiment. This may be a question you are testing; for example, *What conditions are needed for iron to rust?*
- Make sure any tests you do are fair. In a **fair test** you need to change something, measure something and keep everything else the same.
- After identifying all the equipment and chemicals required and designing your procedure, check for any safety requirements. Work out if you can reduce risk to make your experiment as safe as possible.
- Do the experiment, and collect and record data. A computer spreadsheet is useful here.
- Duplicate your tests for more accurate results. For example, if you are germinating seeds, set up three or four pots for each test, not just one.

Remember, there are no incorrect answers when doing an experiment—only *unexpected* ones.

6 You will need to prepare a bibliography as part of your report, so keep careful notes of any books, magazines and websites you use.

Hints on preparing a report

1 Select the most appropriate form for your report. For example:
- a written summary for the teacher
- a display for a science fair or the library
- an entry for a science contest
- a short talk
- a presentation using PowerPoint or other software
- a website to showcase your findings.

2 Include any video, photos, drawings or other multimedia you have.

3 Ensure that your report includes the data you collected and analysed, displayed in tables and graphs where appropriate.

4 Be honest in your reporting. In your discussion include difficulties, errors and results you cannot explain.

5 Make sure that your report has a conclusion. Remember, doing the project may be the most important thing for you, but most people will want to know how successful you were in achieving your aim.

6 Ask your teacher about entering your project in a science contest.

ISBN 978 1 4202 3822 8

Explore the websites below for examples of exciting science projects, as well as awards and competitions!

Double Helix Club

This site has a list of cool experiments you can do yourself. If you join the club, you will receive a magazine six times a year and you can go to regular club events in your local area. You can even work with some of Australia's leading scientists doing real research. For example, in 2007, club members participated in Millipede Madness, in which they mapped the distribution of Portuguese millipedes across Australia. These black bugs have invaded homes in all states except Queensland.

Science Project

The site has a list of ideas for science projects, classified by year level and subject.

BHP Billiton Science Awards

CREST

CREST stands for Creativity in Science and Technology. Look them up for ideas for projects and videos of past projects.

Science Talent Search

Run by the Science Teachers Association of Western Australia

ISBN 978 1 4202 3822 8

IN THIS CHAPTER

Science Understanding

- construct and interpret food chains and food webs to show relationships between organisms in an environment
- investigate the effects of human activity on local habitats, e.g. crown of thorns starfish on coral reefs
- explain how living and non-living factors can affect the survival of organisms in ecosystems
- develop an understanding of biodiversity and the human impact on biodiversity
- present a plan to reduce the carp population in Australian waterways

Science Inquiry Skills

- analyse information and draw conclusions about the relationships between organisms in food webs

CH•6 Ecosystems

ISBN 978 1 4202 3822 8

GET STARTED: *QUESTION*

- You eat food to obtain minerals and energy for growth. But where does food come from? List all the types of food in a hamburger and suggest where they come from.
- Although dogs and hamsters are both common domestic pets, they have very different diets. Research the common foods that dogs and hamsters eat, and suggest why their diets are so different.
- Seals and crocodiles live in different ecosystems and if they swapped the ecosystems they live in they would die. What features does each of these animals have to help it survive in its ecosystem?

6.1 Living in a food web

Food chains

What did you eat for breakfast this morning? The food you eat supplies you with materials for growth, and energy for muscle movement. Let's find out where your breakfast food may have come from. Take eggs, for example. These are made by hens which eat grain to make the eggs, and the grain comes from plants such as corn and wheat.

Plants are able to absorb sunlight as an energy source to make food. So the energy you obtained from your breakfast foods came originally from the Sun's energy, which the corn and wheat plants absorbed.

The diagram below is called a **food chain**. It shows the feeding relationships among organisms. One organism provides the food for the second. The second provides the food for the third, and so on. The food chain can be written simply as:

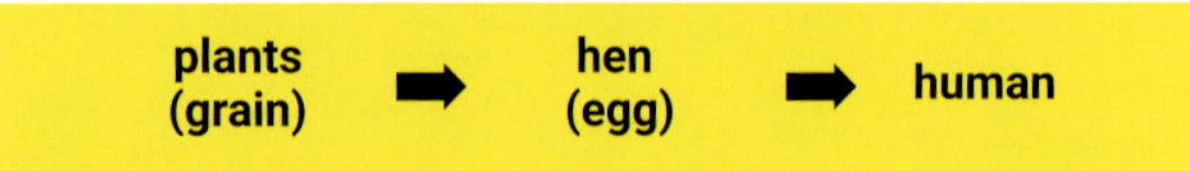

The arrows show the direction in which the food passes.

1 List all the foods you have eaten in the last 24 hours.
2 Beside each food item list the source of the food. For example, if you ate a pie for dinner, the meat would come from cattle and the pastry from plants (grain).
3 Draw a food chain for each food item.

Food chains are usually not very long. Each food chain in the examples on this page contains three organisms. How many organisms did your food chains contain?

Did any of your food chains contain four or more organisms? Can you think of a food chain that might contain four or more organisms?

Other food chains for your breakfast foods can be drawn.

Figure 6.1 A simple food chain

ISBN 978 1 4202 3822 8

Photosynthesis and respiration

Most food chains begin with plants or algae. This is because plants and algae produce their own food from carbon dioxide and water using the energy of sunlight.

You learnt in Chapter 5 that chlorophyll, the green substance in plants, is able to absorb the energy of sunlight. This energy is used to make food from carbon dioxide and water in the process of photosynthesis.

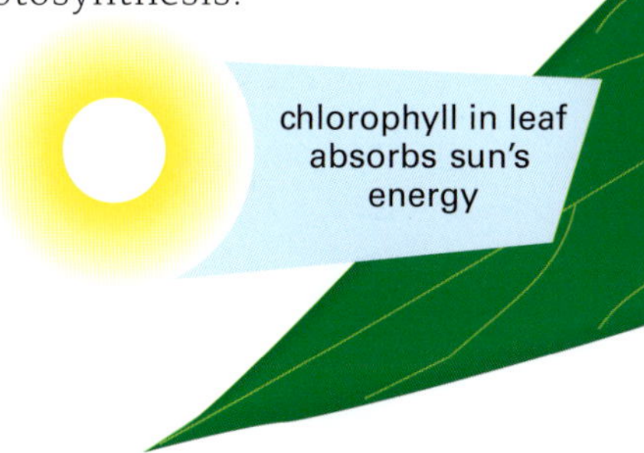

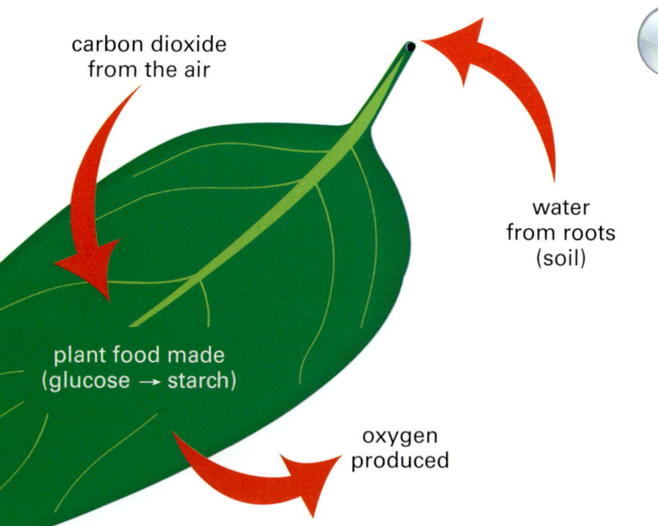

Figure 6.2 Photosynthesis

The main products of photosynthesis are simple sugars such as glucose, which are then converted to starch for storage. The other important product made in photosynthesis is oxygen. This is released into the air, or into the water if the plant lives in water.

PHOTOSYNTHESIS

carbon dioxide + water ➡ glucose + oxygen

glucose ➡ starch

All living things need food as a source of energy for movement, growth and other body functions. The process of getting energy from foods is called **respiration** (RES-pe-RAY-shun). Respiration occurs in cells. Here glucose, the fuel, reacts with oxygen to release energy. Carbon dioxide and water are given off.

RESPIRATION

glucose + oxygen ➡ carbon dioxide + water

ACTIVITY

A bean plant was kept out of sunlight for a number of days. By the end of this time, the plant had lost its bright green colour and looked sick.

Do green plants need sunlight to survive? Work in a group to discuss this question. Design an experiment that you think would be a 'fair test' to answer this question.

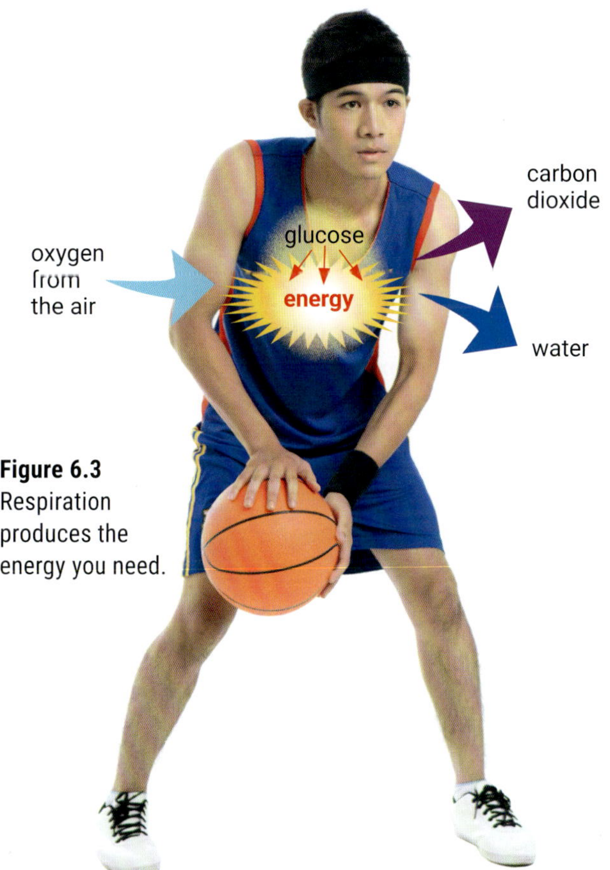

Figure 6.3 Respiration produces the energy you need.

ISBN 978 1 4202 3822 8

Photosynthesis and respiration

Aim

To test for the gases given off in photosynthesis and respiration.

Risk assessment and planning

- Read through the investigation carefully.
- You need at least 40 minutes to do Part A. Work in a group and discuss in which order you will do the various steps in Part A and Part B of the investigation.
- Why have steps 1 and 2 in Part A and Part B been included in this investigation?
- Make a list of all the safety precautions you have to take in this experiment.

PART A Testing for oxygen

Materials

SAFETY GLASSES MUST BE WORN

Corrosive

- small beaker
- methylene blue indicator (in dropping bottle)
- oxygen-removing solution (16 g/L **sodium dithionite/hydrosulfite**, $Na_2S_2O_4$, in water. Prepare the solution just before use.)
- spatula
- test tube and one-holed stopper with a right-angled piece of glass tubing
- hydrogen peroxide solution
- 'pinch' of manganese dioxide powder
- 2 test tubes with stoppers
- small piece of aquarium water plant (e.g. elodea)

Method

1 Half fill the beaker with tap water. Add 4 or 5 drops of methylene blue solution. Add drops of oxygen-removing solution until the blue colour just disappears.

2 To make oxygen, use a spatula to add a rice-grain size quantity of black manganese dioxide powder to a test tube. Pour a small amount of hydrogen peroxide into the test tube, and immediately put the stopper in the test tube. Bubble the gas into the solution in the beaker.

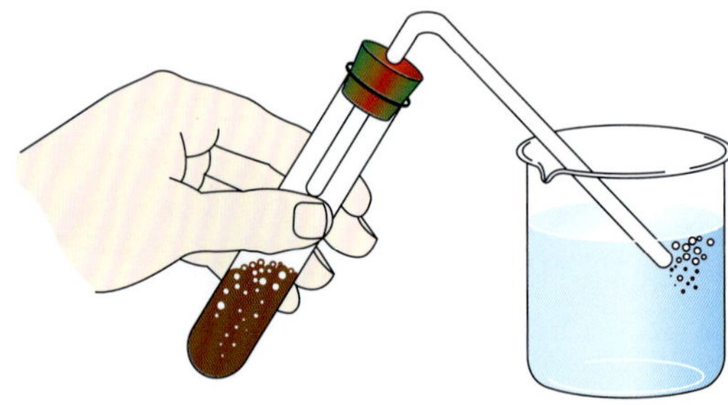

The blue colour indicates that oxygen has been produced.

3 To test that plants produce oxygen, first make up a beaker with water, methylene blue and oxygen-removing solution as in step 1. Then follow the instructions in the diagram below.

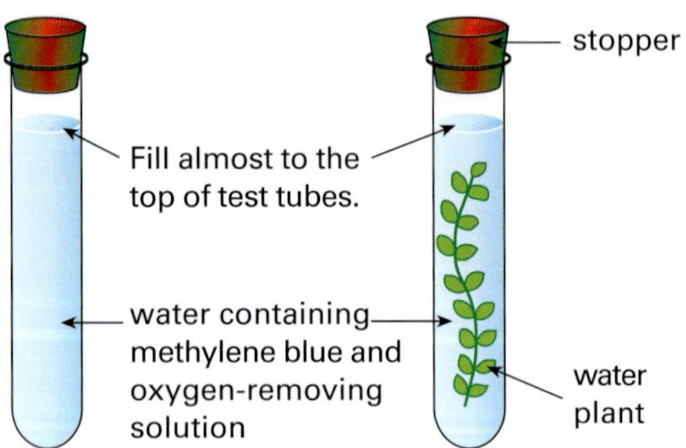

4 Leave your set-up near a window or bright light for at least 20 minutes.
Record your results.

Discussion

1 Which things did you keep the same for both test tubes?
2 Why did you have one test tube without a plant?
3 Discuss ways in which the test could be improved?

ISBN 978 1 4202 3822 8

PART B Testing for carbon dioxide

Materials

- test tube and stopper with a right-angled piece of glass tubing (from Part A)
- bromothymol blue solution
- 2 or 3 marble chips (calcium carbonate)
- dilute **hydrochloric acid** (1 M)
- distilled water
- drinking straw
- 250 mL beaker

Method

1 Half fill the beaker with distilled water. Add 2 or 3 drops of bromothymol blue solution.
2 Put a few marble chips into a test tube. Then add a small amount of hydrochloric acid. Immediately put the stopper in the test tube, and bubble the gas into the solution in the beaker. **Bromothymol blue turns green then yellow when carbon dioxide is present.**
3 Empty the beaker and repeat step 1.
4 Use a drinking straw to blow gently into the blue solution in the beaker.
Record any colour change.
Write a report of your findings.

PART C Inquiry

Design a test to show that plants produce carbon dioxide when they respire.

Discuss the plan with others in your group, and write up a draft. In a well-designed test, you need to compare the test (the container with the plant in it) to an identical container without a plant.

Make a list of the equipment you will need. Then show the draft plan to your teacher for approval before you begin.

Producers and consumers

Organisms such as plants and algae that can photosynthesise and make their own food are called **producers**. The algae are the producers in the food chain on the right.

Organisms that eat other organisms are called **consumers**. Animals are consumers because they do not make their own food and have to rely on other organisms for food.

In the food chain, the mosquito wrigglers, frogs and snakes are all consumers. The mosquito wrigglers are called first-order consumers, the frogs are second-order consumers and the snakes are third-order consumers.

Figure 6.4 Can you identify the producer in this food chain?

ISBN 978 1 4202 3822 8

Food webs

Suppose you were asked to observe the feeding relationships among the organisms in a garden. Using these observations you then constructed the food chains below.

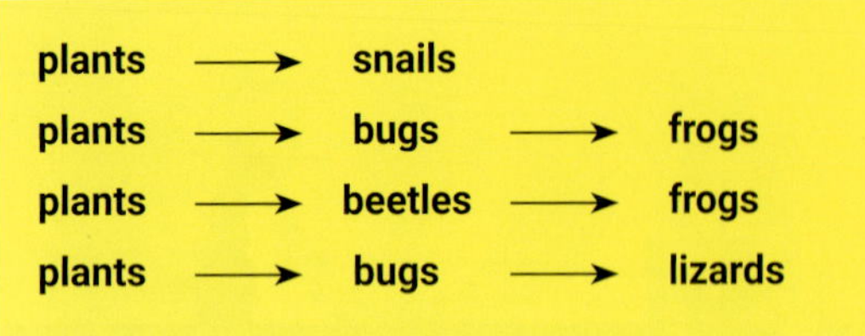

To get a more complete picture of the feeding relationships in the garden, these food chains can be combined to give a **food web**.

Food webs show the feeding habits of all the organisms that live together in a particular place. For example, you can see from the food web that frogs eat beetles as well as bugs, but lizards eat bugs only.

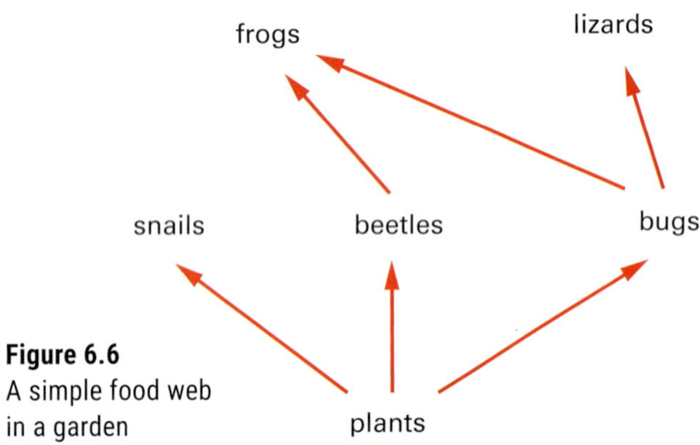

Figure 6.6 A simple food web in a garden

Figure 6.5 Herbivores are consumers that eat only producers.

Food webs can be very complex since animals usually eat many different foods. Biologists have agreed to draw food webs in a standard way.

Look at the forest food web below (Figure 6.7). The producers are placed at the bottom. Next come the consumers that eat the producers. These animals are first-order consumers and are also called **herbivores** (HER-be-vores). After the herbivores come the animals that eat other animals. These are second-order and third-order consumers and are called **carnivores** (CAR-ne-vores)—*carne* is Latin for *meat*.

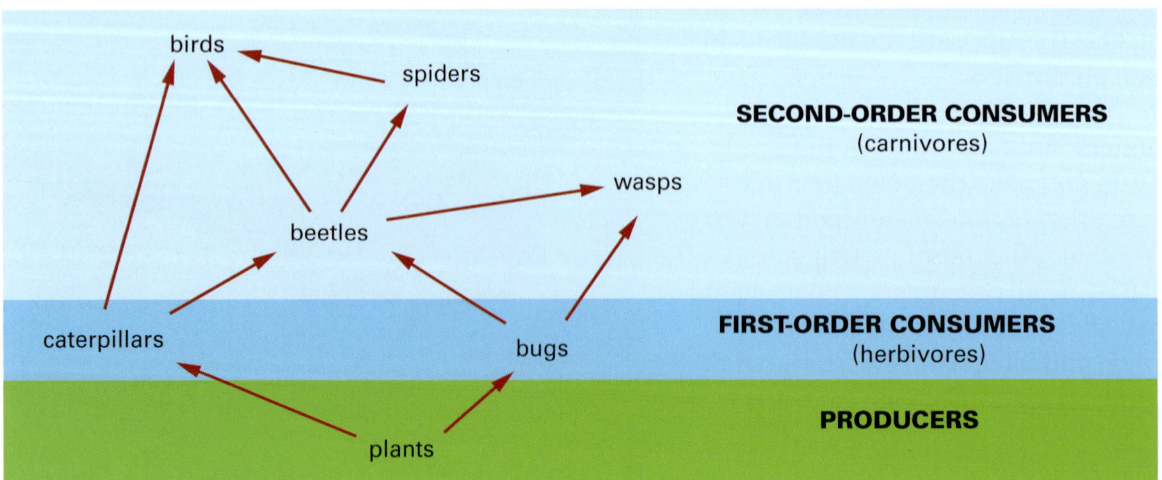

Figure 6.7 A forest food web

ISBN 978 1 4202 3822 8

Figure 6.8 Decomposers break down dead organisms.

When living things die

The photo above shows the remains of a dead wallaby. After a while, only the bones will remain. Animals that eat the flesh and organs of dead animals are called **scavengers**. These include ants, insect maggots, crows and hawks. In aquatic habitats, bottom-dwelling animals such as crabs, lobsters and prawns are very effective scavengers.

The breakdown (decay) of dead organisms is due to microscopic bacteria and fungi. These organisms are called **decomposers**.

Scavengers are classed as consumers in a food web even though they usually eat dead organisms. However, decomposers are not classed as consumers because their method of obtaining food is quite different. Consumers eat food and then digest it internally. Bacteria and fungi, on the other hand, release chemicals that break down the organism's body.

Decomposers are a very important part of a food web. The materials in a dead organism's body are broken down into simple substances that pass into the air, water or soil. Carbon dioxide and other gases such as hydrogen sulfide (rotten egg gas) pass into the air. Some substances pass into the soil and increase the soil fertility, and plants use these substances for growth. In this way decomposers recycle materials in the food web.

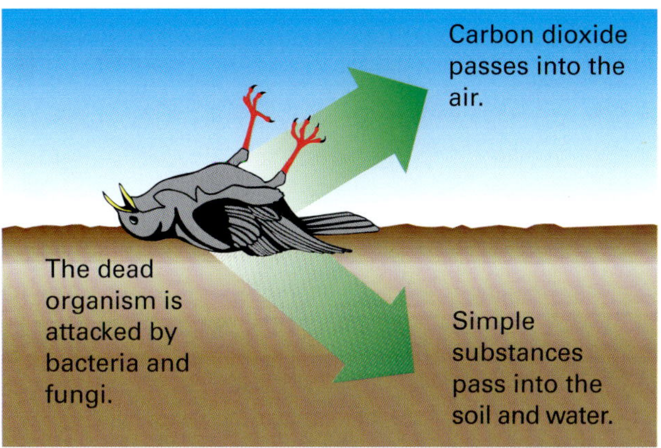

Figure 6.9 Decomposers recycle materials in the food web.

SCIENCE AS A HUMAN ENDEAVOUR

How decomposers can help solve murders

A body has been found just off a forest track. Police investigators call a *forensic entomologist* to help them determine the time since death.

An entomologist (ENT-a-MOL-o-gist) studies insects and other arthropod relatives such as spiders, centipedes, mites and ticks.

Soon after death, certain flies will lay their eggs in body openings such as eyes, nostrils and ears, or in wounds that may be present. These eggs then hatch into maggots that start to decompose the body. Certain beetles also lay their eggs in the body. These eggs hatch into larvae that start to decompose the body.

At the scene of the crime, the forensic entomologist takes samples of any fly maggots or beetle larvae. By knowing which insects attack the body first, and the size and weight of the maggots, the scientist can tell police how long the person has been dead.

By observing the types of insects attacking the body, the forensic entomologist may also be able to infer whether the person died at the scene or was dumped there after death.

ISBN 978 1 4202 3822 8

Food webs exist for different places on Earth where animals and plants live. Here is an example of a complex food web in Antarctica. In this example, the producers are the phytoplankton, which are microscopic plants that float in the ocean. Study this food web and answer the questions that follow.

1 **a** Name the animals that are at the top of this food web.

b What level consumers are each of these animals?

2 **a** Identify the producer in this food web.

b Where does this producer get its energy from?

3 **a** Name all the consumers that feed on squid.

b What would happen to this food web if squid became extinct? Explain your answer.

4 Name two third-order consumers and list the food chains that you used to identify them.

5 If fishing reduced the stocks of fish in this food web, what would be the effect on other organisms? Give some examples.

6 If humans hunted the baleen whale, what level consumer would humans be?

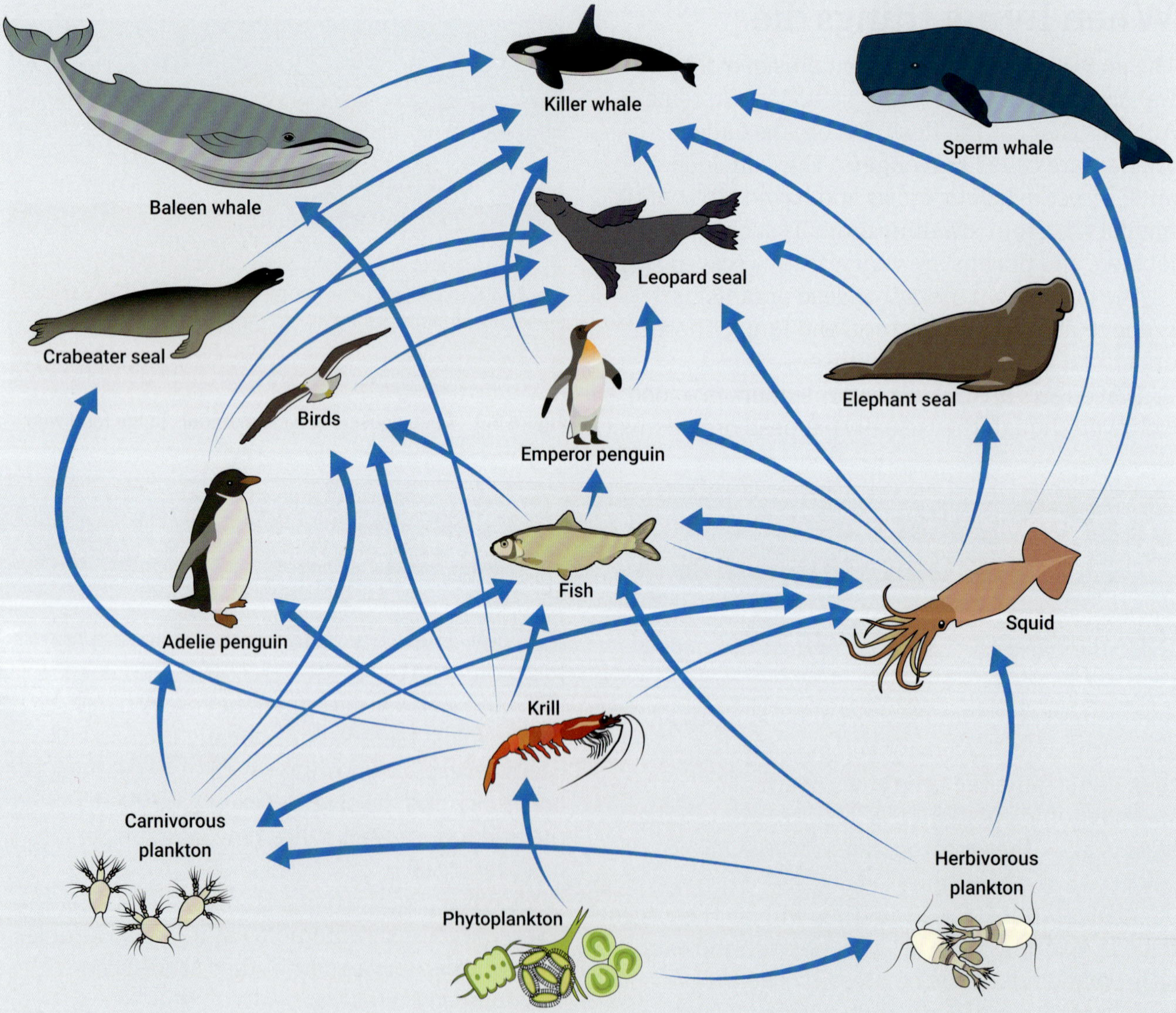

Killer whale
Sperm whale
Baleen whale
Leopard seal
Crabeater seal
Elephant seal
Birds
Emperor penguin
Fish
Adelie penguin
Squid
Krill
Carnivorous plankton
Herbivorous plankton
Phytoplankton

ISBN 978 1 4202 3822 8

ACTIVITY

Karen and Ian are two biologists who observed the feeding habits of animals and plants around a small freshwater creek and pond over four months. Here is a report of some of their observations.

'We found a number of types of water plants growing in the pond and the creek that flowed into it. Many animals such as turtles and waterhens ate these plants. There were also waterlilies growing in the pond. The underside of the leaves contained lots of small animals such as snails and water insects, as well as eggs. We did not see any animals eat the leaves.

The water contained microscopic plants and algae that were eaten by the tiny waterfleas and mosquito wrigglers. The waterfleas could be just seen with your eye. These waterfleas were eaten by small fish, water beetles and shrimp.

Snails moved slowly over the rocks in the creek and ate the green algae on the rocks. Turtles also ate the algae.

Water beetles ate mosquito wrigglers and small fish, as well as shrimp if they could catch them.

At night, we observed frogs eating beetles and mosquitoes. Green snakes ate fish, frogs and beetles. During the day, herons would fly into the pond to feed on the beetles, snails and fish.'

1. Draw a food web from the biologists' observations.
2. Which organisms are producers in the food web? Which are herbivores? Which are second-order consumers?
3. Are there any organisms in the creek and pond that were not eaten by other organisms?
4. From the food web, draw a food chain that includes microscopic plants, snakes and water beetles.
5. Could herons be in a food chain that included water beetles and shrimp? Draw the food chain.
6. What would happen to the numbers of other organisms in the creek and pond if the number of fish suddenly increased? Explain how you arrived at your answers.
7. Why are waterlilies important to some of the animals in the creek and pond?
8. Do you think that this food web would be the same through the whole year? Give reasons for your predictions.
9. What other factors may influence the life in the creek and pond? Which of these factors might have a major effect on the food web? Explain the reasons for your answer.

ISBN 978 1 4202 3822 8

1 Copy and complete the following sentences.
   a Food chains generally begin with _____ or _____.
   b Animals that eat other animals are called _____.
   c In food chains and webs, plants and algae are called _____ because they make their own food.
   d The energy needed by plants and algae to make their own food comes from _____ .
   e _____ and _____, which break down the bodies of dead organisms, are called _____ .

2 Some of the following statements are false. Select the false ones and rewrite them to make them correct.
   a Producers make their food by the process of photosynthesis.
   b First-order consumers are also called carnivores.
   c Organisms that contain chlorophyll are called consumers.
   d A domestic cat could be classed as a first-order consumer.
   e Decomposers break down the bodies of dead organisms and recycle the materials in the food web.

3 Look at the food web below.

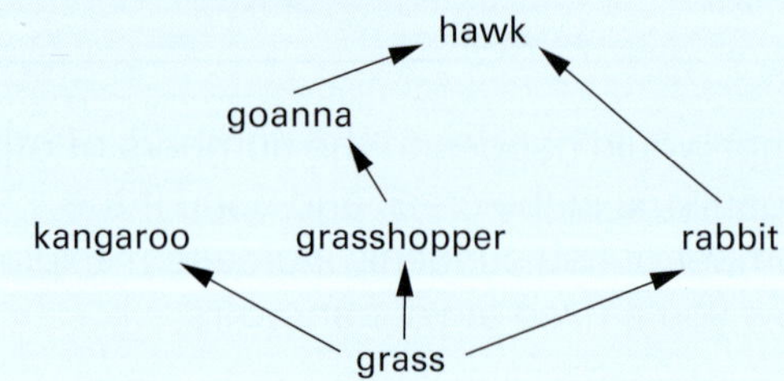

   a Which organisms are consumers?
   b Which of the animals are herbivores?
   c Which are carnivores?
   d Draw a food chain that includes the goanna.
   e Suppose the number of grasshoppers increased. What effect would this have on the food web?

4 Can an organism be a first-order consumer and a second-order consumer at the same time? Give examples with your explanation.

5 From the animals below, make a list of the carnivores and a list of the herbivores. Make another list of the animals that eat both animals and plants. These are called omnivores. For ones you are not sure of, make a separate list and write NOT SURE above it.

   cow, shark, wombat, beetle, moth, tuna, fly, parrot, goldfish, cat, mouse, bee, jellyfish, wallaby, seagull, human, guinea pig, blue whale, seal, tadpole, squid, dog

6 The diagram below shows a food web containing a carnivore, a decomposer, a herbivore and a plant. For each of the letters in the diagram, choose the description that matches it. (One is used twice.)

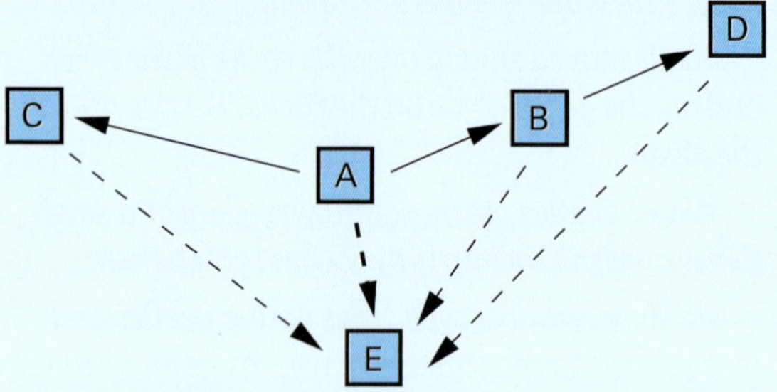

7 Draw food chains to show humans as:
   a herbivores
   b first-order consumers
   c third-order consumers.

ISBN 978 1 4202 3822 8

1 Most of the food chains of the organisms that live in the soil start with the decaying remains of plants. For example, earthworms eat decaying plant remains and microscopic animals that live in the soil.

a Would you expect to find earthworms in desert areas of Australia? Give a reason for your answer.

b Draw a food chain that includes earthworms.

c To a gardener, earthworms in the soil are a good indicator of fertile soil. Suggest a reason for this.

2 A company wants to build a tourist resort on an area now covered with mangroves. Suppose you are a biologist who studies the feeding relationships in mangroves. The local council hires you to make recommendations about the biological importance of the mangroves.

You make the following observations.

The mangrove leaves fall into the mud and are decayed by fungi and bacteria. Microscopic unicellular algae grow rapidly in areas where there is a lot of leaf decay. Small prawns, mud whelks and microscopic animals feed on the algae and the bacteria. Mud crabs feed on the microscopic animals and the decaying leaves. Small fish feed on the prawns and the microscopic animals. Larger fish feed on the small fish and prawns. Many types of birds, including stilts and herons, feed on the small fish, prawns and crabs.

a Draw a food web for the mangrove area.

b The developers of the tourist resort say that it would bring in millions of dollars to the town. What arguments would you put forward to keep the mangroves as they are?

3 The diagrams below show the mouthparts of two types of herbivorous insects. They have been magnified many times. Suggest how each insect uses its mouthparts to obtain food.

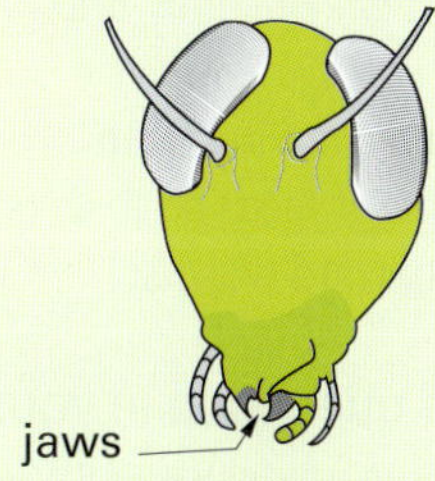

grasshopper's mouthparts

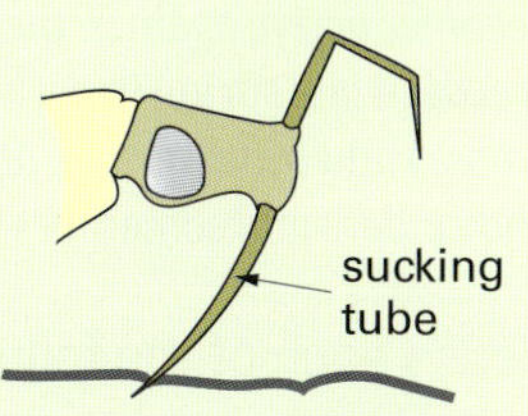

bug's mouthparts

ISBN 978 1 4202 3822 8

6.2 Ecosystems

Why does a particular type of animal or plant live in the place it does?

The frog in the photo is called the motorbike frog. Its scientific name is *Litoria moorei*. The map shows the region of Western Australia in which it is found.

The living place of an organism is called its **habitat**. The photo shows the typical habitat of the motorbike frog in forests and coastal heathland in south-western WA.

EXPLORE ONLINE

Check out the **motorbike frog**. Use the website to see more pictures and listen to frog sounds.

Figure 6.10 The habitat of the motorbike frog

ACTIVITY

Work in groups of 3 or 4 for this activity.

Read the following description of the places in which the motorbike frog is found.

> *The frog lives in wet temperate forests and near creeks within coastal heathland. It hides under stones, logs and in the mud between grasses in late autumn, and becomes active again in the early spring nights. It breeds in late spring to early summer, calling from grasses and beside creeks and dams. It feeds on insects and spiders and sometimes shrimps and worms.*

Discuss each of the following questions in your group. Be prepared to discuss your answers with the whole class.

1. Write a description of the living place of the motorbike frog. In your description, explain what you think 'wet temperate forests' and 'heathlands' are.
2. Why does the frog hide between late autumn and early spring?
3. Why does the frog breed at a certain time of the year? Is the time of year important?
4. Why does the frog become active at night? Suggest what it might do during the daytime.
5. Draw a food web that includes the motorbike frog. Extend your food web to include the animals that might feed on the frog.
6. Look at the colour and texture of the frog's skin. Suggest why it hides under stones, logs and in the mud, rather than in the vegetation around creeks.
7. Why do you think the motorbike frog is found only in this region of Australia?

ISBN 978 1 4202 3822 8

Organisms in ecosystems

The motorbike frog and the other organisms that share its habitat interact with one another and depend on each other for their survival. Organisms are affected by living things called **biotic factors**. Biotic factors include organisms to eat, predators and organisms that compete with them for shelter, food and other resources. The organisms also depend on the non-living things in their habitat (**abiotic factors**).

An **ecosystem** is formed when the organisms in an environment interact with its complex biotic and abiotic factors. An ecosystem may be made up of many smaller habitats which each organism prefers to live in. Usually an ecosystem is balanced so that all organisms have enough of the resources they need to survive, reproduce and thrive.

A terrarium ecosystem

A terrarium can be set up by using a glass aquarium and filling it with sand, soil, rocks, algae, plants, fungi, dead leaves and bark, small logs and even some small animals such as insects, spiders and worms. When water is added to the soil and the glass lid placed on the aquarium to keep the moisture in, you have a terrarium ecosystem. Each of the animals, plants, algae and fungi interact with other organisms and with the non-living parts of the terrarium.

The terrarium can operate for a long period of time without materials from the outside being added to it. All the materials needed for the living things are contained in the ecosystem. No materials are added from the outside—no food for the animals or fertilisers for the plants. The only input is the sunlight, which is needed for the process of photosynthesis.

Figure 6.11 A terrarium is an example of an ecosystem. It contains non-living things such as soil, rocks and logs, and living things such as animals, plants and fungi.

ACTIVITY

Work in a small group to design a terrarium and write instructions on how it might be set up. Use the websites below to help with your write-up.

1 **Designing a terrarium**
Write instructions on how to construct a terrarium so that other people can easily follow them. In the instructions you should:
- list the non-living things you need, such as soil, water, etc.
- list the living things that you think should go in the terrarium, and in which order to add them
- write details and draw sketches of the layout of the terrarium.

2 **How the terrarium functions**
Your terrarium should operate as a functioning ecosystem. For this to occur you need to consider the following questions.
For each question, discuss the answers with other members of your group, come to a consensus and then write the group's answer.
- How would you position the terrarium to ensure there is enough sunlight, without it overheating?
- How are oxygen, carbon dioxide and water used and recycled in the terrarium?
- What happens when organisms die?

3 If you have time, set up the terrarium and test your design. Take care with any animals in the terrarium.

EXPLORE ONLINE

Terrariums
Outlines the basic steps in creating a terrarium.

Terrarium questions and answers
Useful and practical information about setting up terrariums.

ISBN 978 1 4202 3822 8

Why do organisms live where they do?

There are four important reasons why organisms live where they do.

1. For animals, there is ample food. For plants, the soil contains adequate minerals and moisture.
2. There are few animals that feed on the particular organism. These animals are called predators. (For example, the eastern small-eyed snake is a predator of Haswell's frog.)
3. There are few other organisms that need the same type of food, soil nutrients and living space. These other organisms are called competitors because they compete for food and living space. (Another frog called the brown tree frog is a competitor because it lives in the same area as Haswell's frog and eats the same type of food.)
4. The climate and weather conditions are suited to the particular organism. This includes a good supply of water and clean air.

Temperature and activity

The body temperature of frogs changes with the outside temperature. Therefore, in the colder weather of winter, frogs hide away in logs or crevices.

On the other hand, the body temperature of mammals (such as possums, wombats, mice, wallabies, bats and humans) and birds remains fairly constant all year round. These animals can be active all year round.

Most animals that have a changing body temperature live in warm climates. Those that live in regions that become cold in the winter have special ways to survive the lower temperatures: they hide underground or in logs, and become inactive. Some animals (for example, many butterflies and moths) lay eggs as the weather becomes cooler and then die. The eggs last through the winter and do not hatch until the weather becomes warmer. These animals would die without these special methods of survival.

ACTIVITY

Your teacher will do this as a class activity.

You need about 10 ants in a small bottle with a lid, and a plastic container (ice-cream container).

1. Observe the way the ants move about at room temperature.
2. Pour iced water into the container and place the jar of ants into it.

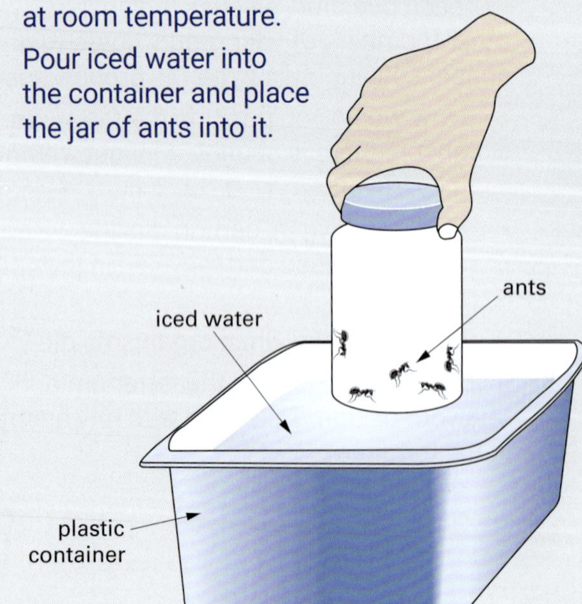

WARNING: You are using living animals. Take care of them, and when finished, put them back outside.

Observe the movement of the ants over the next few minutes. Compare it with their movement in step 1.

3. Pour out the cold water and replace it with warm water. Make sure the water is 'hand hot'. (If you can leave your hand in the water it is at the right temperature.)
4. Place the jar of ants in the warm water.

How did the higher temperature affect the movement of the ants?

Suggest why more insects are found buzzing around lights on a summer evening than in winter.

ISBN 978 1 4202 3822 8

Survival—eat or be eaten

Any animal or plant in a food web can be the food for another animal. To survive in a particular habitat, animals have to find enough food to eat and be clever enough not to be eaten.

To find food, an animal should have:

1. very keen senses—sight, smell, hearing or touch
2. fast muscle reactions—carnivores need to be able to move quickly to catch their prey
3. claws to hold prey and suitable mouthparts to eat the prey.

To avoid being eaten, an animal might:

1. be able to move very quickly
2. avoid being seen by blending in with the colour or texture of the surroundings
3. be poisonous to other animals or have spines, barbs or prickly skin.

The colour and texture of an animal's skin are important for its survival in its habitat. For example, the motorbike frog is brown and has a rough skin texture. When hiding among stones and in the mud banks of creeks, the motorbike frog is not easily seen by predators such as snakes and birds.

Figure 6.12 Echidnas have sharp spines to keep predators away.

In this activity you will make inferences about how animals survive in their particular habitat.

You will need about six preserved animals or coloured photos of animals.

1. Before you begin, draw up a data table like the one shown here. Use at least half a page.
2. Observe each animal carefully. Look for structural features such as shape, size, spines, claws, type of body covering and colour.
3. Write your inferences about survival in the right-hand column of the table. Discuss with your group inferences about the animal's habitat, the foods that it might eat, and how it gets its food.

You could also write inferences about what might eat the animal and how it might avoid being eaten.

What advantages do spines, horns, poisons or stings give animals?

Why do some animals 'play dead' and others roll into a ball shape?

Animals	Observations	Inferences

ACTIVITY

Discovering ecosystems

This activity gives you an opportunity to learn about different ecosystems. You will find interesting and informative websites that you can use to discover more about these ecosystems and the animal and plant life in them.

Produce a report or do a presentation about your chosen ecosystem.

Polar ecosystems—Antarctica

Antarctica is the world's coldest ecosystem (the lowest temperature recorded was −89.6 °C in 1983). The land ecosystem on the Antarctic continent is very simple. Plants and animals have to be able to withstand harsh conditions.

EXPLORE ONLINE

Australian Antarctica Division—Science
This is an excellent website giving information about science in Antarctica.

Use it to find out about:

1 the types of animals and plants in the Antarctic ecosystem, what they feed on, their predators and how they breed
2 how the organisms survive the harsh conditions of the Antarctic ecosystem.

Antarctica webcam
Want to find out what the weather's like at Mawson Station today?

There are many other good websites on Antarctica. Type *Antarctica* into your search engine.

Rainforest ecosystems

Plants that grow in rainforests have to compete for food, water and light in a habitat where there is a dense growth of plants. The leaves of rainforest plants are large to catch as much as possible of the sunlight that comes into the forest.

Use the internet and library to find out where rainforests occur in Australia and what types of conditions are needed for their survival. Find out why rainforest ecosystems have such a diversity and abundance of life as compared with other Australian ecosystems.

Rocky shore ecosystems

Australia is surrounded by sea. The coastline contains sandy beaches, rocky headlands and offshore rocky and coral reefs.

The rocky shores are home to a wide variety of living things. Use the Rocky Reef website below to find out about the organisms on a rocky shore ecosystem.

EXPLORE ONLINE

Rocky Reef
You will find a wealth of photos and information about organisms that live in this marine ecosystem.

MESA
Explore this website by clicking on any interesting links. You can also type your search word in the search frame.

EXPLORE ONLINE

Rainforests Australia
This is a commercial website but it has detailed information about the animals and plants of Australian tropical rainforests and useful links to other sites.

ISBN 978 1 4202 3822 8

1 Consider the following carnivores—cat, hawk, shark and snake.
   a What features do each of these animals have to be able to catch their food successfully?
   b What features do each of the animals have to avoid being eaten?

2 Kim has a pet mouse that she keeps in a terrarium in her house. One day she notices that her little brother has left the lid off the mouse house, and the mouse has escaped.
   a What are the four important factors that will determine whether the mouse survives outside its house?
   b Explain why each of these factors is important.

3 Consider the following herbivores—parrot, green leaf bug, tortoise and horse. What features does each animal have to avoid being eaten by predators?

4 Only mammals and birds are found on the Antarctic continent. Yet in the oceans around the land and icepacks, fish and other animals are found whose body temperatures change. How can these animals live there?

5 The jellyfish in the photo below feeds on small fish and prawns. It does not move very far or very quickly, and usually relies on currents to move it from place to place. Suggest how it might catch its food and avoid being eaten.

6 Explain the terms *biotic* and *abiotic* using examples.

1 The European rabbit was originally found in Spain. It was introduced into Australia in 1788, and on several occasions after this to 'enrich the country'. By 1890, rabbits were in plague proportions in south eastern Australia.

   a Suggest why rabbits spread so quickly in Australia.
   b What do you think 'enrich the country' meant to the early European settlers?
   c Suggest why rabbits were not found in plague proportions in Spain or the rest of Europe?

2 Animals are found in different areas because they can move from place to place. Plants generally cannot move and have to rely on their seeds being scattered over a wide area. Each of the plant seeds below has a different method of being distributed. Suggest what these methods are. (The drawings show the actual sizes of the seeds.)

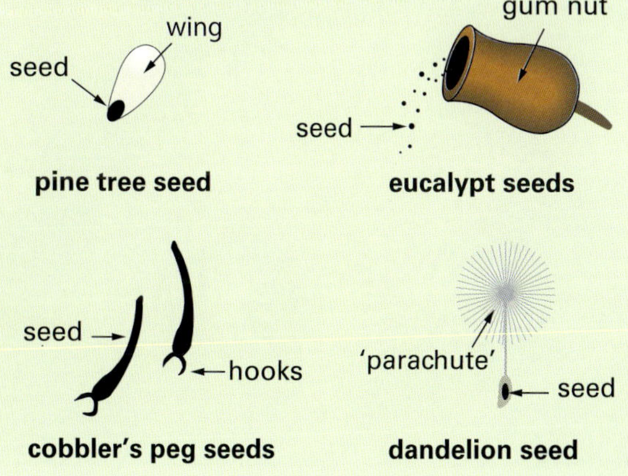

ISBN 978 1 4202 3822 8

6.3 Ecosystems under threat

On a coral reef you will find large numbers of sea anemones, brain coral, brittle stars, anemone fish and coral sharks. All of these organisms interact with each other in various ways, competing for food and living space. Some are producers, others are consumers at various levels, while others are decomposers and scavengers.

Changes in ecosystems

All natural ecosystems change from time to time. Weather changes can cause droughts, floods, cyclones and bushfires. This, in turn, changes the types and numbers of plants and animals that live in particular ecosystems.

If you study a coral reef ecosystem like the one pictured above, you would find the numbers of each type of animal vary from year to year. Sometimes, however, there is a dramatic change.

The crown of thorns starfish is found on coral reefs. It feeds on the small animals that make coral. These are called *coral polyps*. Over the past 40 years, there have been three major outbreaks of the crown of thorns starfish on the Great Barrier Reef. In 2003, there were so many starfish that large areas of reef were destroyed. On one reef alone there were over one million starfish.

Figure 6.13 A coral reef ecosystem

Figure 6.14 A crown of thorns starfish

Why does this happen? First, let's look at some of the features of the starfish.

- A single female can release up to 60 million eggs each year.
- An arm that is severed from a starfish's body can regenerate into a new starfish.
- It can push its stomach outside its body to make eating the polyps more efficient.
- It releases a poison from its spines which can cause the death of small animals.
- It has few predators because its body is thickly covered in sharp spines.

The diagram below shows part of a larger coral reef food web containing the crown of thorns starfish.

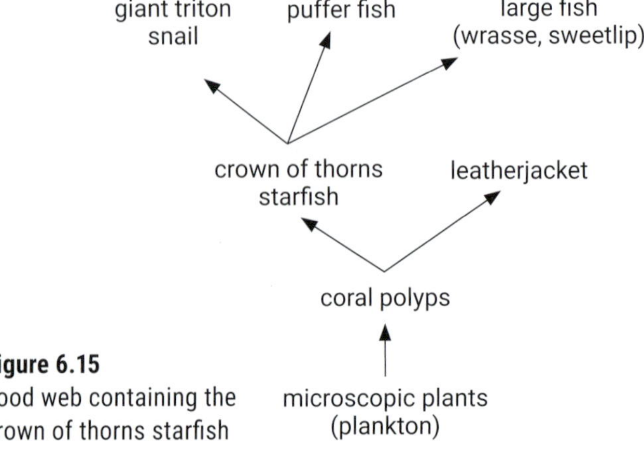

Figure 6.15 Food web containing the crown of thorns starfish

ISBN 978 1 4202 3822 8

What causes the changes?

Scientists are not sure what causes the huge population explosions of the crown of thorns starfish. Some ideas have been put forward.

1 Natural changes

Because one starfish produces so many offspring, small changes in sea water temperature, the amount of salt in the water and the amount of food (microscopic plants) can cause more starfish to survive. This can cause a very rapid increase in the population.

2 Removal of predators

The starfish has very few predators, so if any predators are removed from the food web, the starfish numbers will increase. For many years, triton snails and Maori wrasse, which feed on starfish, were taken from the reef. They are now protected. Over-fishing around the reef could also reduce the populations of other predator fish.

Figure 6.16 The humphead Maori wrasse is a very active predator of the crown of thorns starfish. It was commercially fished up until 2003, when it became illegal to fish or keep the fish.

3 Nutrients from rivers

Rivers carry nutrients into oceans. In times of high rainfall or floods, rivers carry extra nutrients (fertilisers from farms) into oceans. Scientists have found that extra nutrients in the reef waters cause an increase in the microscopic plants which, in turn, provides food for the developing starfish, thus increasing their numbers.

Biodiversity

There is a huge variety of living things on planet Earth. This is called **biodiversity**. It describes the different animals, plants, algae and microorganisms that form all the ecosystems on Earth, such as the coral reef ecosystem in the photo on the previous page.

Why is biodiversity important?

Ecosystems are constantly changing due to factors such as diseases and severe weather conditions. Ecosystems that have a wide variety of organisms tend to be able to resist these changes. On the other hand, in an agricultural ecosystem with low biodiversity such as a wheat field, a disease or a predator has a greater chance of destroying most of the organisms. Diverse ecosystems are also a source of new foods and medicines for humans.

Threats to biodiversity

The world has an estimated 13.6 million species of living things. Australia has more than one million of these, and over twice as many as Europe and North America combined. This means we have a special duty to protect out biodiversity.

The photo below shows a coral reef that has been damaged by silt that was carried into the ocean from agricultural land. Notice there are almost no fish. This indicates that there is now little biodiversity on the reef.

Figure 6.17 Soil carried into the ocean by a river has covered the coral on this reef, killing the coral organisms.

ISBN 978 1 4202 3822 8

There are four main threats to biodiversity.

1 ***Introduced species***

An introduced species describes an organism that is not native to that ecosystem. Its introduction can cause major disturbance to the food webs in the ecosystem. This is almost always caused by humans, sometimes accidentally and other times on purpose.

For example, the cane toad was introduced into Queensland in 1935 to control the cane beetle which was destroying sugar cane crops. It has now spread widely across Australia and is responsible for the reduction in populations of frogs, reptiles, fish and mammals.

Figure 6.18 The cane toad was introduced into Queensland from Hawaii in 1935 to control the cane beetle.

2 ***Destruction or alteration of habitats***

The clearing of land for building new houses and industries destroys the plant populations and also changes the food webs that exist in the ecosystem. It has been estimated that when 100 hectares of native forest is cleared, about 1500 birds die from exposure, starvation and stress.

Figure 6.19 Land being cleared for a housing estate

3 ***Chemical damage (pollution)***

The unwise use of pesticides and herbicides on farms can have a major effect on the populations of native organisms. Most organisms are very sensitive to these chemicals and can be affected if the poisons are carried into creeks, rivers and lakes. Excessive use of fertilisers also affects the plants on the surrounding land and, in turn, changes the balance of the food webs.

4 ***Climate change***

Climate change is a serious world-wide problem and could be the major factor that causes loss of biodiversity. Australia is one of the countries most at risk from climate change. Droughts and bushfires could result from higher temperatures and lower rainfall. Flooding of coastal areas could result from a rise in sea levels due to melting polar ice caps.

On the next page you can investigate some of the factors that can cause changes to the living places of organisms.

ISBN 978 1 4202 3822 8

INVESTIGATION 6.2 Factors that affect organisms

Aim

To model what might happen to ecosystems that are affected by floods or increased water temperatures.

Risk assessment and planning

You can do the two parts of this investigation in any order.

- For each part, read through the Method and make a list of the materials you need. Select the materials for each part from your equipment store or trolley.
- For each part, make a list of the safety precautions you will need to take.
- Prepare data tables where appropriate.

Materials

- methylene blue solution
- oxygen-removing solution (50 g/L **sodium dithionite/hydrosulfite**, freshly prepared)
- 250 mL glass graduated cylinder
- thermometer
- 250 mL beaker
- mud or silt
- marking pen
- burner, tripod and gauze mat
- dropper
- stirring rod
- sheet of white paper

Toxic

Teacher note: Fill a few buckets with tap water a couple of days before use. This will be enough water for the whole class.

PART A Turbidity

Method

Turbidity refers to how clear water is. If water is turbid, it is cloudy. This cloudiness is due to the amount of suspended particles of silt in the water.

1 Use a marker pen to draw an X on a sheet of white paper.

2 Place a graduated cylinder over the paper so you can see the X clearly.

3 Fill a 250 mL beaker with some clean tap water and pour it into the graduated cylinder up to the 250 mL mark.

Is the X clearly visible? Take note of the clarity of the water and how well you can see the X on the paper.

4 Tip out the water in the graduated cylinder.

5 Fill the beaker with tap water and add about a teaspoonful of silt or mud. Stir it thoroughly.

6 Sit the graduated cylinder over the X and gently pour the muddy water into it. Stop when you cannot see the X.

At what mark did you lose sight of the X?

7 Test what happens when you add more silt.

Discussion

1 Account for the difference in the visibility in the water.

2 How does this part of the investigation model the effect of floods on a lake or ocean ecosystem?

3 Explain how turbidity might affect the organisms in a lake, creek or river, or in the ocean.

ISBN 978 1 4202 3822 8

PART B Dissolved oxygen

Method

1 Pour exactly 200 mL of water from the bucket into a beaker.

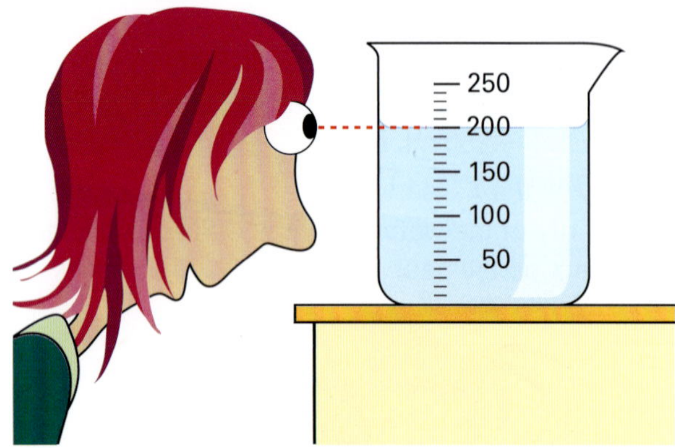

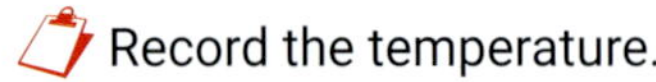
Make sure there is exactly 200 mL of water.

2 Use a thermometer to find the temperature of the water.

Record the temperature.

3 Add three drops of methylene blue to the water and stir very gently.

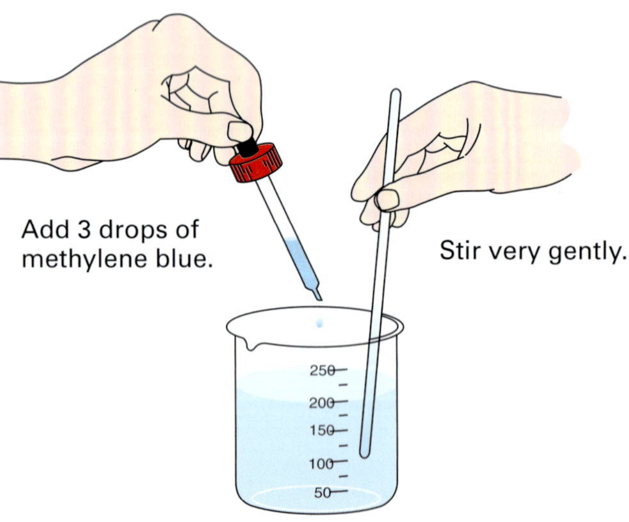

4 While stirring, add oxygen-removing solution a drop at a time until the blue colour just disappears.

Record the number of drops added.
This is a measure of the amount of dissolved oxygen in the water.

5 Wash out the beaker and pour another 200 mL of water into it.

6 Set up a burner, tripod and gauze mat, and heat the water to about 50 °C.

7 Turn the burner off and repeat steps 3 and 4.

Record the number of drops added.

Discussion

1 Use your results to write an inference about the effect of temperature on the amount of oxygen that dissolves in water.

2 Suggest what might happen to the organisms in a lake if the water temperature increased during prolonged hot weather?

3 An electricity power plant was built beside a river. It used the river water in its cooling towers and released the heated river water back into the river. Suppose you are an environmental engineer. What recommendations would you give to the plant owners? Make sure you explain your recommendations.

PART C Inquiry

How does silt in water affect photosynthesis?

Use the procedures from this investigation to design an experiment to answer the question.

Note: Use an aquarium plant such as elodea in large glass jars or small aquariums. See Investigation 6.1 on page 136.

Make sure you are able to compare the amount of photosynthesis in turbid water with that in clear water.

ISBN 978 1 4202 3822 8

SCIENCE AS A HUMAN ENDEAVOUR

Carp—an environmental problem

Have you ever wondered what happens to goldfish that are released into lakes or rivers? Goldfish are a type of carp and have become a pest in Western Australian waterways. They grow to 40 centimetres long and weigh up to 2 kilograms.

Carp are a pest across Australia. Since being introduced to Victoria and New South Wales over 100 years ago, carp are now the most abundant fish in some freshwater areas, where they cause similar problems to the goldfish in Western Australia.

Why are carp such a problem?

Carp have been successful in Australian rivers because they can tolerate a wider variety of conditions including low oxygen levels, high water turbidity and pollutants than most native fish. They also eat a wide range of food—snails, native fish eggs, insect larvae, seeds and water plants.

But it's the carp's destructive feeding habits that cause problems for other organisms. Because they stir up the bottom of the river while searching for food, they increase the silt in the water. The turbid water reduces the light for water plants and makes it unsuitable for native fish. Carp also uproot plants when they feed. It is possible that the uprooting of plants causes erosion and the collapse of river banks.

Carp breed in large numbers. A female carp can produce more than 1 million sticky eggs. The murky waters aid in the survival of eggs and baby fish.

Solving the problem

Carp are now found in very large numbers in the river systems in Australia. Scientists have made suggestions for managing the carp problem and the government is funding programs to get rid of carp.

1. Clean up waterways to reduce stagnant and murky water and restore natural water flows.
2. Revegetate river banks and surrounding land to stop the flow of mud and silt into the waterways.
3. Encourage commercial harvesting of carp to supply fish shops both here and abroad, and also to make high-quality fertiliser.
4. Kill the fish with poisoned food pellets in isolated waterways such as farm dams.
5. Introduce viruses specifically to kill carp.

Questions

1. What sort of water conditions can carp tolerate, and how do they make the waterways unsuitable for native fish?
2. Use the information on this page to draw a food web that includes carp.
3. How has land clearing made conditions in waterways more suitable for carp?
4. Scientists are reluctant to introduce poisons and viruses into the waterways to kill carp. Suggest why.
5. Draft a plan that you could submit to the federal government that would reduce the carp population in Australian waterways.

ISBN 978 1 4202 3822 8

1 Look at the food web (Figure 6.15) on page 150.
 a Which organisms are producers?
 b What type of consumer is the giant triton snail?
 c Which organism is a competitor of the crown of thorns starfish?

2 The crown of thorns starfish is found in very low numbers on a certain coral reef. All the other organisms in the food web on page 150 are found there. Suppose conditions change and the number of starfish rapidly increases. Infer what might happen to the other organisms.

3 Use this food web to answer the questions below.

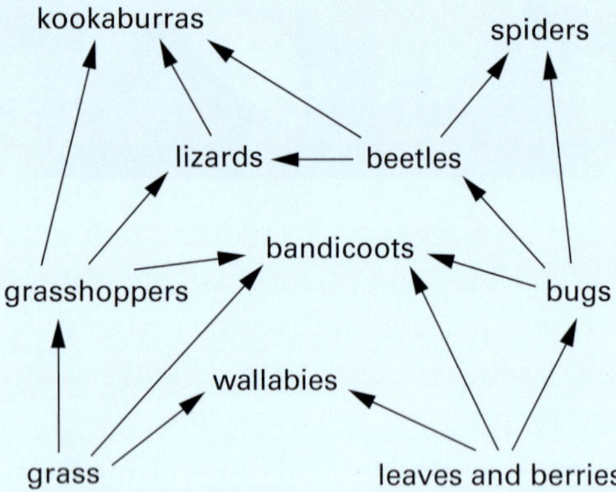

 a Which organisms are predators of bugs?
 b Which organisms are competitors of beetles?
 c Which organisms might affect the population of bandicoots? Explain your answer.
 d Suppose feral cats killed many of the lizards in this community. Infer what might happen to the other organisms in the food web.
 e Pesticide being sprayed on a nearby farm drifted into the area of this food web and killed many insects and spiders. Suggest how the organisms in the food web could have been affected.

4 a What is biodiversity?
 b Explain briefly the factors that can threaten biodiversity.
 c Which of the factors are caused by the actions of humans? Give a reason for each.

1 Scientists were studying the populations of four types of frog that lived in the same location in southern NSW. The frogs all ate a variety of insects including crickets, beetles, bugs and flies. The frogs were preyed upon by snakes and owls.

The bar graph shows populations of the four different frogs in 2005.

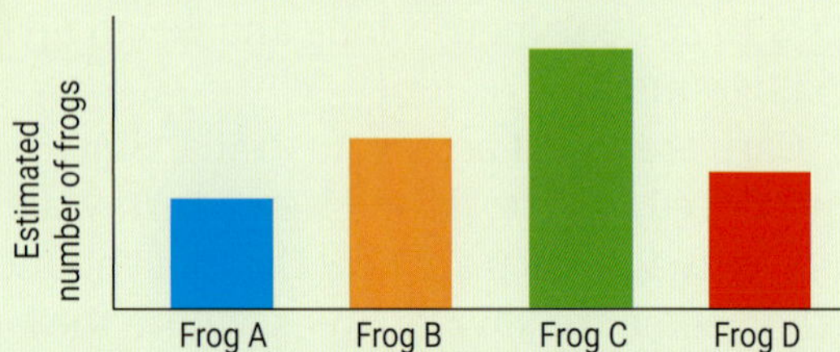

In a survey four years later, the scientists noticed the presence of the introduced fish gambusia (gam-BOO-see-a) in many of the creeks. This fish feeds on the eggs and tadpoles of frogs.

In 2010, the scientists repeated their count of the frogs. The new data is shown below.

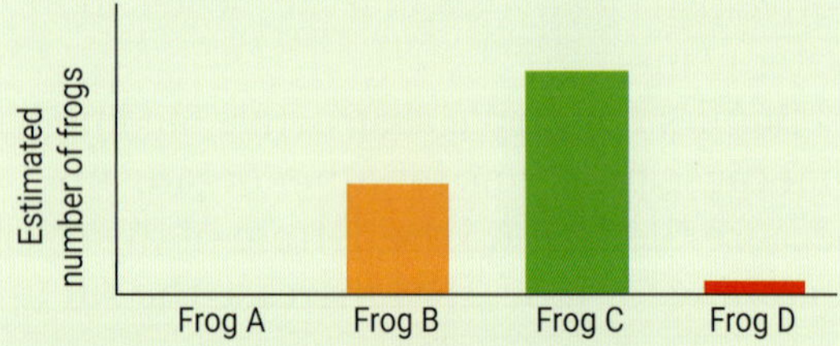

 a Compare the results of the two studies. What can you infer from the results?
 b Draw a food web for the organisms without the gambusia. Suggest how the introduction of the gambusia might affect this food web.
 c How is the biodiversity of the ecosystem affected by the introduction of the gambusia?
 d Are the scientists justified in saying that the gambusia is the cause of the change in biodiversity? Give reasons for your answer.

ISBN 978 1 4202 3822 8

MAIN IDEAS

Copy and complete these statements to make a summary of this chapter. The missing words are on the right.

1 A food chain is a simple way of showing how animals and plants depend on each other for food. A _____ consists of a number of food chains linked together.

2 Respiration is a process in which all living things use _____ and _____ to produce _____, and give off carbon dioxide and water.

3 Plants and algae are called _____ because they photosynthesise and make their own food using the energy from the _____; carbon dioxide and _____ are used in the process.

4 Certain_____ and _____ are called decomposers because they break down the bodies of dead organisms into simple substances.

5 The living place of an organism is called its _____; a system of interrelationships between living things and their surroundings is called an _____.

6 Biodiversity describes the _____ of all the living things on Earth. There are a number of threats to biodiversity, including the _____ of habitats and climate change.

7 The non-living factors that affect organisms in their habitat are called _______.

bacteria
habitat
abiotic
destruction
variety
sun
water
ecosystem
glucose
food web
oxygen
energy
producers
fungi

CH•6 REVIEW

1 Which statement is *not true* for food chains?
  A They show which organisms feed on which.
  B They mostly begin with plants or algae.
  C They show the number of living things in each step.
  D They show the order in which organisms are fed upon by others.

2 Which of the following food chains is correct?
  A grass → snake → frog → grasshopper
  B grasshopper → grass → frog → snake
  C grass ← grasshopper ← frog ← snake
  D grass → grasshopper → frog → snake

3 Read the paragraph below.
  a Use this information to draw a food web.
  b Give a name to the type of community to which you think these organisms belong.

Emily and Max were observing the organisms in a small area of open eucalypt forest. They noticed bugs sucking the sap from the eucalypt leaves and honeyeaters and butterflies eating the nectar from the eucalypt flowers. Grasshoppers ate the new grass shoots and mice ate the grass seeds. Beetles and butcher birds fed on bugs. The butcher birds also ate grasshoppers, beetles and butterflies. Hawks were seen to attack honeyeaters and butcher birds and to eat mice. Snakes were seen to eat mice.

4 In which ways are decomposers and scavengers similar? How are they different?

ISBN 978 1 4202 3822 8

5 The diagram below shows some of the animals and plants that live in a pond.
a Which organisms are producers?
b Which organisms are second-order consumers?
c Which organisms are competitors of small fish? Which are their predators?

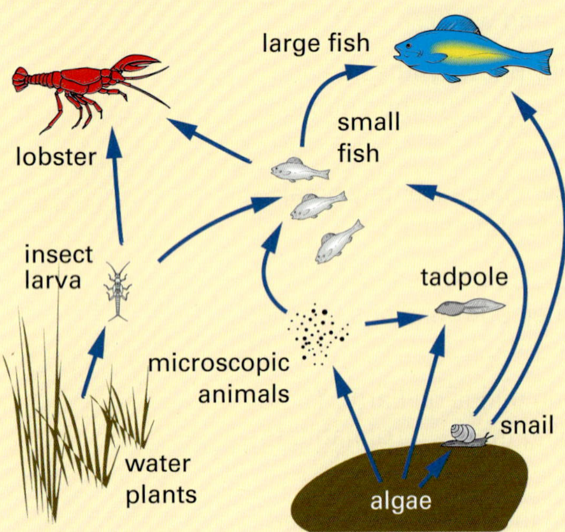

6 Explain why an ecosystem would not survive without decomposer organisms.

7 Matthew and Jessica drew this diagram to show some interactions in a particular ecosystem.

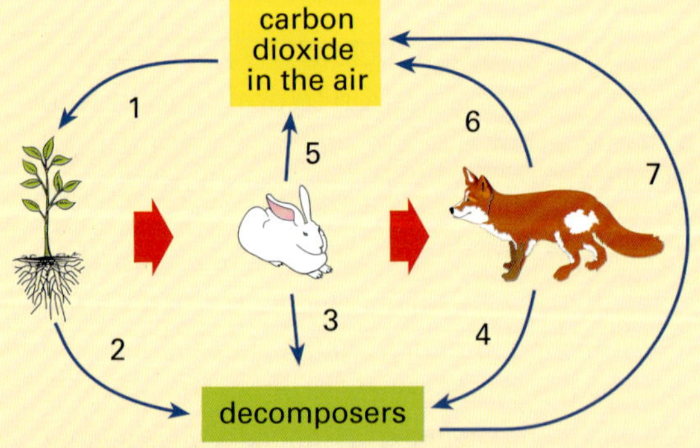

a What is the process labelled 5?
b Describe what happens in 2, 3 and 4.
c What do the red arrows mean?
d Why is process 1 important to this food web?
e Which path shows carbon dioxide being taken out of the air? Which shows it given off into the air?

8 When house cats escape into the bush and feed on native animals they are said to be feral. Feral cats are powerful and skilful hunters and will eat mammals, birds, lizards and insects.
Why are feral cats a threat to the biodiversity in an ecosystem? In your answer explain what biodiversity means.

9 Tom designed an experiment. He set up five tubes with each containing the same amount of water and the same-sized piece of water plant. Tube 1 contained clear water. Tube 2 contained water with a small amount of silt. Tubes 3, 4 and 5 contained water with an increasing amount of silt.

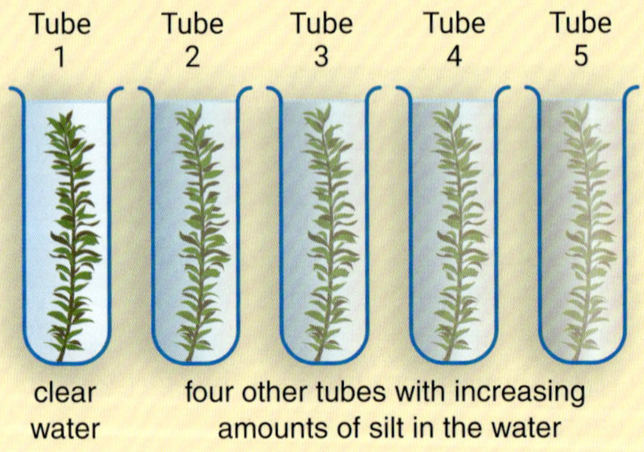

Tom placed all tubes under a large aquarium light. After 24 hours he measured the amount of oxygen in each tube. He expressed the amount of oxygen in tube 1 as 100%. His results are shown in the table.

	Tube 1	Tube 2	Tube 3	Tube 4	Tube 5
% of oxygen	100%	94%	88%	80%	73%

ISBN 978 1 4202 3822 8

a What was the aim of Tom's experiment?
b Which factor did Tom change in his experiment? Which things did he keep the same?
c Write a conclusion for Tom's experiment.

10 a The following diagram shows photosynthesis in a plant. Label parts A to D.

b Name the green substance in plants that absorbs sunlight.
c Which substance that stores energy is produced during photosynthesis?
d In your own words, describe the process of photosynthesis.

11 The following diagram shows the process of respiration. What do labels A to D represent?

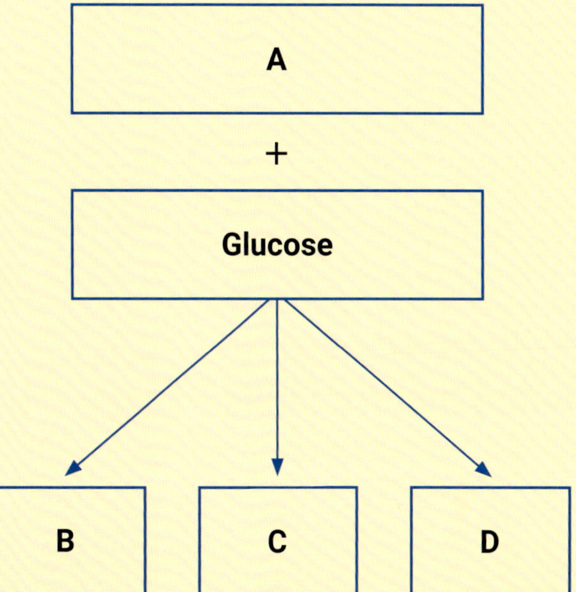

Check your answers on page 241.

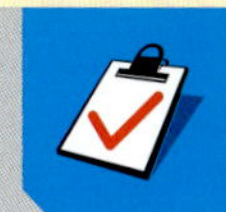

ISBN 978 1 4202 3822 8

IN THIS CHAPTER

Science Understanding

- compare times for the rotations and orbits of the Earth, sun and moon
- use a model of the Earth to answer questions about day and night
- explain the seasons as a result of the tilt of the Earth's axis as it orbits the sun
- use models to understand phases of the moon and eclipses
- relate the tides to the positions of the Earth, moon and sun
- investigate how advances in telescopes and space probes have provided new evidence about space

Science Inquiry Skills

- give examples of different types of models and how they are used in science

CH•7 Earth, moon and sun

ISBN 978 1 4202 3822 8

GET STARTED: *QUESTION*

In a small group, discuss each of these questions:

- This photo of the Earth was taken from the moon. What do you notice about the surface of the moon? Why can you see only half of the Earth?
- Who was the first person to walk on the moon? When did this happen? Why was a spacesuit needed?
- The photo below was taken by pointing a camera directly south in the night sky and leaving the shutter open for many hours. How can you explain the circular star trails?

ISBN 978 1 4202 3822 8

7.1 How the Earth moves

People once thought the Earth was flat. Today we know that it is approximately spherical—round like a ball. The Earth is divided into two *hemispheres* (half-spheres) by an imaginary line called the equator. The part that includes Australia is called the Southern Hemisphere, and the other half is called the Northern Hemisphere. Astronomers have found that the diameter of the Earth is about 13 000 km. They have also found that it is slightly flattened at the poles.

Each day you see the sun rise in the east and set in the west. Does this mean that the sun moves? For thousands of years people thought so. For instance, people in India believed the Earth was a circular disc surrounded by the ocean. In the centre of the world was a great mountain. The sun went around the mountain once a day. At sunset the sun went behind the western side of the mountain. It travelled behind the mountain during the night and came out on the eastern side at sunrise. We now know that it is the Earth that moves. It only appears as though the sun moves around the Earth.

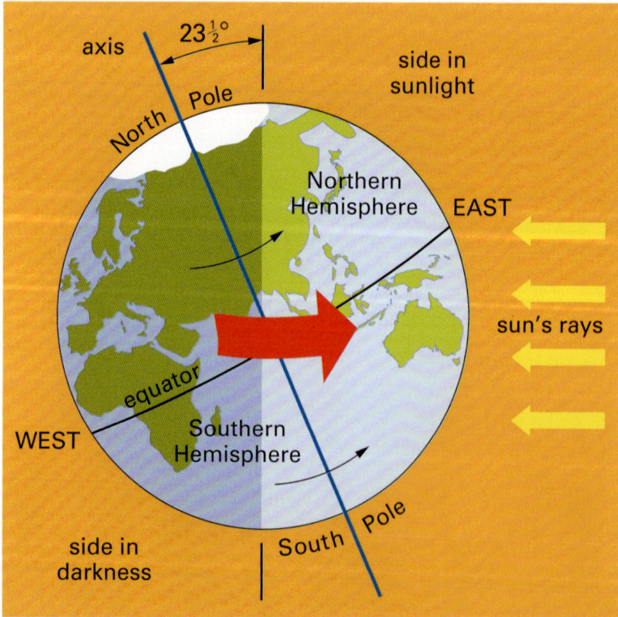

Figure 7.1 The Earth rotates on its axis in an anticlockwise direction when viewed from space above the North Pole.

Earth's rotation

The **axis** of the Earth is an imaginary line through the centre of the Earth from pole to pole. This axis is tilted at an angle of 23½ degrees. The Earth **rotates** or spins on this axis, rotating once every 24 hours. This means that people on the equator are moving at 1700 km/h! You don't feel or see movement because everything else around you moves at the same speed.

As the Earth rotates from west to east, the sun, moon, stars and planets all seem to move the other way—from east to west. This is why you get the circular star trails in the photo on the previous page.

It is because the Earth rotates on its axis that we get night and day. As the Earth rotates, only one half of it faces the sun at any one time. While this half is in sunlight, the other half is in darkness.

Figure 7.2 Australia has just moved into day as the Earth rotates towards the right.

ISBN 978 1 4202 3822 8

Earth's revolution

As well as rotating on its axis, the Earth travels through space around the sun. This is why the stars appear to change position in the sky throughout the year. For example, Orion (the Saucepan) appears in the north-east in summer and disappears in the north-west in autumn. We say the Earth *revolves* around the sun. The path it follows is called its **orbit**. This orbit is almost circular, but slightly oval.

The time taken for one complete **revolution** of the sun is one year. During this time the Earth rotates 365¼ times. This means there are 365¼ days in a year. This is very difficult to divide into equal parts for our days and weeks. So we consider each year as having just 365 days and every fourth year, or *leap year*, has 366 days.

The Earth stays in its orbit because of the gravitational force of attraction between it and the sun. In the activity below you can use a model to help you understand this force.

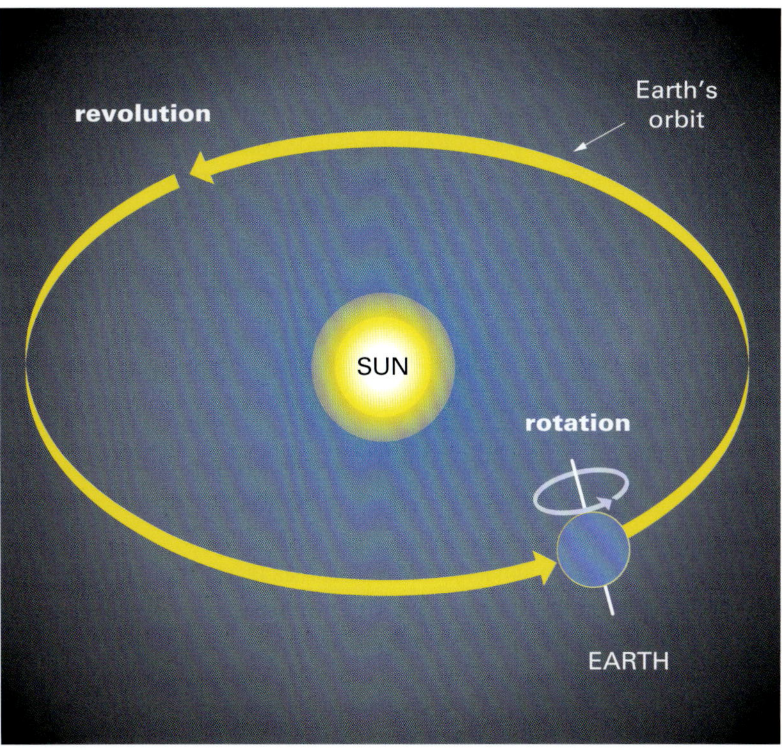

Figure 7.3 The Earth rotates on its axis and revolves around the sun, both in an anticlockwise direction (viewed from space above the North Pole).

ACTIVITY

It is best to do this activity outside where you have plenty of room.

1 Fasten a piece of thread to a styrofoam ball. Pass the thread through an empty pen case or piece of plastic pipe. Tie the other end to a small rubber stopper or similar weight, as shown.

2 Hold the pen case and whirl the ball in a circle so that it stays the same distance from your hand.

In your model, what force keeps the ball in orbit?

What will happen if you cut the string while the ball is moving? Try it. (Cut between the pen case and the weight.)

What force keeps the real Earth in orbit around the sun?

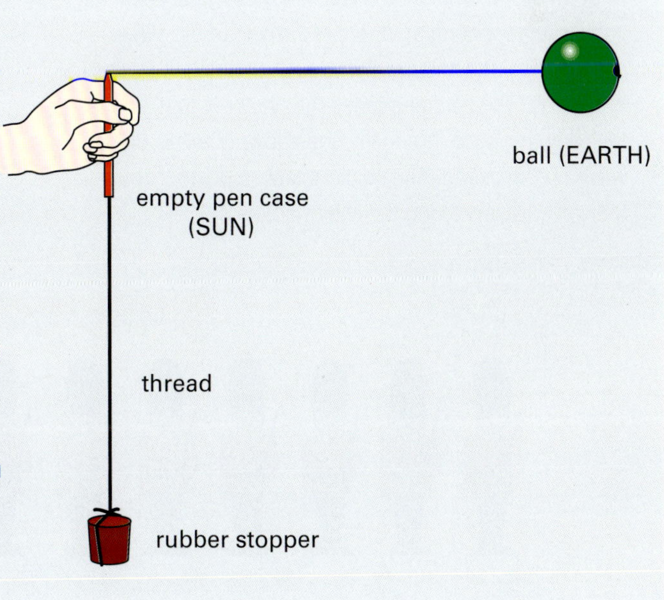

ISBN 978 1 4202 3822 8

SKILLBUILDER

Models

A **model** is a way of representing something that is too small to be seen, or too large or too complicated to be studied easily. For example, in primary school you might have made a model of a volcano. The model helps you to understand how a volcano works.

You can't look from space and see the Earth and the planets moving around the sun. However, if you build a model you can see the relative sizes of the planets and how far apart they are. In this chapter you will use models to represent or show how scientists infer that the sun, Earth and moon move in relation to each other.

Models of aircraft are tested in wind tunnels before being built. Models of buildings can be tested in earthquake simulation machines. As a result of these tests, the plans may need to be modified.

Scientists use computer models to make predictions about the future. If these predictions are accurate, then the model is a good one. If the predictions are not accurate, then the model needs to be modified or discarded. The illustration at top right shows a computer model prediction for global warming in the year 2030.

Sometimes you can think up your own model to represent something you don't understand very well. For example, Jack wanted to explain how sound travels through air. He imagined a line of toy soldiers, each soldier representing an invisible particle of air. When a soldier is pushed sideways it springs back up again—it vibrates. The soldier pushes the one next to it, and in this way the 'push' gradually goes down the line, like a wave on the ocean. The soldiers stay in position, but the wave travels along the line of soldiers.

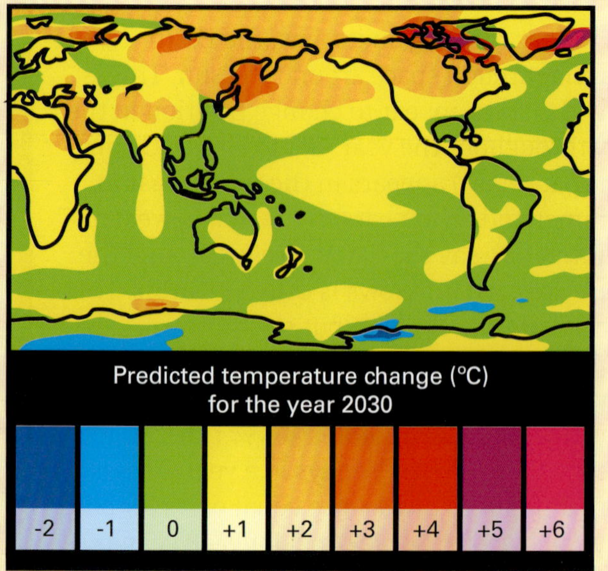

Figure 7.4 Computer model predicting global warming.

Questions

1 a In the activity on the previous page, what does the ball represent? What does your hand represent?
  b Is it a good model of how the Earth orbits the sun? Explain your answer.

2 Why do scientists make models of atoms and molecules?

3 Why would high-rise buildings need to be tested in wind tunnels?

4 Your teacher will show you some classroom models. What do they show that you can't see with the real thing?

5 a What was Jack trying to explain with his toy soldiers?
  b Why did he need to use a model?

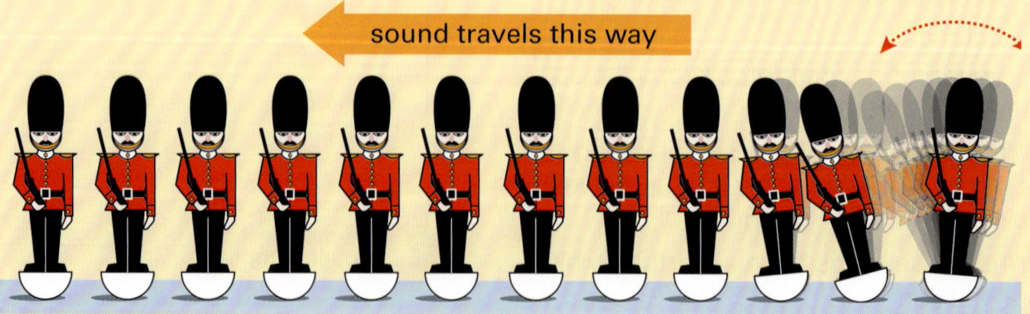

Figure 7.5 Jack's model to explain how sound travels

ISBN 978 1 4202 3822 8

What causes the seasons?

We divide the year into four seasons—spring, summer, autumn and winter. Summer is much warmer than winter, and the days are longer. Seasons are caused by a combination of the tilt of the Earth's axis and the Earth's revolution around the sun. The particular season depends on whether the Earth's axis is tilted towards the sun or away from it.

In the model in Figure 7.6, the Southern Hemisphere is tilted towards the sun. This means the Northern Hemisphere is tilted away from the sun. Let's represent sunlight by torch beams. Torch beam A hits the Southern Hemisphere square-on, and shines over a small area. Beam B hits at an angle and is spread out over a larger area of the Earth. Beam A warms the Earth more than beam B. So, where beam A hits the Earth, it would be summer, and where beam B hits it would be winter.

Summer in Australia is when the Southern Hemisphere is tilted towards the sun, as on the left in the diagram below. In this position there is more of the Southern Hemisphere than the Northern Hemisphere in sunlight. Six months later we are on the other side of the sun. The Southern Hemisphere is now tilted away from the sun, and it is winter.

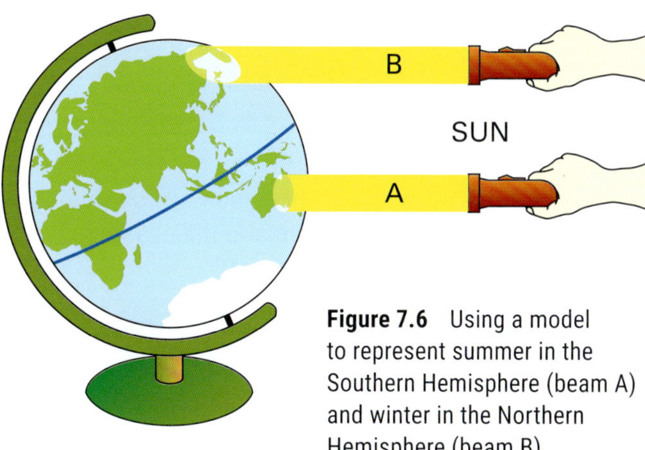

Figure 7.6 Using a model to represent summer in the Southern Hemisphere (beam A) and winter in the Northern Hemisphere (beam B)

The seasons in the Southern Hemisphere are the opposite of those in the Northern Hemisphere. If the Southern Hemisphere is tilted towards the sun (summer), then the Northern Hemisphere is tilted away from the sun (winter).

Investigation 7.1 should help you to understand day and night and the seasons.

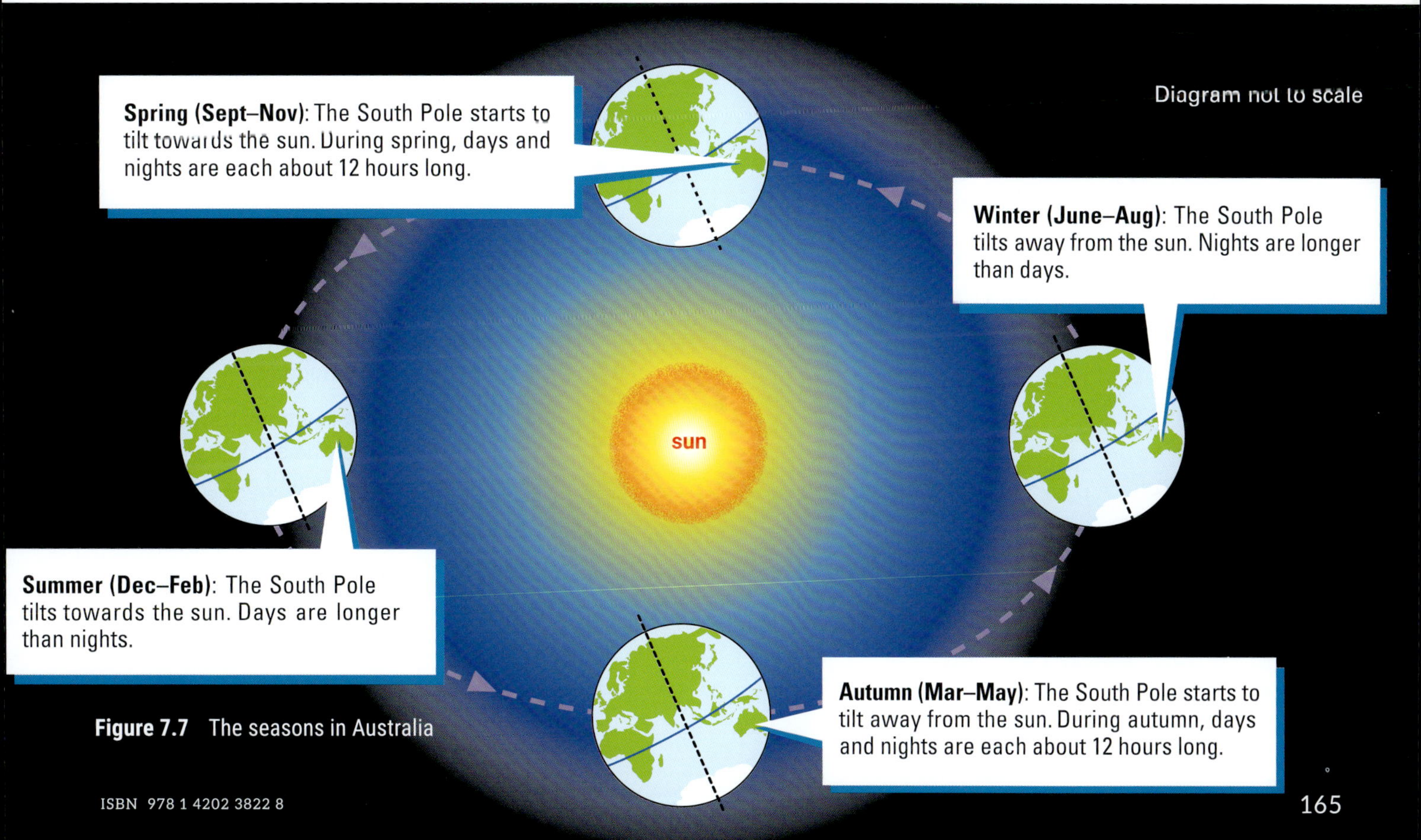

Diagram not to scale

Spring (Sept–Nov): The South Pole starts to tilt towards the sun. During spring, days and nights are each about 12 hours long.

Winter (June–Aug): The South Pole tilts away from the sun. Nights are longer than days.

sun

Summer (Dec–Feb): The South Pole tilts towards the sun. Days are longer than nights.

Autumn (Mar–May): The South Pole starts to tilt away from the sun. During autumn, days and nights are each about 12 hours long.

Figure 7.7 The seasons in Australia

ISBN 978 1 4202 3822 8

INVESTIGATION 7.1 A sun–Earth model

Aim

To use a model to explain how the Earth's motions in space cause day and night and the seasons.

Materials

- polystyrene ball (about 7 cm in diameter)
- wooden cooking skewer or knitting needle
- projector (overhead projector or slide projector)
- 2 pins

> **Risk assessment and planning**
> Read both parts of the investigation carefully before you start. You will need to make accurate observations, so work in a group and make sure you know who is doing what.

PART A Day and night

Method

1 Carefully push the skewer through the centre of the polystyrene ball. Be careful—it has a sharp point.

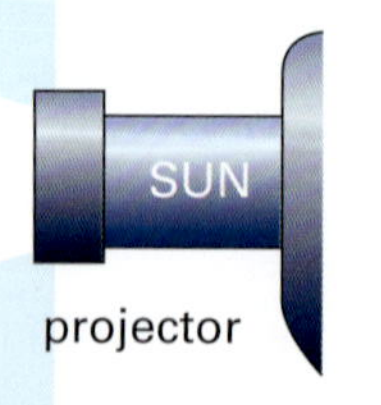

You now have a model of the Earth. Of course there is no real axis going through the centre of the Earth. The model simply helps you understand how the Earth rotates.
Note: Instead of the ball and skewer you could use a geography globe.

2 Draw a rough sketch of Australia in the bottom half (Southern Hemisphere) of the ball. Also draw in the directions north, south, east and west.
Represent yourself by a pin stuck in the ball where you live.

3 Turn on the projector. Hold the ball in the light, and observe from the side.
What does the lit half of the ball represent? What does the dark half represent?

4 Tilt the top of the ball away from the projector as shown, keeping the pin representing you on the light–dark line.
What happens at this time of day?
Turn the ball *slowly* from west to east. Notice that the sun rises in the east.

5 Keep turning slowly until your pin reaches the light–dark line on the other side of the ball.
What happens at this time of day?
In which direction do you look to see the sun set?

6 Watch your shadow (from the pin) as you move from sunrise to sunset.
How does its length and position change?
How can you tell from your shadow when it is midday?

7 Do all places in Australia have day and night at the same time? Use your model to find out.

Discussion

Do you think this is a good model to explain day and night? Explain.

ISBN 978 1 4202 3822 8

PART B The seasons

1 Stick another pin in the ball to represent a person in the Northern Hemisphere. Place it so that the other person is directly north of you—between you and the North Pole, as shown.
With the axis tilted this way, is it summer or winter in Australia?

2 Keeping the ball tilted, rotate it from the sunrise position to the sunset position.
Where does the sun rise first—at your place, or in the Northern Hemisphere?
Where does the sun set first?
Was the day the same length for both people? Explain.
Is there any place on the Earth at this time of year where the sun doesn't set? Is there any place where it is dark for 24 hours?

3 Repeat step 2, but this time tilt the top of the ball *towards* the projector.
In this position is it summer or winter in Australia?
What is the relationship between the seasons and the length of the day?

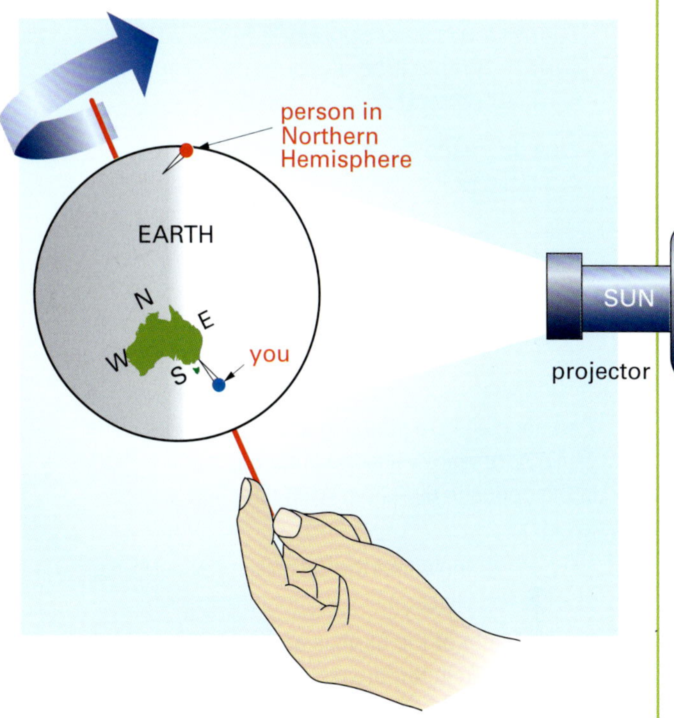

4 If the Earth's axis was not tilted, would the day be the same length for both people? Use your model to find out.

CHECK

1 Match each of these terms with the definitions below.

axis	orbit
hemisphere	revolution
model	rotation

a a way of representing something that cannot be observed directly
b the spinning of a body on its axis
c the path followed by an object in space as it revolves around another object
d imaginary line about which an object in space spins
e half of a sphere
f movement of one body around another body

2 Copy and complete these sentences.
a The Earth takes _____ hours to rotate once on its axis.
b We have seasons because the Earth's axis is _____.
c In summer the _____ are longer than the _____.
d The Earth's axis is tilted at an angle of _____.
e When it is spring in Australia, it is _____ in the Northern Hemisphere.
f The Earth stays in orbit around the sun because of a _____ force.
g In December, the South Pole is tilted _____ the sun.

3 How long does it take the Earth to revolve around the sun? How many times does the Earth rotate on its axis in this time?

4 Write a description of how the Earth moves by comparing it to a sideshow ride.

ISBN 978 1 4202 3822 8

5 Suppose you lived at point X on the Earth's surface. Would it be day or night? Would it be summer or winter?

6 Why do shadows in the southern half of Australia point south at midday?

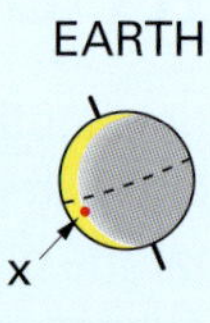

CHALLENGE

1 Two penguins are standing at the South Pole. One walks northwards, and the other turns and walks in the opposite direction. What direction is this? Explain.

2 How do we know that the Earth is spherical? Give at least two reasons.

3 The diameter of the Earth is about 13 000 km. Calculate its circumference.

4 In winter the sun is more to the north (lower in the northern sky) than it is at the same time in summer. Why is this?

5 Find out the meanings of the terms *solstice* and *equinox*.

6 Where is the 'Land of the Midnight Sun'? Why is it called this?

7 In Darwin, the Earth's surface is travelling at about 1500 km/h. In Melbourne, it is travelling at only 1000 km/h. Why is this?

8 The planet Mercury is not tilted. It rotates slowly on its axis once every 59 Earth days, and revolves around the sun in 88 Earth days. What would the days, nights, years and seasons be like?

9 Look back at the photo of the star trails on page 161. If you were at the equator and pointed the camera straight up, the trails would be straight lines. Try to explain this.

EXPLORE

By making a few measurements you can calculate the distance to the sun (if you know its diameter).

1 Obtain two small pieces of cardboard. Use a pin to put a small hole in one of them.

2 Attach the two pieces of cardboard to a metre ruler 50 cm apart, as shown.

3 Tilt the ruler until the spot of light from the pinhole is circular.

4 Carefully measure the diameter of the spot (in millimetres).

5 Use the formula below to calculate the distance to the sun.

$$\frac{\text{distance to sun}}{\text{diameter of sun (1 400 000 km)}} = \frac{\text{distance from pinhole to spot (500 mm)}}{\text{diameter of spot (in mm)}}$$

$$\text{distance to sun (in km)} = \frac{500 \times 1\,400\,000}{\text{diameter of spot (in mm)}}$$

ISBN 978 1 4202 3822 8

7.2 Phases, eclipses and tides

Phases of the moon

The moon produces no light of its own. We see it only because it reflects light from the sun (Figure 7.8). If any part of the moon's surface is not in sunlight, we cannot see that part because space (the background) is black too. Because the moon revolves around the Earth, we see different amounts of its surface lit up during its 29½-day cycle.

Look at Figure 7.9 below. When the moon and the sun are on opposite sides of the Earth (position 5), you see a **full moon**. The entire side of the moon facing Earth is lit up by the sun. When the moon and sun are on the same side of the Earth (1), there is a **new moon**—the side facing the Earth is in darkness. You see quarter moons when the moon is one-quarter (3) and three-quarters (7) of the way from the new moon position. In between are **crescent moons** (2 and 8) and **gibbous moons** (4 and 6).

The changing shapes are called **phases of the moon**. On the next page you will use a model of the sun, Earth and moon to help you understand what causes these phases.

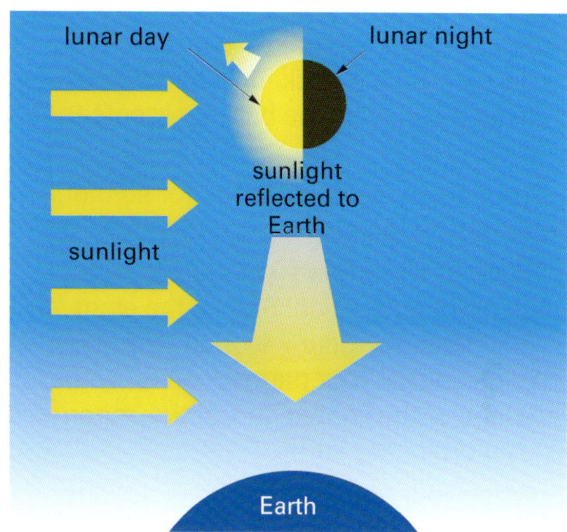

Figure 7.8 The moon produces no light of its own. We see it only because it reflects light from the sun.

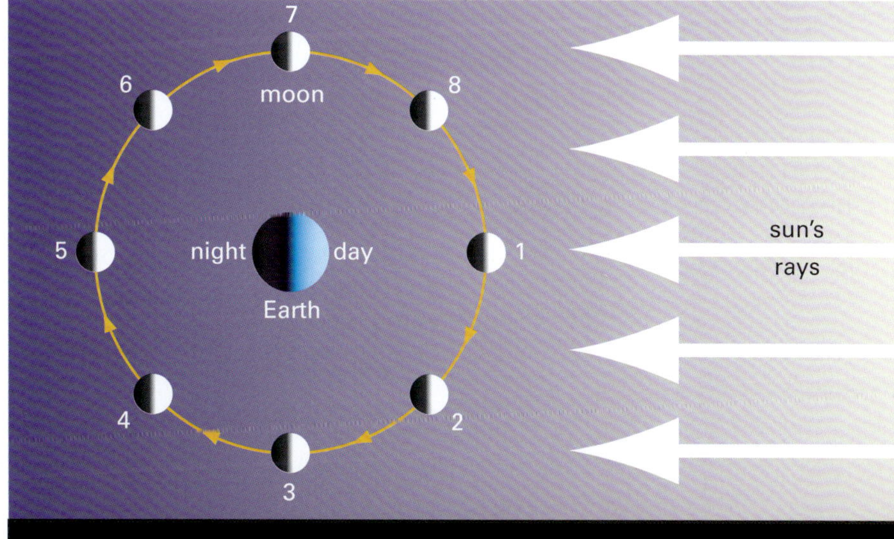

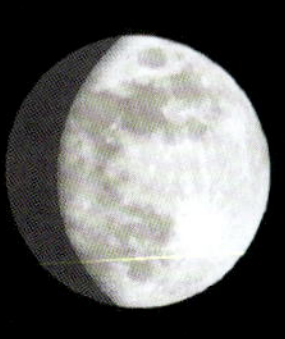
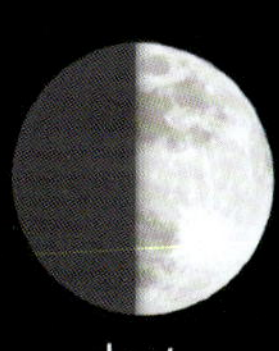
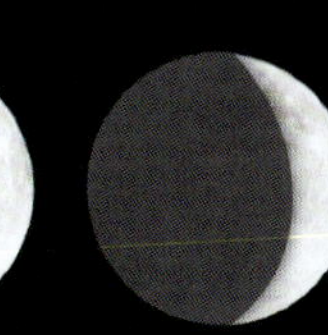

Figure 7.9 We see different parts of the moon's surface lit up during its 29½-day cycle. These changing shapes are called phases.

Note: When viewed from the Southern Hemisphere the moon appears to orbit the Earth in a clockwise direction.

ISBN 978 1 4202 3822 8

ACTIVITY

Moon phases

For this activity you will need to work with a partner. You could also do it as a class activity.

1 Ask your partner to sit in the middle of a darkened room with a notebook and pencil. This person represents the Earth.
2 Stand about two paces away from your partner. Hold a basketball above your head as shown. The ball represents the moon.
3 Turn on an overhead projector so that it shines on the ball.
4 Walk slowly around your partner in a clockwise direction. Stop in the eight positions of Figure 7.9 on the previous page so that your partner can draw the shape of the lighted part of the 'moon' as it orbits the Earth.
5 When your partner has completed the diagrams, swap jobs and repeat the activity.

6 Compare your diagrams with Figure 7.9. Where was the ball when you saw a full moon? Where was it when you saw a new moon?

ACTIVITY

Why do we see only one side of the moon?

You and two partners can make a model of the Earth, moon and sun. Ask your partners to help you as shown. Partner A represents the moon, and partner B the sun. Ask partner A to walk slowly around you (the Earth) while you rotate on your chair in the same direction.

1 How many sides of A's body do you see?
2 How many times does A turn around (rotate) in one revolution? (Ask partner B.)
3 How many times does the moon *revolve* in 29½ days?
4 How many times does the moon rotate in 29½ days?
5 How many sides of the moon (A) can you see from the Earth as it revolves? Why?
6 For how long did B (the sun) see all or part of A's front?

7 How long does the sun shine on one particular place on the moon as it revolves around the Earth?
8 Does the same side of the moon always face the sun?

ISBN 978 1 4202 3822 8

Solar eclipses

Because the Earth and the moon are so much smaller than the sun, they always cast cone-shaped shadows into space, as shown. Once a month the moon passes between the Earth and the sun, as shown. This is the new moon phase. Normally the moon's shadow misses the Earth. Sometimes, however, the shadow touches the Earth. When this happens, the light from the sun is blocked. This is called an eclipse of the sun, or a **solar eclipse**. Eclipses of the sun do not happen very often. They can be seen only over a narrow strip of the Earth.

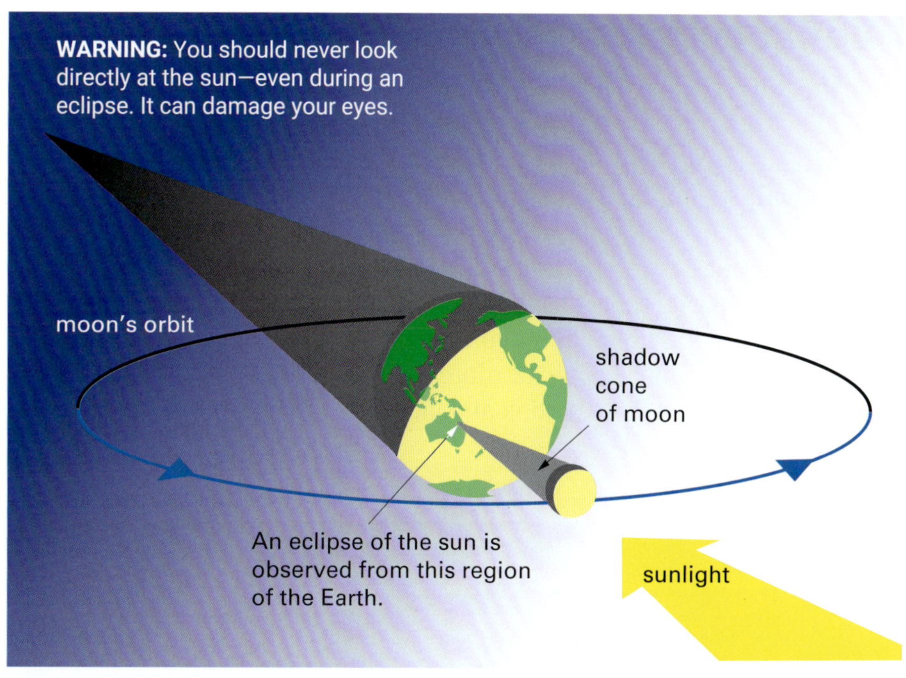

Figure 7.10 A solar eclipse

Figure 7.11 A composite image of a solar eclipse as seen from Earth.

ISBN 978 1 4202 3822 8

Lunar eclipses

At the full moon position the Earth's shadow sometimes falls across the moon's orbit, causing an eclipse of the moon, or a **lunar eclipse**. As the moon passes through the Earth's shadow, it grows darker. It turns not black, but a reddish colour. The Earth's shadow is so wide that it sometimes takes the moon several hours to move through it. Each year there may be one or more lunar eclipses. It is quite safe to look at a lunar eclipse.

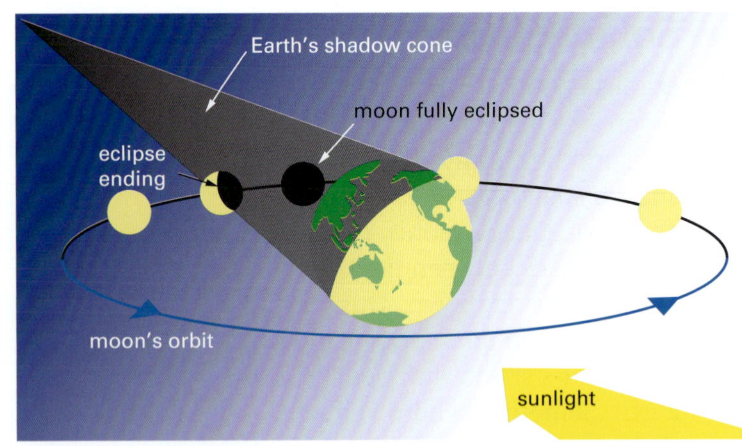

Figure 7.12 A lunar eclipse

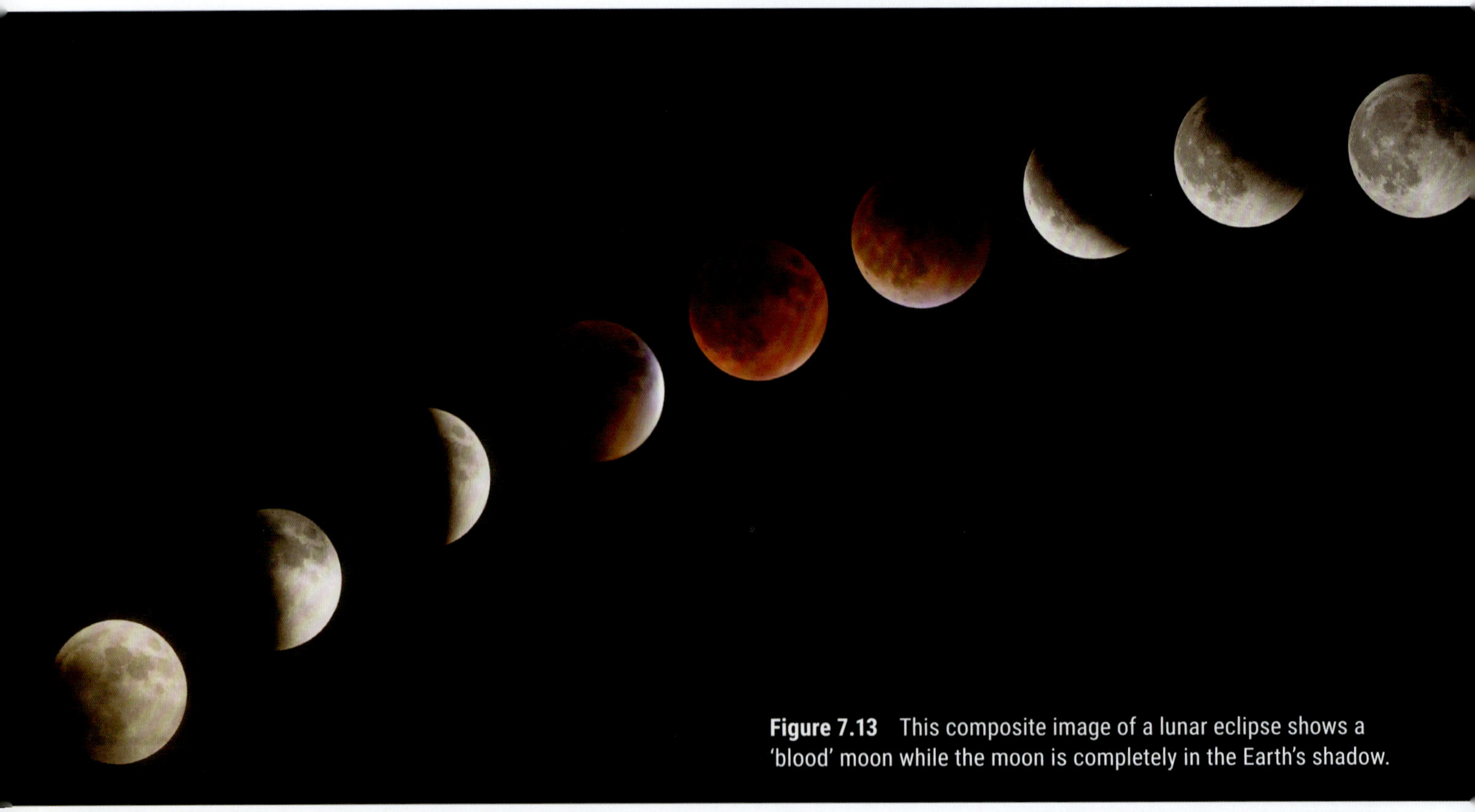

Figure 7.13 This composite image of a lunar eclipse shows a 'blood' moon while the moon is completely in the Earth's shadow.

Use a projector, ball and globe to show how eclipses occur. For a solar eclipse, move the ball (moon) between the projector (sun) and the globe (Earth). For a lunar eclipse move the ball behind the globe.

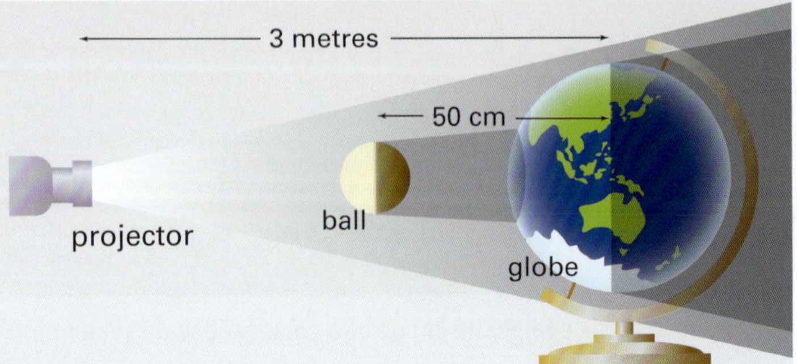

ISBN 978 1 4202 3822 8

In 1969, the *Apollo 11* astronauts left a special mirror on the moon's surface. When a laser beam is fired at the mirror from a telescope on Earth, it bounces straight back. The laser beam travels at the speed of light. So by measuring the time it takes the beam to travel from the Earth to the moon, the distance to the moon can be calculated. This distance can be measured very accurately. Scientists have found that the moon is moving away from the Earth at about 3.8 cm each year—about the rate at which your fingernails grow.

Tides

If you fish or swim at any coastal beach you will be familiar with tides—the periodic rise and fall of the ocean. But what causes these tides?

Scientists explain tides as being caused mainly by the gravitational pull of the moon. Look at Figure 7.14, which shows the Earth's oceans (not to scale). The moon's gravity pulls water towards it, away from B and D, so that it collects at A and bulges out. A second smaller bulge forms at C. This is because the Earth is pulled towards the moon and the water in the oceans is left behind. The overall effect is to create high tides at A and C and low tides at B and D.

As the Earth rotates, the tidal bulges stay roughly in the same place—since the Earth rotates much faster than the moon revolves. The Earth rotates under the tidal bulges, and therefore each place on Earth passes through two high tides each day. So the tide goes in and out twice each day, with one high tide (the one at A) slightly higher than the other.

The sun also has an effect on the tides, but because it is further away, it has less effect than the moon does. When the sun and the moon are lined up (new moon and full moon), extra high tides called *spring tides* occur. When the sun and moon pull at right angles to each other (quarter moons), you get low high tides called *neap tides*.

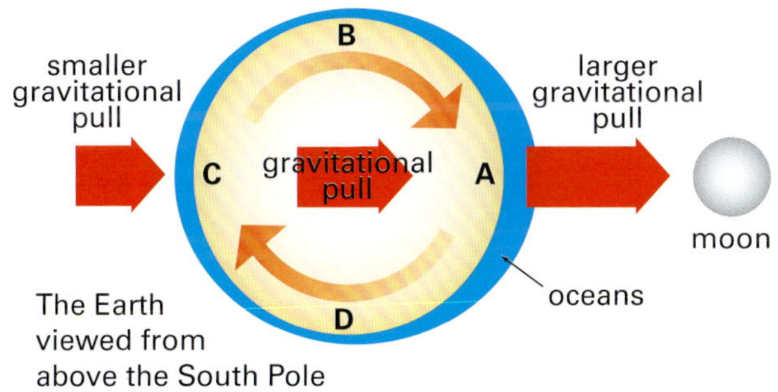

Figure 7.14 How tides are formed by the gravitational pull of the moon

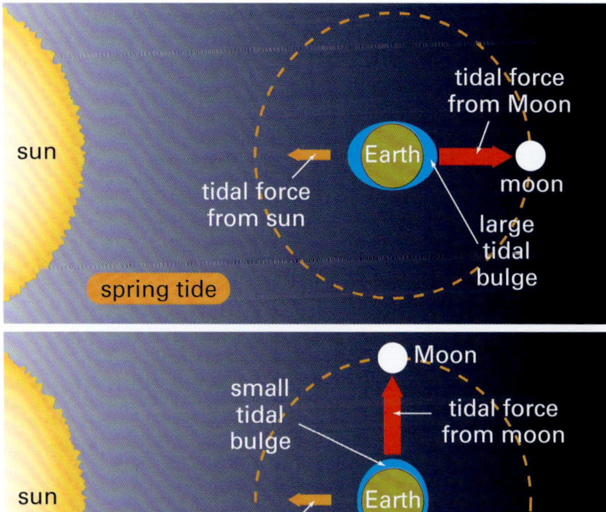

Figure 7.15 How spring tides (top) and neap tides (bottom) are formed

ISBN 978 1 4202 3822 8

1 Why are we able to see the moon even though it does not give off its own light?

2 Use a labelled diagram to show the positions of the sun, Earth and moon when a person on Earth sees a full moon.

3 Why is the new moon dark?

4 Explain the difference between a solar eclipse and a lunar eclipse.

5 Explain why there are usually two high tides and two low tides each day.

6 Most fishers like to fish at the time of the month when there is a full moon. What reasons could there be for this?

7 The tide track graph below shows high and low tides for two days.
   a How many high tides are there each day (over 24 hours)?
   b What is the time and the height of the high tide on Monday morning?
   c What is the time and the height of the high tide on Monday night?
   d Which is the bigger of the two high tides on Monday? What causes this difference?
   e What is the difference between the height of the low tide on Monday afternoon and the high tide on Monday night?
   f What is the time of the first high tide on Tuesday? How much later in the day is this than on Monday?

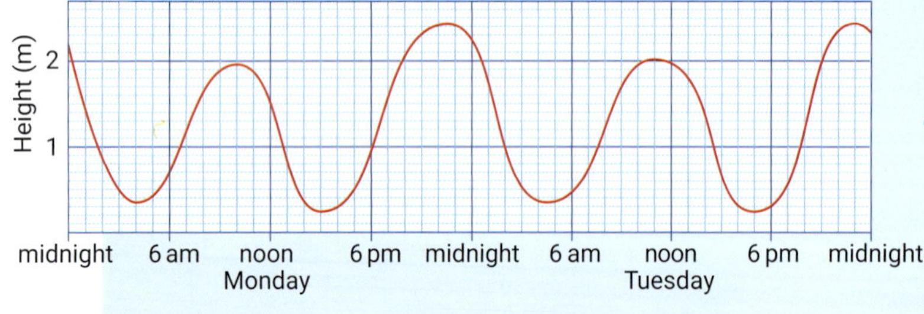

8 Copy the diagram below.

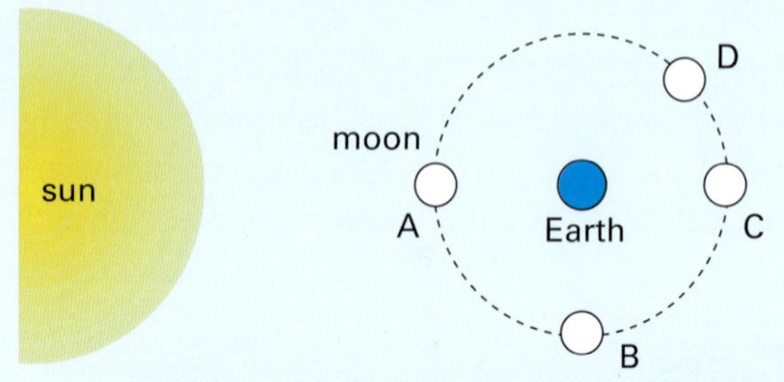

   a For each of the positions A, B, C and D shade those parts of the moon on which light from the sun does not shine.
   b What is the phase of the moon in each position?

9 If the sun is so much bigger than the moon, why doesn't it have a greater influence than the moon on the tides?

10 How would the Earth's tides be affected if the moon was further away?

1 The round outline of the moon can sometimes be seen during the new moon or crescent phase. Why?

2 Explain why the heights of the high tide vary throughout the year.

3 The highest high tides (spring tides) occur when it is a new moon or a full moon, while the lowest high tides (neap tides) occur when it is the first quarter or the last quarter. Use Figure 7.15 to explain this.

4 New moons usually occur during the day. Why?

5 If you were on the near side of the moon, what would you observe when the people on Earth were having a:
   a lunar eclipse?
   b solar eclipse?

ISBN 978 1 4202 3822 8

7.3 Discovering space

Humans have discovered much about space ever since the invention of the first telescope. Since then, technology has continually been invented and improved, providing new evidence about space, our solar system and planets, and the universe.

Telescopes

Telescopes are devices that use lenses and mirrors to focus light and create an image of objects far away that cannot be seen with the unaided eye.

The first telescopes were used around the late 1500s, over 600 years ago. The first to apply for a patent for a telescope was Hans Lippershey, a Dutch eye glasses maker who invented a telescope that magnified things to three times their size, using lenses. Soon after, in the early 1600s, it was Galileo Galilei who made a telescope powerful enough to observe the rings on Saturn, discover the moons of Jupiter, identify sun spots, and see the mountains and craters of the moon, among other things.

In 1668, Isaac Newton invented a telescope that used mirrors instead of lenses, and these reflecting telescopes are the most widely used type today.

Figure 7.16 Galileo Galilei using his telescope in about 1610

Earth-based telescopes

Earth-based telescopes have become more powerful over time, as technology and imaging systems have improved. Earth-based telescopes are usually located at high altitudes where the atmosphere is thinner and there is less light pollution.

As well as traditional telescopes that use light to create images, there are also radio telescopes. Radio telescopes measure electromagnetic radiation in space and have helped scientists explore the big bang and other phenomena.

Figure 7.17 The Anglo-Australian telescope is located at Coonabarabran in NSW and is Australia's largest, with a mirror 3.9 m across.

Figure 7.18 Australian radio telescope compact array, near Narrabri, NSW

ISBN 978 1 4202 3822 8

Space-based telescopes

Space-based telescopes have the advantage of seeing into space without the Earth's atmosphere getting in the way. The atmosphere can distort images, and background light from Earth can also affect image quality. This means a space-based telescope can produce much better images of the universe and phenomena.

One of the most famous is the Hubble Space Telescope that was launched in 1990 and remains in use today. The Hubble telescope has a 2.4 m mirror and can detect visible, infra-red and ultraviolet light, taking amazing pictures of the universe and leading to many incredible discoveries.

The James Webb Space Telescope (JWST) is scheduled for launch in 2018. The JWST will be the largest and most powerful space telescope ever built. The JWST will have a massive 6.5 m gold mirror and will be situated further away from Earth than Hubble. It will be used to study the solar system, including taking images of the planets. It will also be able to study the history of the universe, taking images of other galaxies, identifying new planets and exploring the black hole at the centre of the Milky Way.

Space probes

A **space probe** is a robotic spacecraft that is sent into space, beyond Earth's orbit, to explore space. Probes have been sent on missions to explore the sun, planets, asteroids and comets, and other phenomena in our solar system. The first space probe was launched in 1960, and since then there have been many successful attempts as well as many failures in sending out space probes. Countries including the USA (NASA), Russia, Japan, India and the European Space Agency (ESA) have all sent probes.

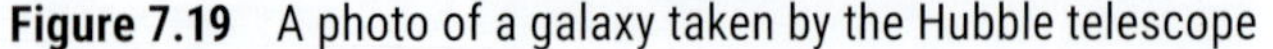

Figure 7.19 A photo of a galaxy taken by the Hubble telescope

Figure 7.20 The James Web Space Telescope will be launched in 2018.

ISBN 978 1 4202 3822 8

What do probes measure?

Space probes are designed to do many things and are becoming increasingly sophisticated. Probes may contain the following instruments for measuring, recording or detecting:

- cameras and telescopes to record visible light images and video
- infra-red sensors to measure temperature
- radars to see planet surfaces through clouds and to map terrain
- ultraviolet sensors to measure atmospheric make-up
- magnetometers to measure magnetic fields
- gravitometers to measure gravity
- radiation sensors
- sensors to measure the solar wind and solar radiation from the sun
- (if a probe has a lander) drills, scoops and sensors to collect and analyse soil, rock and the composition of atmospheric gases, and to look for the presence of water and for signs of life.

What is a probe made of?

A probe needs many parts to survive and travel in space and to complete its mission. A typical space probe needs more than just the sensors and scientific equipment it carries. It also needs an energy source (usually solar or nuclear), a propulsion system, protection from radiation and magnetic fields in space, communication systems to send and receive data, systems and sensors to monitor that the probe's components are all working properly, and a computer to control everything.

The Cassini–Huygens probe

To help us learn more about how a space probe works, we will take a look at the Cassini–Huygens probe. Cassini–Huygens is a joint project of NASA, the European Space Agency (ESA) and the Italian space agency Agenzia Spaziale Italiana (ASI).

Cassini–Huygens is unique as it consists of an orbiter (Cassini probe) that goes around Saturn, and a lander (Huygens probe) that was designed to land on Saturn's moon Titan.

Cassini probe

Cassini was launched in 1997 and reached Saturn's orbit in July 2004. The Cassini probe has eyes that can see much better than human eyes. Its instruments are super sensitive and can detect things that could not be detected from Earth, or even by humans if they were at Saturn.

Figure 7.21 This photo of the Cassini probe (being constructed here) gives you an idea of its 6.7 m height and 5712 kg weight.

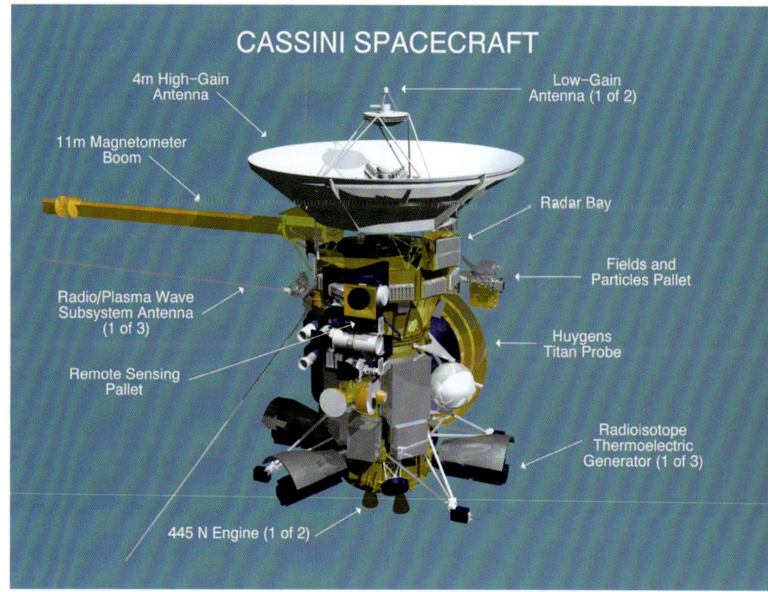

Figure 7.22 The Cassini probe, showing its main instruments

ISBN 978 1 4202 3822 8

Huygens probe

The Huygens probe is the only probe to ever land on a planet in the outer solar system. The probe contained various instruments to image and measure Titan as it descended. These include an aerosol collector pyrolyser, a descent imager and spectral radiometer, a Doppler wind experiment, a gas chromatograph and mass spectrometer, an atmospheric structure instrument and a surface science package.

The Huygens probe descended from Cassini into the Titan atmosphere, landing on its surface in a sandy area. The descent was controlled by a series of parachutes and took about 2.5 hours.

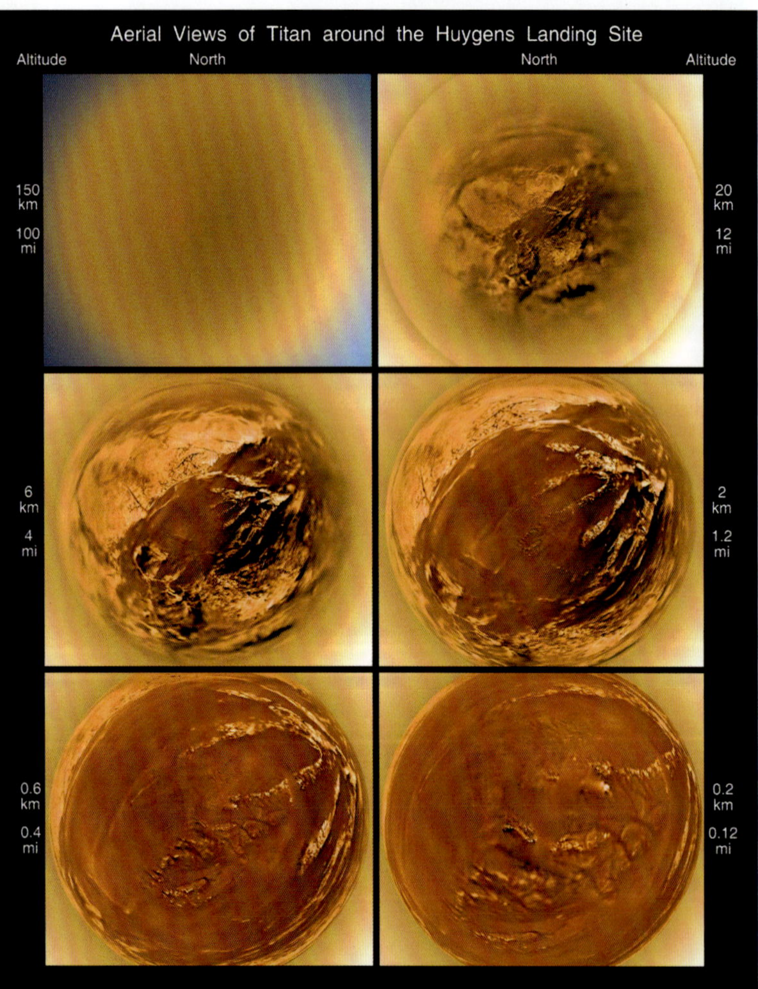

Figure 7.23 This series of photos shows that as the Huygens probe descended to Titan, the surface features became clearer as it went through the cloudy atmosphere.

The lander remained in operation for about 90 minutes after landing, collecting and sending back data the entire time.

What has been discovered at Saturn?

The Cassini probe has been at Saturn for over 12 years and has discovered many things about the planet Saturn and its moons. Some of these include:

- new moons identified
- information about Saturn's rings and how they behave
- mapping the surface of many of Saturn's moons
- measuring conditions on Saturn and many of its moons.
- signs of active subsurface water oceans on Enceladus, Saturn's sixth largest moon, and massive plumes, including water, that erupt from the surface
- mapping the surface of Titan, including its oceans, although these oceans are not water; the surface of Titan (another moon) is active and a lot like Earth
- images of Titan's surface and an understanding of the wind and gases that make up the atmosphere

Find out more about Cassini–Huygens by doing some research to answer the following questions.

1. Who are the Cassini and Huygens probes named after and why?
2. Select and describe three of Cassini's top discoveries about Saturn and its moons.
3. List each of the instruments on the Huygens probe mentioned above and find out what they do.
4. What did the Huygens probe discover while landing on Titan?
5. Describe and compare Saturn's moons Titan and Enceladus.
6. When will the Cassini mission end, and how?

ISBN 978 1 4202 3822 8

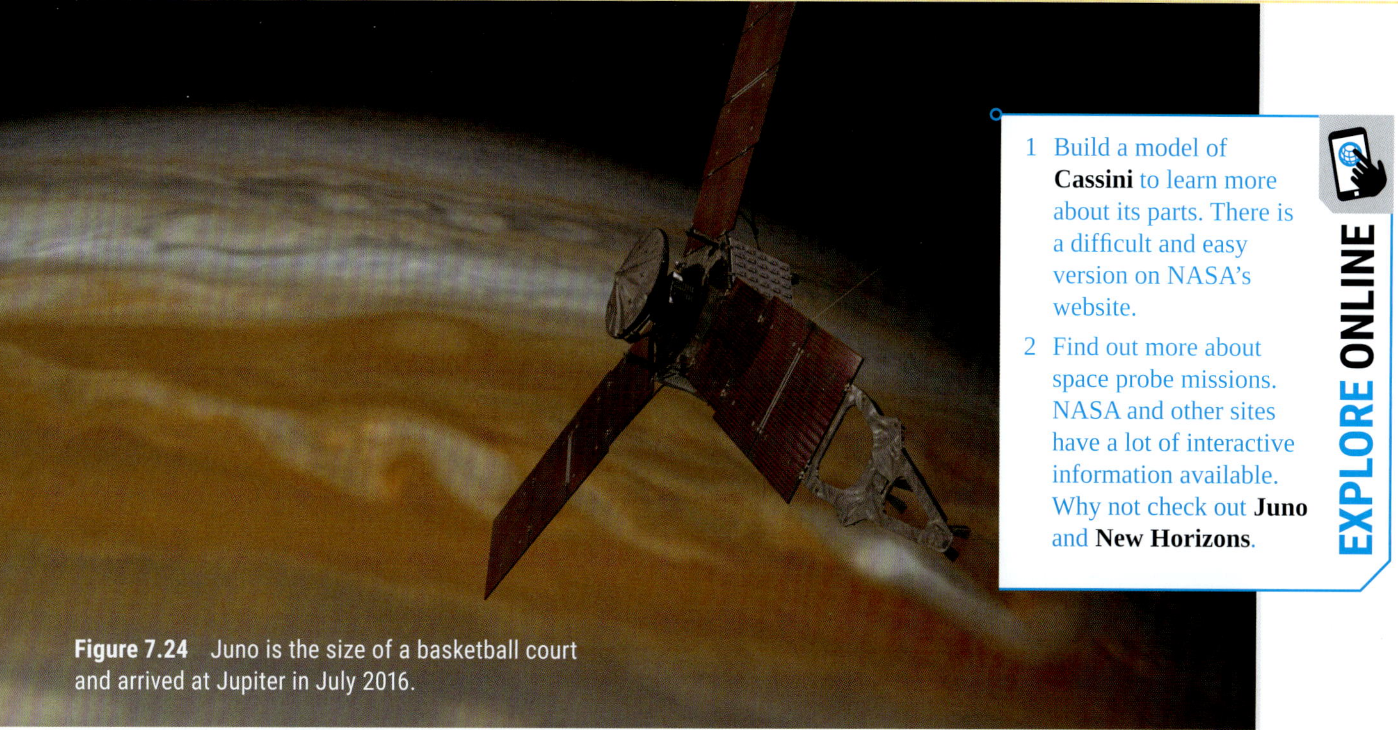

Figure 7.24 Juno is the size of a basketball court and arrived at Jupiter in July 2016.

EXPLORE ONLINE

1. Build a model of **Cassini** to learn more about its parts. There is a difficult and easy version on NASA's website.
2. Find out more about space probe missions. NASA and other sites have a lot of interactive information available. Why not check out **Juno** and **New Horizons**.

SCIENCE AS A HUMAN ENDEAVOUR

Heading to Mars

Mars is one of the most explored planets in the solar system. Humans have walked on Earth and the moon, but not Mars—not yet. NASA, working with many partners, has launched an ambitious plan to put people on Mars around 2030. There are three phases to the mission.

1 *Earth-reliant exploration*

Experiments on board the International Space Station are increasingly focused on new technologies and how to stay healthy in space in preparation for deep-space missions. Other areas of Earth-reliant investigation include:

- advanced communications systems
- extravehicular operations
- Mars mission level environmental control and life support systems
- 3D printing in space to manufacture parts as required.

2 *Proving ground*

The proving ground is the phase in which NASA will test and refine many new techniques and technologies by conducting deep-space operations around the region of the moon:

- a series of exploration missions starting with the test of a new space launch system and the Orion crew capsule in 2018
- the Asteroid Redirect Robotic Mission in 2020, which will collect a large boulder from a near-Earth asteroid and bring it back near Earth, followed by a crewed mission that will allow astronauts to investigate and sample the boulder
- testing a deep-space habitat system that can sustain life in space for long periods
- ideas to reduce resupply needs through reduction, reuse and recycling of consumables, packaging and materials
- all of this requires new propulsion systems, larger rockets than ever before and a variety of other new technologies that will need to be proven.

3 *Earth-independent activities*

Earth-independent activities will build on the knowledge developed during the 'proving ground'

ISBN 978 1 4202 3822 8

phase, and prepare the deep-space technology to eventually get people to the moons of Mars, and to Mars itself without support from Earth. These activities will develop and test technology that can enable:

- living and working in habitats on the way to Mars and in surface habitats; these must support life for years with little maintenance
- collecting resources on Mars to create fuel, water, oxygen and building materials
- developing more advanced communication systems to relay data with a minimum of delay.

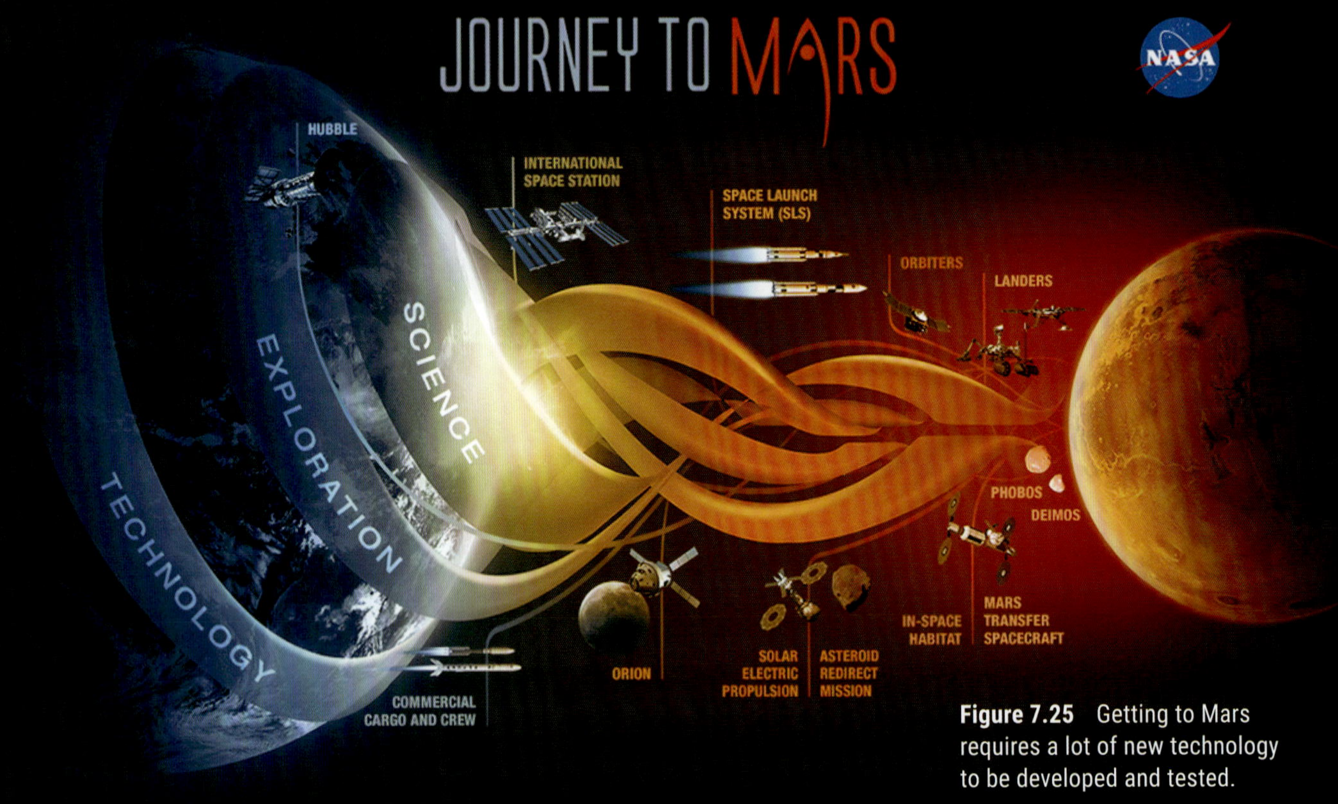

Figure 7.25 Getting to Mars requires a lot of new technology to be developed and tested.

ACTIVITY

Split into groups, each group researching one of the following topics. Produce an interactive report that you can share with the class about your topic.

1. Investigate the environmental conditions on Mars, including temperature, weather, landscape, availability of resources, radiation, atmosphere and other factors. Outline these conditions, and list the obstacles to be overcome in order to live on Mars. Propose technologies to help overcome each of these obstacles.
2. Staying in space for long periods of time is not good for human health. Outline the problems associated with living in zero gravity and spending long periods in space. Produce a plan and design technologies for astronauts that can help them remain healthy on a long journey to Mars.
3. Mars is a long way away and it will take an extended trip just to get there. Investigate the technology required to push a spacecraft out of Earth's atmosphere and then to Mars. What type of propulsion

ISBN 978 1 4202 3822 8

systems will be required and where will the fuel come from? How will the spacecraft slow down and stop? How will it reach the surface of Mars? Design the systems for propelling people to Mars.

4 Living on Mars will require life-support systems that allow people to stay alive for long periods without supplies from Earth. Research what resources will be required to survive on Mars, and design a habitat for the astronauts to live in that allows them to make use of these resources.

5 Mobility on Mars will be important, because people will need to get around. This will require a new generation of space suit, as current suits are very bulky. It will also require vehicles for transport. Design a Mars car, and a new space suit that will be useful on Mars.

6 A transit habitat will be required for a journey to Mars. This will stay in orbit around Mars, and be used by astronauts on the way to and from Mars. Investigate what this habitat will need to be able to do, and then design it.

EXPLORE ONLINE

Start your research by visiting the NASA website. Under the 'Topics' menus is a section called 'Journey to Mars'.

CHECK

1 Describe the function of a telescope.

2 a Who is believed to have invented the telescope?
  b Which famous scientist then produced one of the first telescopes? Outline some of this scientist's discoveries.

3 Explain the advantages of space-based telescopes compared to Earth-based ones.

4 What is the name of the new space telescope being launched in 2018?

5 a What is a space probe?
  b List four things that a space probe may measure or investigate.
  c List some advantages of sending a space probe rather than a manned mission.
  d Identify four things that a space probe needs to function in space.

6 Name two space probes and their destinations.

7 a Explain the difference between the Cassini and Huygens probes.
  b List two discoveries made by the Cassini–Huygens probes at Saturn.

CHALLENGE

1 Design your own space probe.
  a Decide where your probe will go, and what its mission is.
  b List the parts your probe will need to function and keep working.
  c List the equipment and sensors your probe will need to complete its scientific mission and tests.
  d Draw your space probe design and label the parts you have included.

2 Investigate the difference between a reflecting telescope and a refracting telescope. Use diagrams to help explain your answer. You will need to do some research.

ISBN 978 1 4202 3822 8

Copy and complete these statements to make a summary of this chapter. The missing words are on the right.

1 Astronomical observations can be explained in terms of the positions and _____ of the sun, Earth and moon. For example, as the Earth orbits the _____ you see different stars at different times of the year.

2 The Earth _____ on its axis once every 24 hours, causing day and night. It takes one year to _____ around the sun.

3 A gravitational force of attraction keeps the Earth in _____ around the sun.

4 The _____ are caused by the revolution of the Earth and the tilt of its axis.

5 We always see the same side of the _____. It has no water and no air, and has many _____ on its surface.

6 We see the moon because it _____ light from the sun. The changing shapes of the visible moon as it revolves around the Earth are called _____ of the moon.

7 An eclipse of the sun (solar eclipse) occurs when the moon casts a shadow on the _____. An eclipse of the moon (_____ eclipse) occurs when the Earth casts a shadow on the moon.

8 A space probe is a _____ spacecraft sent to explore deep space.

9 Probes and _____ have provided new knowledge about space.

moon
craters
lunar
movements
rotates
phases
seasons
orbit
reflects
Earth
revolve
sun
telescopes
robotic

CH•7 REVIEW

1 The planet Mercury has a very short year (only 88 Earth days). This is because (compared to Earth) it:

A revolves faster around the sun.
B is further away from the sun.
C is a much smaller planet.
D rotates faster on its axis.

2 Which of the following statements about telescopes are true?

A The first telescope was invented by Galileo Galilei.
B Telescopes use mirrors and lenses to create images of objects that we can't see with our eyes.
C Telescopes in space have an advantage over those on the Earth.
D Telescopes can only use light to create images of space.

3 Suppose that over a period of time an astronaut who has made a forced landing on a planet records the following observations:

- Day and night on the planet are always of equal length.
- Daylight lasts about 14 hours.
- The sun rises in the west and sets in the east.

From this information, it is most reasonable to conclude that the planet rotates from:

A east to west once every 14 hours.
B west to east once every 14 hours.
C east to west once every 28 hours.
D west to east once every 28 hours.

ISBN 978 1 4202 3822 8

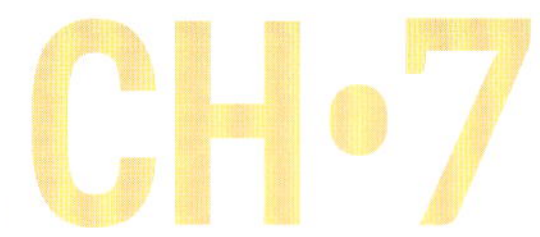

4 The planets Frankenstein and Dracula, which orbit the star Horror, are in the positions shown.

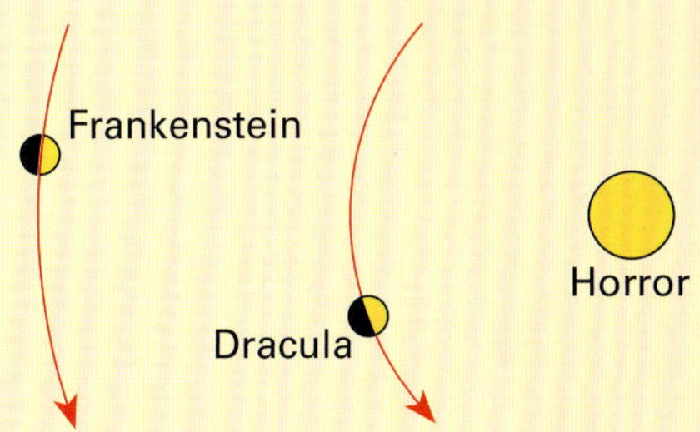

Which one of the choices below shows the appearance of Dracula as seen by an observer on the day side of Frankenstein?

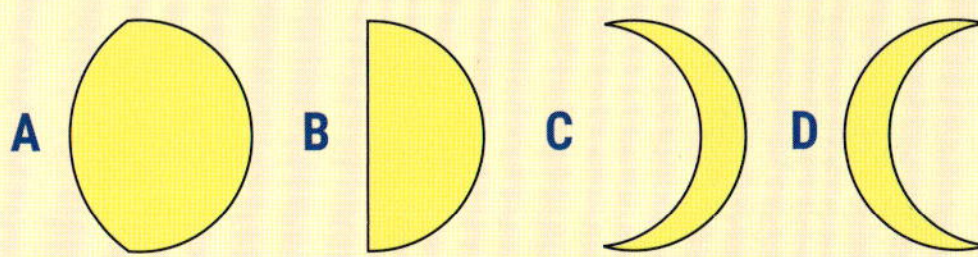

5 The drawings below show a few of the many possible positions of the sun, Earth and moon.

a Which one shows the position necessary for a solar eclipse?

b Which one shows the position necessary for a lunar eclipse?

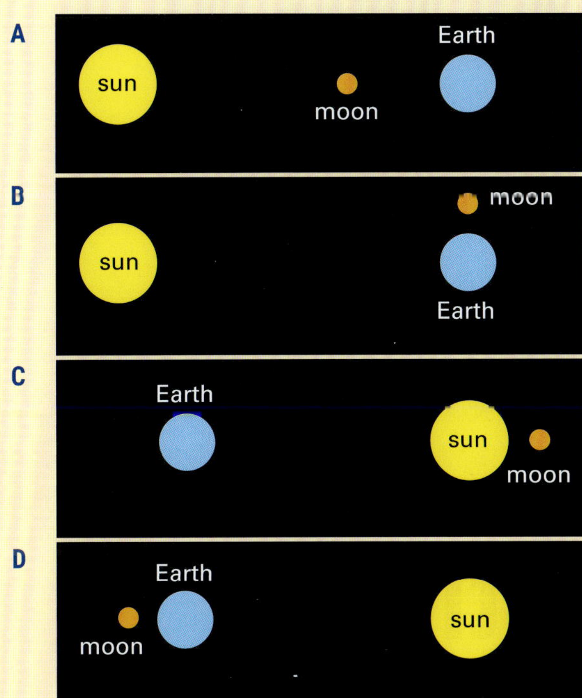

6 What is the difference between rotation and revolution? Give an example to illustrate your answer.

7 Sarah said: 'If the moon's gravitational pull causes the tides, then high tides can only occur at night.' Write a short paragraph to convince her she is wrong.

8 In science you often give the reason for one thing leading to another. In other words, you relate the cause (the reason) to the effect (what occurs). Draw up two columns. Call the left-hand column 'Cause' and the right-hand one 'effect'. Consider the sentences below. For each one, write the cause part of the sentence in the left-hand column and the effect part in the right-hand column.

a The tilt of the Earth's axis causes the seasons.

b A lunar eclipse occurs when the Earth moves between the sun and the moon.

c Night follows day as the Earth rotates.

d When the moon completely blocks the sun, a total eclipse of the sun occurs.

e The same side of the moon always faces Earth, because the moon revolves in the same time that it rotates.

f As the Earth orbits the sun you see different stars at different times of the year.

9 A space-based telescope is better than an Earth-based one because:

A it is above Earth's atmosphere, making it closer to objects in space.

B it is above Earth's atmosphere, making images clearer.

C in space it will not corrode in the atmosphere.

D in space there is less air to blow it around.

10 Which of the following is not required for space probes to function in space:

A a control system to maintain operations

B an energy source

C a communications system

D a map of the planets and solar system

Check your answers on page 242.

ISBN 978 1 4202 3822 8

IN THIS CHAPTER

Science Understanding

- recognise the differences between pure substances and mixtures, and identify examples of each
- perform different separation techniques, including decanting, filtration, distillation, evaporation, crystallisation and chromatography
- identify the solvent and solute in a variety of solutions
- explore and compare separation methods used in the home
- investigate how chromatography is used in forensic science

Science Inquiry Skills

- work in a group to find a way to purify a sample of impure creek water
- reflect on the methods used to separate mixtures

CH•8 Separating mixtures

ISBN 978 1 4202 3822 8

GET STARTED: *QUESTION*

Work in a small group to solve one or more of these problems.

- Your four-wheel drive has broken down in the middle of the Simpson Desert and you have no water to drink. You find a damp patch of sand near the base of a cliff. How can you get drinkable water from this damp sand?

- Your uncle has given you a large bottle of 5c, 10c, 20c and 50c coins. Can you design a device to separate the coins?

- Your science teacher is very angry with the class. Someone has poured sand, salt and iron filings into one jar! Until they are all back in separate containers, no one can go to lunch. Design an experiment that would allow you to separate these substances.

ISBN 978 1 4202 3822 8

8.1 What's a mixture?

Different substances have different **properties**. They can be solids, liquids or gases. But there are many other properties that allow you to tell one substance from another. For example, you can detect sugar by its sweet taste. You can detect kerosene by its smell. Glass is transparent (you can see through it). Diamond is extremely hard. Beetroot is a purple-red colour. A piece of lead is very heavy.

The materials around you can be grouped into pure substances and mixtures. **Pure substances** contain only one substance. They always have the same properties, no matter where they come from. Examples are sugar, gold, pure water and helium gas. However, most materials around you are **mixtures**—several different substances mixed together. Examples are air, soft drink, concrete and lipstick.

The amounts of each part of the mixture (called their *proportions*) can vary widely. This changes the properties of the mixture. For example, concrete is a mixture of cement, sand, gravel and water. Mixing these four substances in different proportions will change the properties of the concrete.

Lipstick normally contains:

- castor oil
- beeswax
- carnauba (to stop it melting)
- esters (to make it slippery)
- antioxidant (to stop it going off)
- aloe vera (to stop lips becoming dry)
- mineral oil (to make lips glossy)
- red dye No. 21
- perfume

Figure 8.1 Lipstick is a complex mixture.

The parts of mixtures can be solids, liquids or gases. For example, soft drink is a mixture of water and carbon dioxide gas, plus sweetener, flavouring and colouring.

Examples	Type of mixture	Main parts of mixture
black coffee	solid in liquid	coffee powder in water
air	gas with gas	nitrogen and oxygen
soft drink	gas in liquid	carbon dioxide in water
smoke	solid in gas	tiny bits of soot, dust, etc. in air
wine	liquid in liquid	alcohol in water
brass	solid with solid	copper and zinc

CHECK

1 Which of the following are mixtures, and which are pure substances?
   - a air
   - b petrol
   - c polluted water
   - d gold
   - e orange juice
   - f sugar
   - g helium gas
   - h concrete

2 Copy and complete these sentences.

   The features by which a material can be identified are called _____. Materials that always have the same properties are called _____ substances. Materials that are made up of different substances are called _____. The properties of a mixture can _____.

3 Explain why concrete is a mixture and not a pure substance.

4 In your notebook, match up the following types of mixtures with the examples.

a	mixture of gases	smoke
b	mixture of solids	air
c	mixture of gas in liquid	soil
d	mixture of solids in liquid	muddy water
e	mixture of solids and gases	lemonade

ISBN 978 1 4202 3822 8

8.2 Solutions

When you stir sugar in a glass of water, it disappears into the water. We say it **dissolves** in the water. The sugar and water have mixed to form a **solution**.

A solution is a special mixture that looks and behaves like a single substance. It consists of a liquid and the dissolved substance, which is spread evenly throughout it. Consider what happens when instant coffee dissolves in hot water. The substance that dissolves (the coffee) is called the **solute**. The substance that does the dissolving (the water) is called the **solvent**. So the solute dissolves in the solvent, forming a solution.

Figure 8.2 Forming a solution

Solutions are very important to you. The food you eat is digested and dissolved in water. It is then carried around your body in the blood plasma, which is a solution consisting of about 90 per cent water. The wastes produced by your body are also carried away in this solution.

Two liquids can also form a solution. For example, wine is a solution of alcohol (solute) in water (solvent). Fuel for two-stroke motor mowers and outboard engines is a solution of oil in petrol.

A particular substance may not dissolve in every solvent. For example, salt is soluble in water, but insoluble in alcohol. Water is an excellent solvent, but to dissolve some things you have to use other solvents. Some commonly used solvents are shown in the table at the top of the page.

Solute	Solvent (dissolves the solute)
nail polish	nail polish remover
ballpoint pen stains	methylated spirits
grease marks on clothes	eucalyptus oil
oil-based paint	turpentine
tar on car paintwork	kerosene

Figure 8.3 Oil is insoluble in water.

Figure 8.4 A solution will appear clear, and may be coloured.

When substances don't dissolve

A substance that dissolves is said to be *soluble*. A substance that will not dissolve is *insoluble*. Some insoluble substances sink in water (settle out), and others float on top. If you shake up an insoluble solid (such as chalk dust) with water, it may seem to dissolve at first. However, if you look closely you will see that the liquid is cloudy and the chalk settles when you let it stand for a while. Such a mixture is not a solution, but a **suspension**. Muddy water is another example of a suspension, because the mud settles to the bottom when you let it stand.

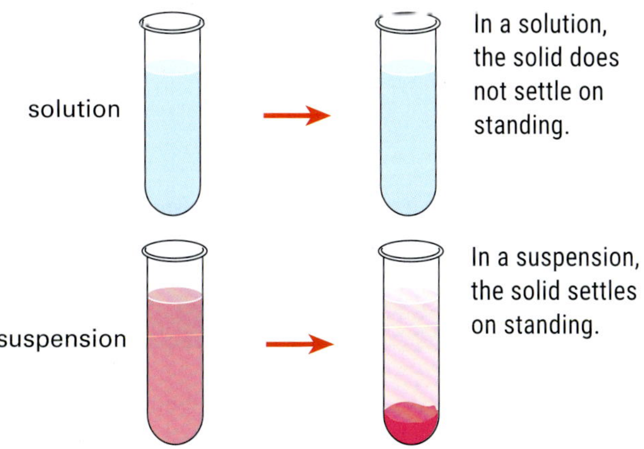

Figure 8.5 The difference between a solution and a suspension.

ISBN 978 1 4202 3822 8

INVESTIGATION 8.1 Soluble or insoluble?

Aim

To test whether various substances are soluble in water and in alcohol.

Materials

- test tubes (at least 6)
- rubber stoppers to fit test tubes
- test tube rack
- spatula
- marking pen
- **alcohol** or **methylated spirits** (in a dropping bottle)
- samples of:
  - salt
  - sugar
  - coffee
  - flour
  - **iodine** (solid)
  - jelly crystals
  - grass (ground up)
  - ballpoint pen spot on small piece of paper

Risk assessment and planning

Before you start, check that you know the safety rules for your laboratory.

Read through both parts of the investigation, then prepare a data table like the one below.

Substance	Soluble in water?	Soluble in alcohol?	Observations
salt			
sugar			
coffee			
flour			

Why do you have to be careful when using iodine? How do you dispose of leftover iodine?

Warning: Do not touch iodine with your fingers. It is poisonous. Your teacher will tell you how to dispose of any leftover iodine. Do not wash it down the sink.

PART A Is it soluble in water?

Method

1 Use the spatula to pick up a small amount of salt—about the size of a grain of rice. Place this salt in a test tube. Use the marking pen to label the tube 'salt'.

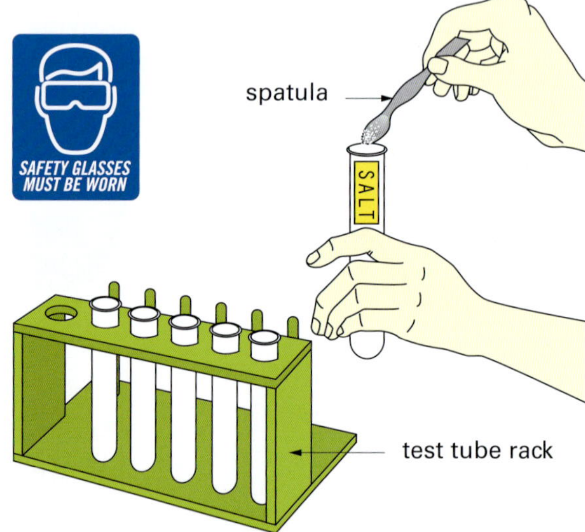

2 One-third fill the test tube with water. Shake the tube using the following method. Hold the test tube firmly between your thumb and index finger. Then tap the bottom of the test tube sharply with the index finger of your other hand. (You may need to practise this.)

Record whether the substance is soluble, slightly soluble (a bit dissolves) or insoluble.

Record any other observations as well. For example, if a solution was formed, what colour was it? Was a suspension formed?

3 Repeat steps 1 and 2 for each of the other samples.

ISBN 978 1 4202 3822 8

PART B Is it soluble in alcohol?

Repeat Part A, using alcohol or methylated spirits instead of water.

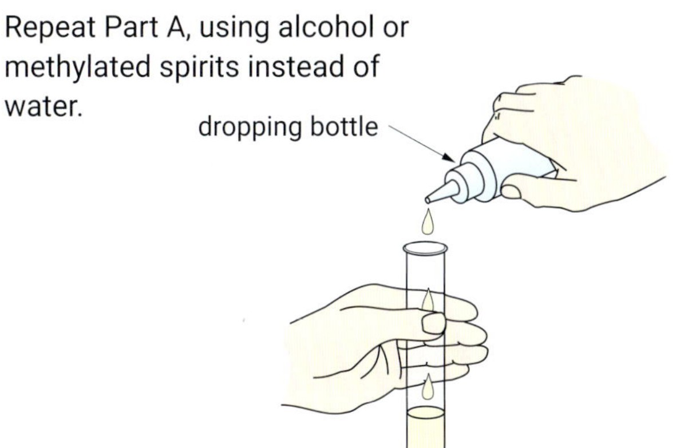

Discussion

1 How can you tell whether a substance has dissolved or not?

2 Which substance dissolved most easily in water?

3 Compare the solubilities of the substances in water and in alcohol.
   a Which substances were soluble in water but not in alcohol?
   b Which substances were soluble in alcohol but not in water?
   c Which substances did not dissolve in either water or alcohol?

4 You have a ballpoint pen stain on your school uniform. How could you remove it?

Solubility

A cup of coffee is like any liquid solution. It comes in many different strengths. If you like your coffee stronger, add more coffee granules. If you like it weaker, add less coffee.

We use the terms **dilute** (dye-LOOT) and **concentrated** (CON-cen-TRAY-ted) to help us compare solutions. A dilute solution contains only a small amount of solute in a given volume of solvent. A concentrated solution contains a large amount of solute in the same amount of solvent. You may have used the terms *weak cordial* or *strong coffee*—but the correct scientific terms are *dilute cordial* and *concentrated coffee*.

Figure 8.6 Dilute and concentrated solutions

The colour of a solution gives you some idea of its concentration. The darker the colour, the higher the concentration; or a more dilute solution would be lighter in colour. These statements are generalisations.

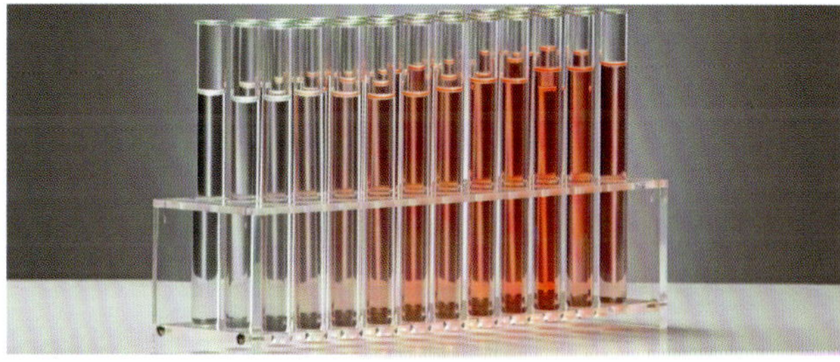

Figure 8.7 Which solutions are more concentrated?

Saturated solutions

There is a limit to the amount of solute that will dissolve in a solution. When a solution will dissolve no more solute, it is said to be **saturated**. Until it reaches this point, it is *unsaturated*. If you add more solute to an unsaturated solution, it will dissolve.

The amount of solute needed to saturate a solution depends on the temperature. For example, at room temperature (around 20 °C) you can dissolve about 2 kg of sugar in a litre of water, but when the water is boiling (100 °C) you can dissolve almost 5 kg. Most solids are more soluble in warm water than in cold water. We say that their **solubility** increases as the temperature increases. This is another generalisation.

ISBN 978 1 4202 3822 8

Colloids—solutions or suspensions?

The Swan River in Perth is known for its brown colour. This is because of clay and other impurities in the water that are so fine that they don't settle to the bottom, as they would normally do in a suspension. Instead, the clay particles are spread evenly throughout the water, forming what is called a **colloid** (COL-oid). A colloid has properties that are in between those of a solution and a suspension.

The particles in a colloid may be tiny bits of solid, liquid droplets or gas bubbles. The colloid may also be a solid, a liquid or a gas. The table on the right lists the common types of colloids.

A liquid-in-liquid colloid is called an **emulsion** (ee-MULL-shun). A common example is ordinary homogenised milk, in which tiny globules of milk fat are spread throughout water. It is processed by forcing the milk through small holes to break up the larger fat globules in the cow's milk. This is why the cream (the fat) doesn't come to the top when standing.

Although it is easy to see fogs, foams and emulsions, it is often hard to tell the difference between solutions and some types of colloids. One way to do this is to shine a beam of light through them. The beam can be seen in the colloid because it bounces off the tiny particles. This is why you can see the headlights of a car in fog. However, the beam cannot be seen in a solution.

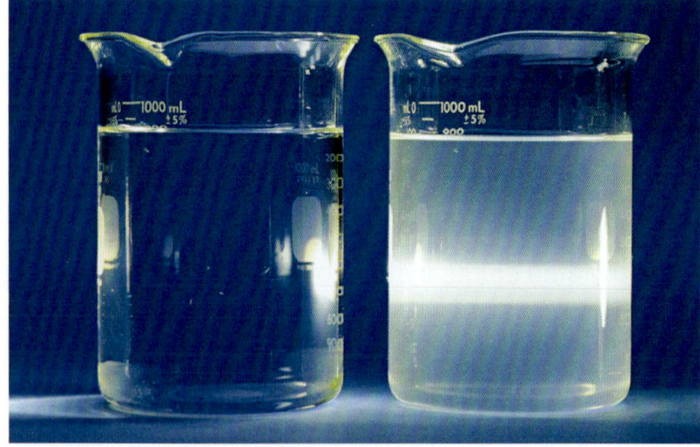

Figure 8.9 Can you tell the colloid from the solution?

Colloid	Type	Examples
sol or gel	solid in liquid	most paints, starch in water, clay in water, jelly
liquid aerosol	liquid in gas	fog, clouds, sprays from spray cans
liquid emulsion	liquid in liquid	mayonnaise, milk
solid emulsion	liquid in solid	cheese, butter, face cream
foam	gas in liquid	whipped cream, beer froth, soap suds
solid foam	gas in solid	pumice, marshmallows, meringues

ACTIVITY

A Forming an emulsion

Shake up some olive oil with vinegar in a stoppered test tube, and let it stand for a while. Do the olive oil and vinegar mix?

Now add a pinch of mustard powder and shake. Allow the mixture to stand again. What do you observe now? (Salad dressing is made this way.)

Write an inference to explain your observations.

B Solution or colloid?

Dissolve a few crystals of hypo (sodium thiosulfate) in a beaker of water and shine a strong beam of light through it. When looking from the side can you see the beam in the solution?

Now add a few drops of dilute hydrochloric acid and observe what happens to the beam.

Try to explain what has happened.

Figure 8.8 The water in the Swan River looks brown because it is a colloid of clay in water.

ISBN 978 1 4202 3822 8

SKILLBUILDER

Concentration

The **concentration** of a solution is often given as a percentage. For example, a 5% hydrochloric acid solution is dilute. A 30% solution is concentrated.

If you have a dog at home, you may sometimes wash it in a dog shampoo or flea-killing liquid. These chemicals can be dangerous, and have to be mixed with water in the correct proportions. Suppose you have to make up a 5% dogwash solution. This means you need 5 parts of dogwash dissolved in enough water to make up a total of 100 parts. That is, you mix 5 parts of dogwash with 95 parts of water. This is a 5% solution.

Questions

1 How could you tell the difference between a 1% red food colouring solution and a 5% solution?

2 The label on a bottle of cleaner says it contains 15% ammonia. If the bottle contains 200 mL of cleaner, how much ammonia is in the bottle?

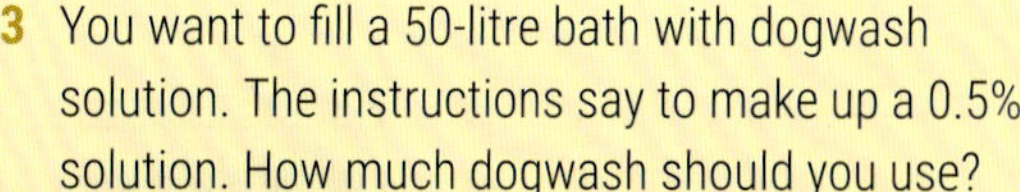

3 You want to fill a 50-litre bath with dogwash solution. The instructions say to make up a 0.5% solution. How much dogwash should you use?

CHECK

1 Copy and complete these sentences.
   - a When sugar is mixed with water, it ______. This shows that sugar is ______ in water.
   - b Sand is ______ in water.
   - c In salt water, the salt is the ______, and the water is the ______.
   - d The solute in a solution does not settle out, but the solid in a ______ does.
   - e A mixture with properties in between a solution and a suspension is a ______.
   - f A solution that can dissolve no more solute is ______.
   - g Most solids are more ______ in hot water than in ______.
   - h A ______ solution is one that contains a small amount of solute. When more solute is added, the solution becomes more ______.

2 Which of the following statements are true, and which are false?
   - A Water dissolves everything.
   - B Solutions are always coloured.
   - C Some gases dissolve in water.
   - D Emulsions settle out on standing.
   - E A solute can be a solid, a liquid or a gas.
   - F More solute can be dissolved in an unsaturated solution.
   - G Adding more water to a dilute solution will make it more concentrated.

3 Explain in your own words the difference between a solute and a solution.

4 How can you tell the difference between a solution and a suspension?

5 Imagine that while doing Investigation 8.1 (page 188) you noticed the following:
   - Nathan put his thumb over the mouth of a test tube to shake it.
   - Donna used her fingers to put a pinch of salt into a test tube.

   Explain to Nathan and Donna why their methods were unsafe.

6 In your notebook, complete the table below by putting in the names of the solute and solvent for each solution.

Solution	Solute	Solvent
a sea water		
b hot chocolate		
c turpentine in which you have just cleaned a paint brush		
d bath water		
e soft drink		

ISBN 978 1 4202 3822 8

7 The photo below shows three different solutions of Condy's crystals in water.
   a What clue in the photo suggests that the solutions contain Condy's crystals?
   b How do the three solutions differ?
   c How can you explain this difference?

8 The instructions on jelly crystal packets say to dissolve the crystals in boiling water. Suggest a reason for this.

CHALLENGE

1 You have painted something with an oil-based paint. Why can't you clean the brushes with water?

2 Is fog a solution, a suspension or a colloid? Explain your answer.

3 Katy shone a beam of light through some muddy water. She could see the beam. When she tried this again the next day, she could not see the beam. Explain her observations.

4 The following solutions vary in concentration. Arrange them from the most concentrated to the least concentrated.
   A a glass of milk with one teaspoon of flavouring
   B a glass of milk with half a teaspoon of flavouring
   C a glass of milk with two teaspoons of flavouring
   D half a glass of milk with two teaspoons of flavouring

5 A jug contains four glasses of milk. You want to make flavoured milk with the same concentration as solution C in Question 4. How much flavouring would you need to add?

6 Which one of the following generalisations is the most general?
   A The hotter the solvent, the more solute it dissolves.
   B The hotter the water, the more sugar dissolves.
   C Sugar dissolves better in hot water than in cold.
   D The hotter the water, the more a substance dissolves.

   Explain your choice.

7 Describe how you would make a saturated sugar solution. If someone asked you to check their solution to see if it was saturated, how would you do it?

8 Design your own experiment to investigate the factors that affect how quickly sugar will dissolve in water.

9 The oceans at the poles contain 2.9% salt, but the oceans at the equator contain about 3.5%. Suggest a reason for this.

 ISBN 978 1 4202 3822 8

8.3 Separating mixtures

Separating suspensions

You are in the kitchen and have boiled some peas in water, but you don't want the water. You gently tip the saucepan so that the water runs out, leaving the peas in the saucepan. Pouring off the liquid like this, while keeping the solid in the container, is called **decanting**. It is a way of separating the liquid part of a suspension from the solid part.

If a suspended solid settles very slowly, you can speed up the separation by using a *centrifuge*. This is a machine designed to separate mixtures by a spinning motion. A spin-drier is one type of centrifuge.

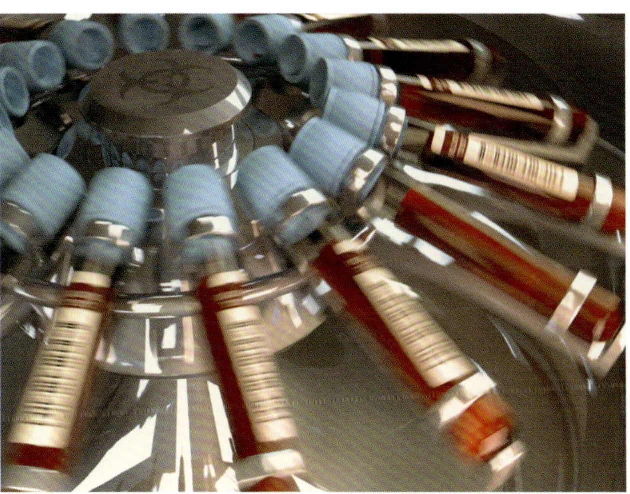

Figure 8.10 A high-speed centrifuge is used at the blood bank to separate the components of blood.

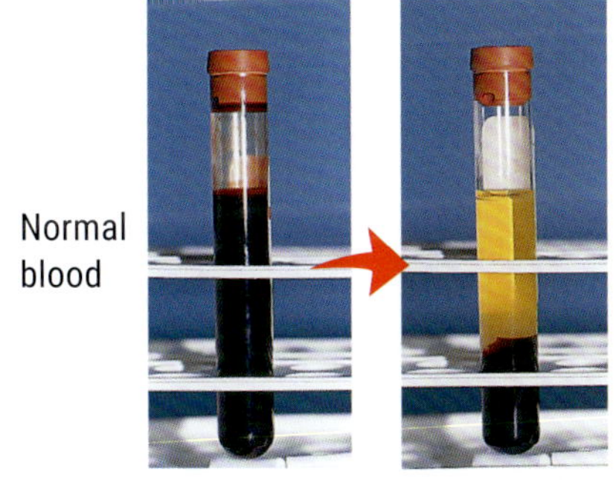

Figure 8.11 What happens when a blood sample is centrifuged.

Figure 8.12 Decanting

A centrifuge is used to separate cream from milk, and at a blood bank a centrifuge is used to separate blood cells from *plasma*. When test tubes of blood are spun in a centrifuge (Figure 8.10), the heavier red blood cells settle to the bottom, leaving the pale yellow liquid plasma on top. The plasma and red blood cells can then be separated by decanting.

Decanting is not a very good method for complete separation. Some liquid is usually left behind. Unless you are very careful, you are likely to pour off some solid with the liquid. A better way of separating suspensions is by **filtering**.

Suppose you have a suspension of chalk in water. The chalk can be separated from the suspension using *filter paper*. The filter paper has microscopic holes in it. The water can pass through these holes, but the suspended chalk cannot. This is similar to separating sand and gravel using a *sieve*. The small sand particles pass through, but the larger gravel particles do not.

The solution that passes through the filter paper and collects in the beaker is called the *filtrate*. The solid material that remains in the filter paper is called the *residue*.

ISBN 978 1 4202 3822 8

In our day-to-day life we use *filters* to separate solids from liquids and gases. For example, vacuum cleaners have special bags that filter dust and dirt from the air that is drawn in. The hairs in your nose filter the dust from the air you breathe. There are filters in a car to clean the petrol, air and oil. Filters are used to purify the water we drink, and to clean the water in swimming pools.

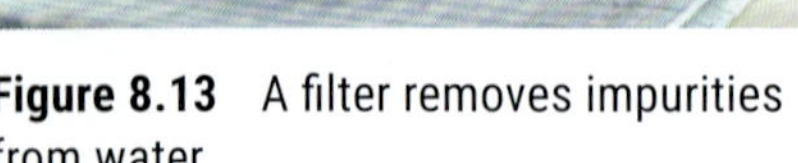

Figure 8.13 A filter removes impurities from water.

Figure 8.14 Dirty and new air filters from a car engine

INVESTIGATION 8.2 Filtering and decanting

Aim

To separate a soil–water mixture by filtering and by decanting.

Materials

- soil
- three 250 mL beakers
- 2 or 3 pieces of filter paper
- filter funnel
- retort stand and ring clamp
- glass stirring rod
- teaspoon
- wash bottle

Risk assessment and planning

Read through Part A and describe to your partner what you have to do. Your partner will then describe Part B to you.

PART A Filtering

Method

1 Make a suspension by stirring about 4 teaspoons of soil in a beaker of water. Pour half of this suspension into a second beaker and let it stand for about a day.

2 Set up the filtration apparatus as shown below. Adjust the height of the stand so that the spout of the funnel touches the inside wall of the beaker. This allows the water to flow out evenly, without splashing.

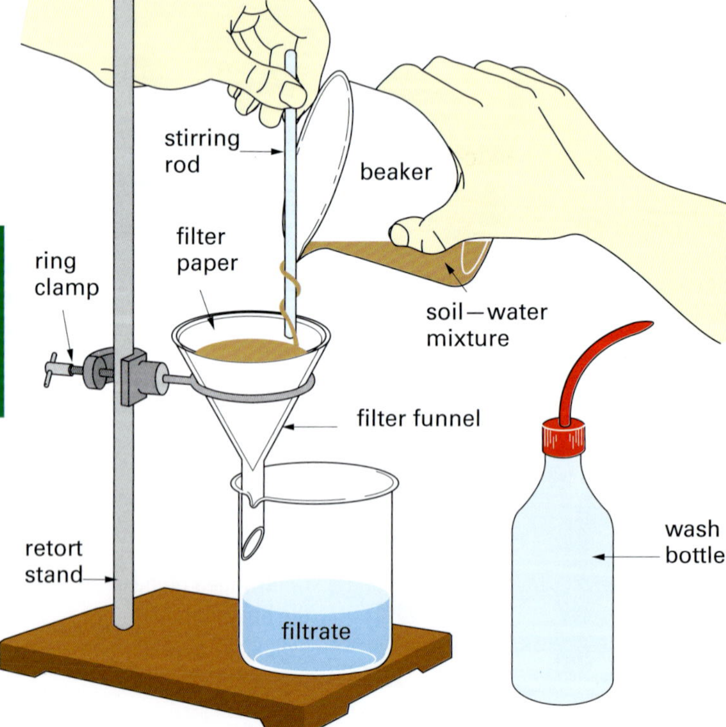

Figure 8.15 Filtration apparatus

ISBN 978 1 4202 3822 8

3 Fold the filter paper and open it out into a cone as shown (as you did in Chapter 1).

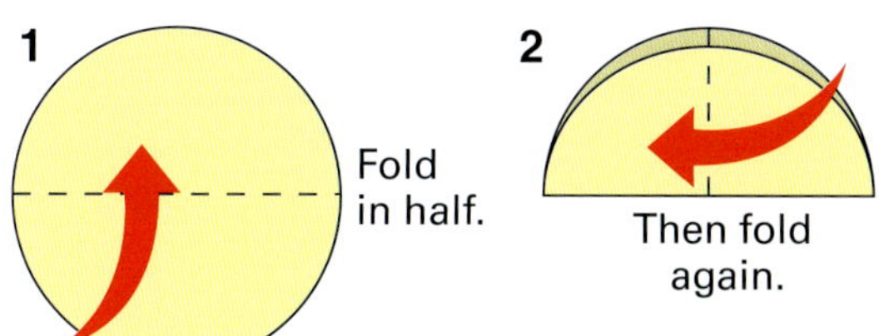

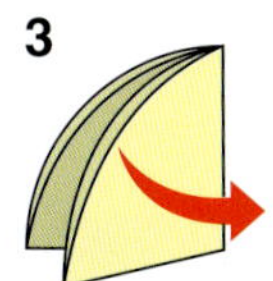

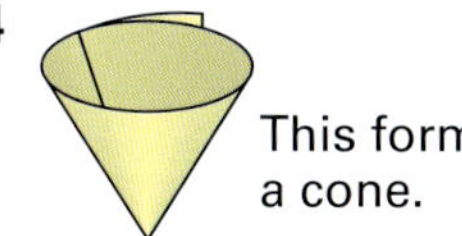

4 Place the cone into the funnel. Use the wash bottle to wet the paper so that it sticks to the sides of the funnel.

5 Hold the stirring rod as shown in Figure 8.15, with its lower end almost touching the filter paper. This will allow the water to flow gently into the filter paper.

Carefully pour some of the soil–water mixture down the rod into the funnel. Don't let the water level reach the top of the filter paper.

6 Use the wash bottle to rinse the remaining soil from the beaker into the filter funnel. Keep the filtrate for Part B.

7 Draw a diagram of the filtration apparatus. Draw a simple two-dimensional view as shown on the right. Notice how much simpler it is than the three-dimensional view in Figure 8.15. For example, there is no line across the top of the beakers, and the ring clamp has been simplified.

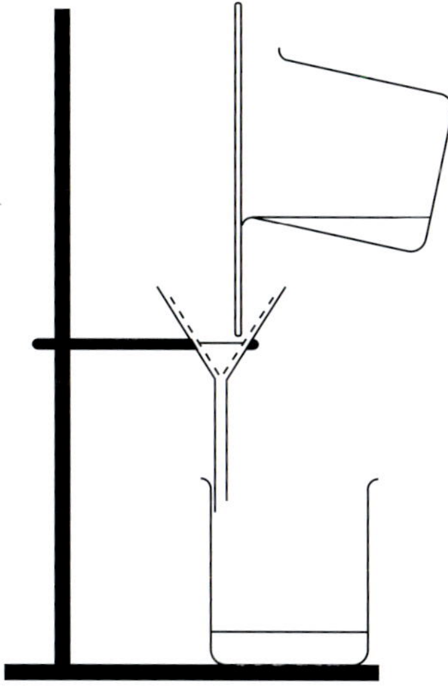

Neatly add the following labels:

- filter funnel
- filter paper
- filtrate
- residue
- stirring rod

PART B Decanting

1 Look at the beaker containing the soil–water mixture that has been standing for a day.

What do you notice?

2 Carefully decant the water into a second beaker. To do this, hold a stirring rod over the mouth of the beaker as shown below. This way the liquid runs down the rod without splashing.

Decanting

Compare the decanted water with the filtrate from Part A. Is it as clear?

3 Filter the decanted water.

Discussion

1 How easy was it to filter the decanted water, compared with the original soil–water mixture? Suggest a reason for this.

2 Explain why you:

a wet the filter paper in Part A step 4

b used the wash bottle in step 6

c poured the suspension down a stirring rod when filtering and decanting.

ISBN 978 1 4202 3822 8

Figure 8.16 At this salt plant, sea water is run into large ponds. Heat from the sun causes the water to evaporate, leaving the salt behind.

Separating solutions

Once a solute has dissolved in a solvent to form a solution, you cannot separate it by filtration. The solution simply passes through the filter paper in the same way that water does.

If a solution consists of a solid dissolved in water, you can separate the solute and the water by heating. The water *evaporates*—turns into a vapour and seems to disappear into the air—leaving the solid behind. Salt can be obtained from sea water by this method.

If you want the liquid, you must somehow trap it as it evaporates and *condense* it back to a liquid. This process is called **distillation**. In a solar distillation plant, the sunlight passes through glass plates and the heat causes salty bore water (from underground) to evaporate. The water vapour condenses on the glass roof, and the water droplets run down the inside of the glass plates into the collection gutter. The water is pure, because the salty solutes have been left behind.

Distillation can also be used to separate two or more liquids with different boiling points, for example, water and alcohol. This process is used in the making of whisky and brandy, and in the separation of crude oil into petrol, kerosene, diesel fuel and lubricating oil.

EXPLORE ONLINE

Go online and find out how to make a simple solar still, by following the links to **Solar still**.

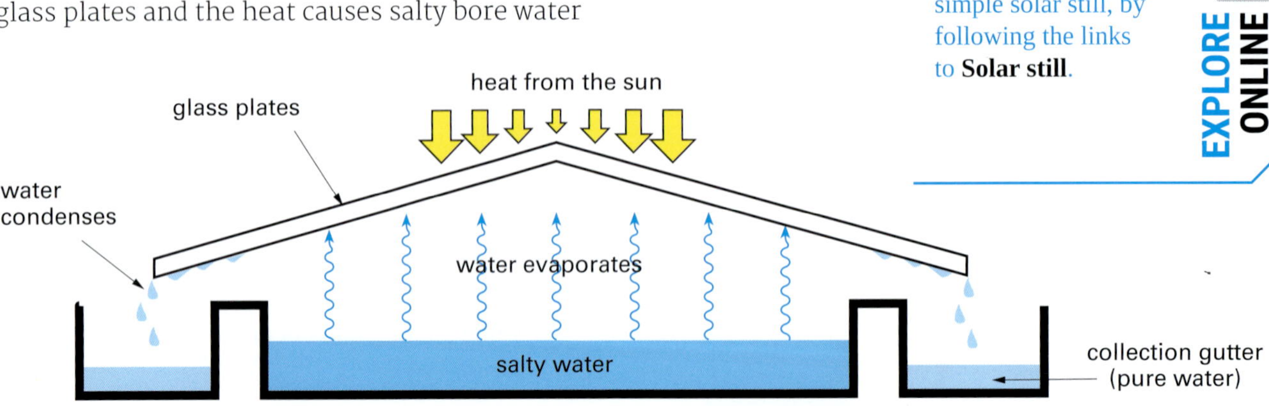

Figure 8.17 How a solar distillation plant works

ISBN 978 1 4202 3822 8

INVESTIGATION 8.3 Evaporating and distilling

Aim

To separate the solute and the solvent in a solution by evaporation and by distillation.

PART A Evaporating

Materials

- copper sulfate solution (0.5 M)
- boiling chips (broken porcelain)
- 250 mL beaker
- Bunsen burner
- heatproof mat
- watch glass
- gauze mat
- tripod

Risk assessment and planning

Read the Method for Part A.

- Suggest why you put the watch glass on top of the beaker of boiling water, instead of directly on the gauze mat over the burner.
- Suggest why you don't evaporate the copper sulfate solution completely over the burner.

Rules for safe use of a burner

1. Keep the burner away from books, and away from the edge of the bench.
2. Use a heatproof mat under the burner.
3. Always light the burner with the air hole closed.
4. Switch to a yellow safety flame when not heating.
5. The barrel of the burner gets very hot. If you have to move the burner, turn it off first. Move it by holding the base or the gas hose—not the barrel.
6. Check that the gas is off properly when you have finished.

Method

1. Put the tripod and gauze mat on a heatproof mat as shown below.
2. Half fill the beaker with water. Add some boiling chips to prevent *bumping* (violent eruption of bubbles from the bottom of the beaker).
3. One-third fill the watch glass with copper sulfate solution. Place the watch glass on top of the beaker as shown.
4. Put on your safety glasses. Light the burner and adjust to the blue flame (see page 12). Then put it under the tripod and boil the water in the beaker. The copper sulfate solution will evaporate slowly.
5. When almost all the copper sulfate solution has evaporated, turn off the burner and let the apparatus cool. (If you heat the solution any longer, it will start to splutter.)
6. Leave the remaining solution in the watch glass in a warm, protected place to finish evaporating. This process is called *crystallisation* and may take a day or two.

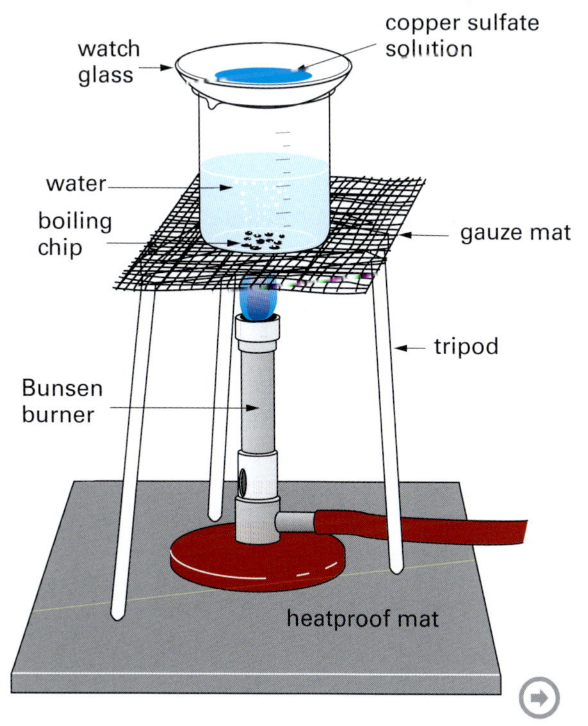

ISBN 978 1 4202 3822 8

Discussion

1 What was left in the watch glass after a day or two?

2 In your own words, explain how evaporation caused the solute to be separated from the solvent.

3 What was the purpose of the gauze mat when heating?

4 Why is it essential to wear safety glasses for this investigation?

PART B Distilling

Materials

SAFETY GLASSES MUST BE WORN

Same as for Part A, plus:

- conical flask
- one-holed stopper to fit flask
- length of glass tubing (at least 40 cm long and bent as shown at right)
- 2 retort stands and clamps
- test tube

Risk assessment and planning

Read the instructions and study the diagram.

- What do you think is the purpose of the glass tubing?
- What safety precautions will you need to take?

Method

1 Set up the distillation apparatus as shown.

2 One-quarter fill the flask with copper sulfate solution. Add some boiling chips.

3 Put on your safety glasses. Light the burner, adjust it to the blue flame, and heat the solution in the flask.

4 As the water boils, observe the water vapour:
   a rising in the flask and moving through the glass tubing
   b condensing back to liquid and dripping from the glass tubing into the test tube.

5 Collect a sample of distilled water in the test tube, then turn off the burner.

Discussion

1 Explain what happened in the:
   a conical flask
   b test tube.

2 The liquid you collected in the test tube is called the *distillate*. Why is it clear, not blue?

3 The glass tubing is called an *air-cooled condenser*. Suggest a reason for this name.

4 Design a water-cooled condenser.

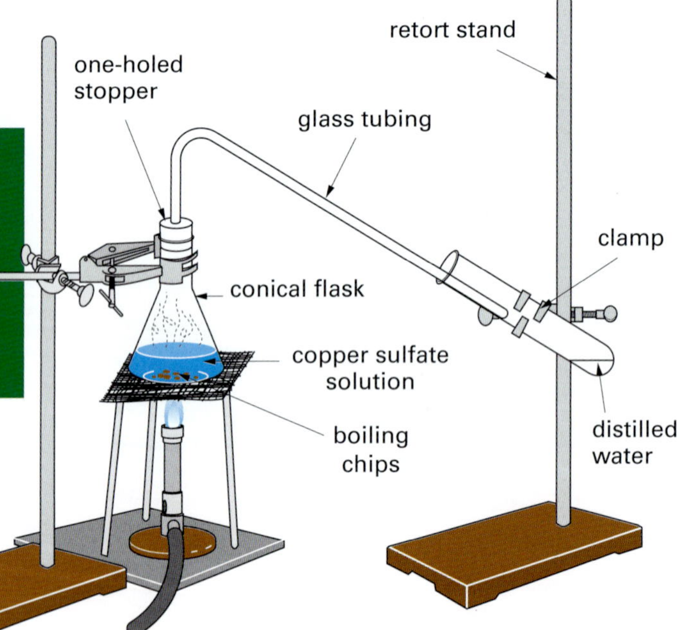

Figure 8.18 Distillation apparatus

ISBN 978 1 4202 3822 8

Separating solids

Sometimes we need to separate a mixture of solids from each other. The following four methods all depend on differences in the properties of the solids.

1 **If one solid is soluble in water and the other is insoluble, you can add water.** When you filter the mixture, the residue is the insoluble solid. The filtrate contains the soluble solid in solution. It can be recovered by evaporation. The process can be summarised in a flowchart.

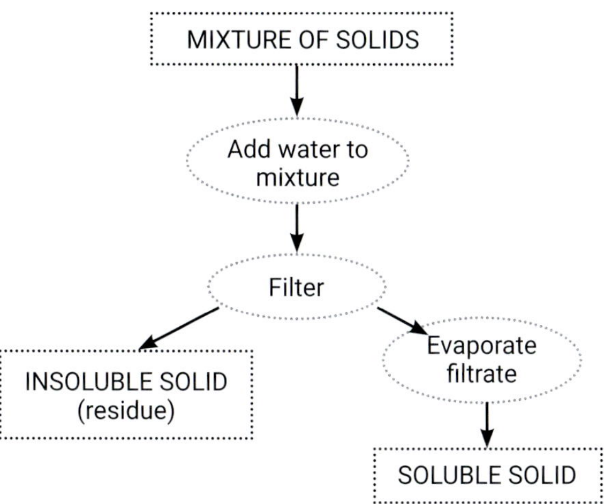

2 **If one solid is attracted to a magnet and the other is not, you can use magnetic separation.** This method is used in industry to separate the magnetic minerals in mineral sands.

Figure 8.19 A magnet will separate the magnetic minerals in sand.

3 **If one insoluble solid floats on water and the other sinks, you can add water to the mixture and skim off the floating solid.** For example, you can separate sawdust and sand this way. Sometimes this method can be used even if both solids normally sink in water. A special chemical is dissolved in the water, and air is bubbled into it. A froth of bubbles floats to the top, taking one of the solids with it. This method is called *froth flotation*. It was invented in Australia, at Broken Hill, and is often used to separate valuable minerals from rock.

4 **If one solid is heavier than the other, you can use gravity separation.** A good example of this is gold panning. Here the water is swirled about in the pan, allowing the heavy gold to sink and the lighter mud and sand to be washed off the top. This is like decanting.

Figure 8.20 When you pan for gold, you use gravity to separate heavy gold particles from 'lighter' sand.

ISBN 978 1 4202 3822 8

EXPERIMENT 8.1 Water purification

The problem to be solved

The normal water supply has broken down. The only water available is creek water, which is greenish in colour, smells, and has all sorts of things in it; for example, twigs and mosquito wrigglers. How can you make this water pure enough to drink?

Method

1 Form a group with other students. Your teacher will give you a sample of about 200 mL of impure creek water. Your task is to recover as much pure water as possible.

2 Observe the creek water and record what impurities are in it.

3 In your group, discuss ways of purifying the water.

Which of the separation techniques you have learnt in this chapter could you use?

The flow diagram below shows how water is purified in a water treatment plant. Could you modify this for use in the laboratory? How?

4 Decide how you will attack the problem.
- Which technique(s) will you use?
- What equipment will you need?
- Who will do what?
- How much time will you need?

5 When you and your teacher are happy with your plan, put it into action.

Keep a record of what you did.

What was the water like after you purified it?

How much purified water did you recover?

6 If your technique isn't successful, try another. You may need to discuss the problem with your teacher. You may also need to use the library.

Writing your report

Write a report describing what you did, for someone else to read. You could prepare a poster for presentation to the rest of the class. Include a discussion of how successful your method was.

- Is your method practical?
- How long did it take?
- Would your method work for larger volumes of water?

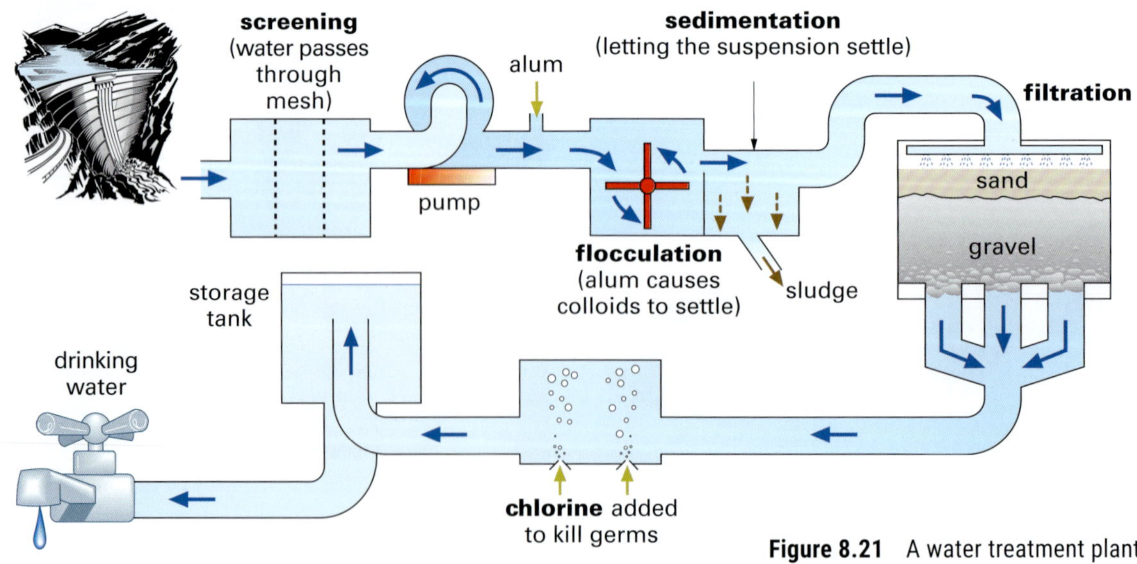

Figure 8.21 A water treatment plant

ISBN 978 1 4202 3822 8

Separating colours

Chromatography (CROW-ma-TOG-ra-fee) can be used to separate a mixture of coloured substances. (*Chromos* is the Greek word for 'colour'.) For example, this method will separate the coloured substances in black ink, as shown at the right.

Gas chromatography

Gas chromatography is used in industry and in scientific research to detect very small amounts of chemicals in mixtures. It is used to test the purity of medicines and to see if harmful pollutants are being released into the air.

Forensic scientists use it to detect poisons and drugs in blood or traces of chemicals at crime scenes. The peaks on the graph on the monitor in the photo are the different chemicals in the sample being tested.

Figure 8.22 Gas chromatography is used by anti-doping agencies to test for drugs in sport.

ISBN 978 1 4202 3822 8

Paper chromatography

Aim

To plan and carry out an investigation to separate the different coloured substances in inks or food colourings using paper chromatography.

Materials

- various coloured inks from ballpoint pens, felt pens or marking pens (Indian ink works well)
- food colourings
- 250 mL beaker
- filter paper or blotting paper
- dropper
- scissors
- adhesive tape
- jelly beans, Smarties or similar sweets
- small paint brush

Risk assessment and planning

- You can use one or more of the three methods listed.
- Black and dark colours usually give good results.
- For some inks, e.g. ballpoint pen, you may need to use alcohol or methylated spirits instead of water as the solvent.
- To remove the food colourings from a jelly bean or similar, put it in a watch glass and add three drops of water. Brush the jelly bean with a small paint brush until the colouring dissolves in the water.
- Allow the filter papers to dry, then label them and stick them in your notebook.
- When you have finished the investigation, write a report describing in a few sentences what you found out. For example, which ink or jelly bean contains the most colours?

Method

Use the 'Risk assessment and planning' box and the diagrams below to plan what you are going to do and how you are going to do it.

Method A

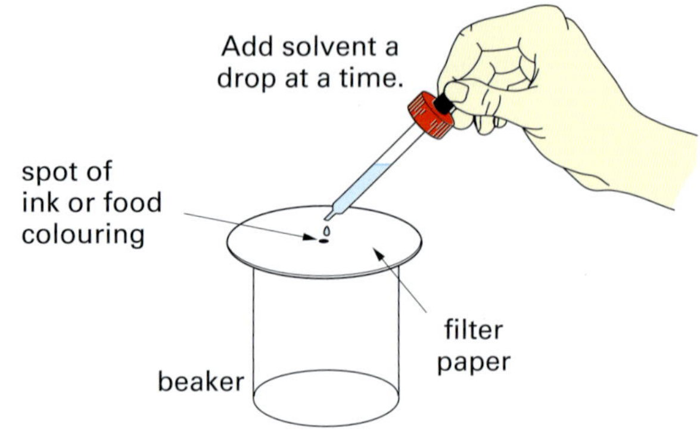

Method B

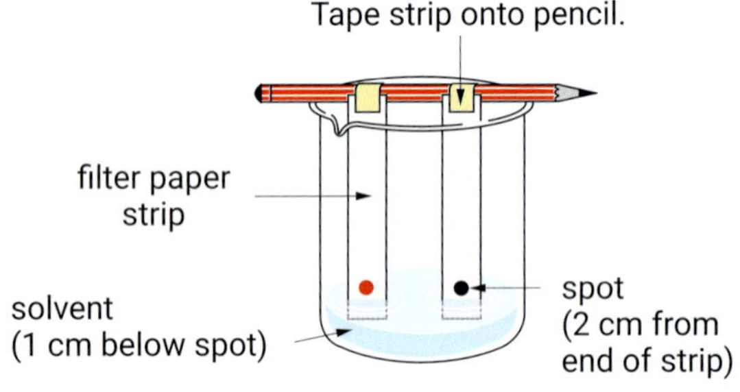

Method C

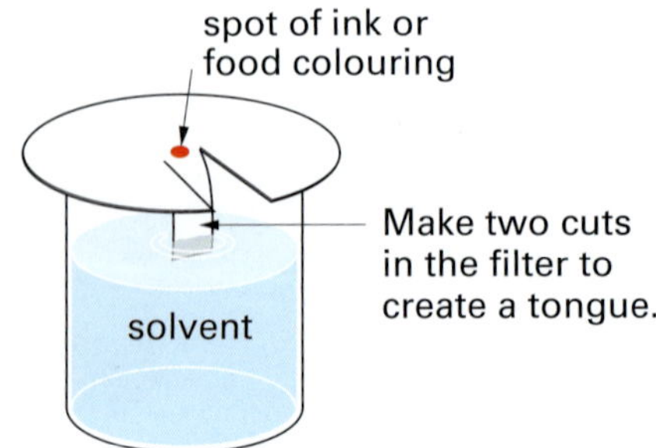

ISBN 978 1 4202 3822 8

SCIENCE AS A HUMAN ENDEAVOUR

Desalination

Western Australia has two desalination plants, which together can provide more than half the fresh water required in WA.

Since opening in 2006, the Perth desalination plant has been taking ocean water and removing salt and impurities from it. The desalination plant produces around 45 billion litres each year, almost 20% of the water supply that Perth requires. The second desalination plant is located in Binningup, near Bunbury, and has the ability to purify 100 billion litres per year. This desalination plant is powered by renewable energy sources from the Greenough River solar farm and the Mumbida wind farm.

Desalination plants work by removing the salt and other impurities from ocean water using a separation technique called reverse osmosis. Reverse osmosis is done by 'pushing' the water through tiny holes in a membrane. The water can fit through the holes, but the salt cannot, meaning that the water (the solvent) becomes purified by ending up on the opposite side of the membrane to the salt.

Use online resources such as the Water Corporation website to answer the following questions.

Questions

1 Where did Western Australia get fresh water from before these desalination plants were operating?
2 How do desalination plants work? Design a flowchart or diagram that shows each step of the process.
3 What happens to the waste produced during the desalination process?

Figure 8.23 The waste from a desalination plant can be stockpiled on land or discharged back into the ocean.

4 Why is it important to use renewable energy sources when using new technologies?
5 Extension question: Research the history of desalination and present a timeline of the process.

Present your answers in a digital format such as a PowerPoint, video, website or Prezi.

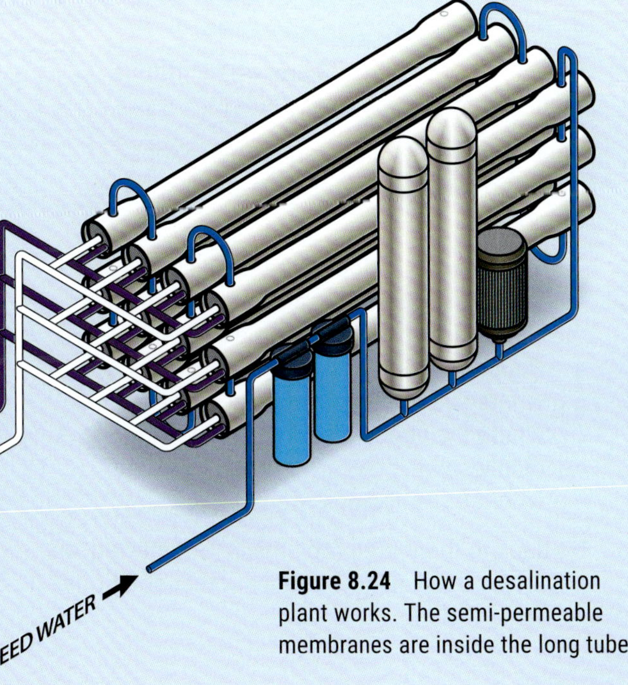

Figure 8.24 How a desalination plant works. The semi-permeable membranes are inside the long tubes.

ISBN 978 1 4202 3822 8

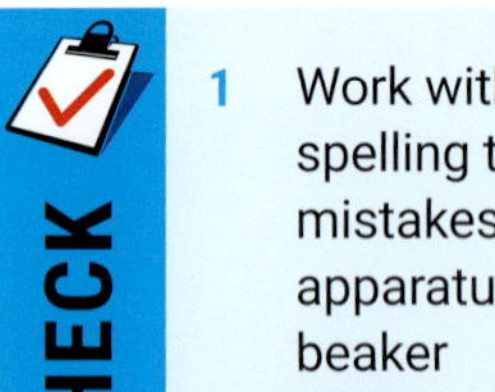

1 Work with a partner and give each other a spelling test of these words. Correct any mistakes.

apparatus	filter funnel
beaker	laboratory
dilute	solubility
distillation	solute
evaporation	solution

2 Look at the food strainer below. Explain how it works.

3 Why is filtering usually a better method of separation than decanting?

4 Suppose you filter river water that contains mud, sand, dissolved salt and some plant materials. Which of these materials will be present in the:

a residue?

b filtrate?

5 What is a centrifuge? Where is a centrifuge found in most homes?

6 Write a sentence or sentences using these words correctly: condensation, distillation and evaporation.

7 The diagram at the top of the page shows the apparatus used to distil salt water. Write down the correct letter for each of the following:

a Bunsen burner

b where evaporation takes place

c where the vapour changes to a liquid

d distilled water

e where the salt stays

8 Go back to the three problems in 'Get started' on page 185. Can you now suggest other solutions?

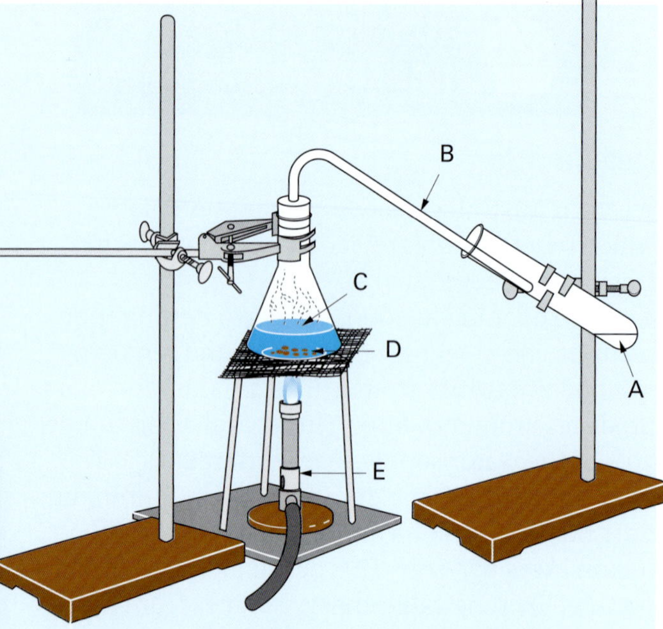

9 Which method would you use to do each of the following?

a separate iron filings from sawdust

b make fresh water from sea water

c remove the water from wet clothes

d remove the dust from the smoke going up a factory chimney

e separate the coloured dyes in an ink

f separate cream from milk

g separate a mixture of salt and pepper

10 A dye is known to be a mixture. When a spot of the dye was put on a strip of filter paper and the strip placed in alcohol, three coloured spots appeared, as shown.

a Why have the parts of the mixture separated?

b Which coloured substance do you think is the most soluble in alcohol? Why?

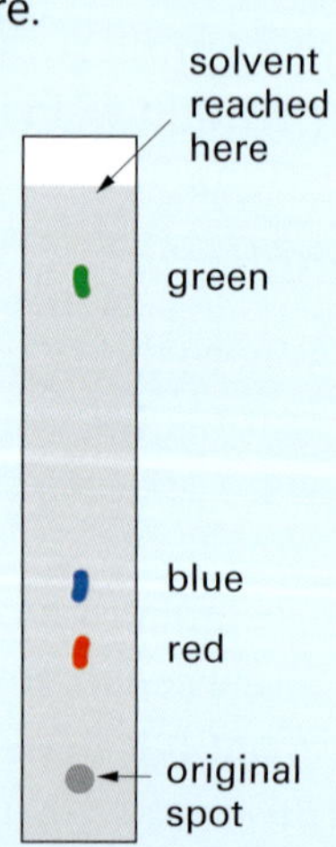

11 Draw a simple diagram of the apparatus needed to separate a mixture of sea water and sand so that you obtain clean sand. On your diagram, label at least three pieces of equipment, and show where the salt and sand end up.

ISBN 978 1 4202 3822 8

CHALLENGE

1 In Investigation 8.2 (page 194) you used filtration apparatus. What is the difference between equipment and apparatus?

2 Why is it important to replace the filters used in cars from time to time?

3 Using Figure 8.10 on page 193, explain how a centrifuge works.

4 A chemist used paper chromatography to investigate some ink. Her results are shown below.

a Which different coloured dyes did the ink contain?

b Infer the colour of the ink.

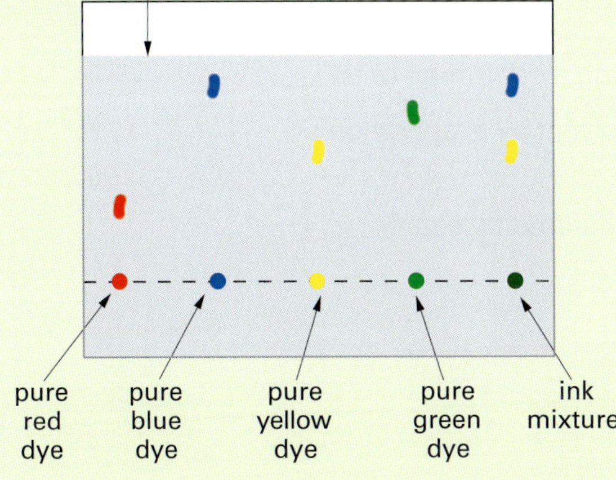

5 Kirk, Nathan, Patsy and Jade each had a mixture to separate. The four mixtures (not in order) were:

i mud and water

ii mud and salt

iii salt and water

iv mud, salt and water.

Patsy's first step in her separation was to add some water to her mixture. Kirk separated his mixture by decanting it. Jade had more steps in her experiment than Nathan did.
Which student separated which mixture?

6 After using an electric kettle with hard water for some time an insoluble substance builds up inside the kettle. Infer where this substance comes from.

7 Look at the photos of the filters on page 194. Describe how you think these work.

8 The photo below shows a separating funnel. It can be used to separate two liquids that do not mix; for example, oil and water. Explain how you think it works.

9 Imagine you are a waste management engineer. You have been supplied with a mixture that contains sand, sawdust, iron filings and lead shot. Your task is to separate as much of each component as you can, so that they can be recycled.

10 Suppose you own a lolly shop. You suspect your supplier is selling you a cheap, inferior brand instead of Smarties, but charging you for the real thing. How could you check this?

ISBN 978 1 4202 3822 8

Copy and complete these statements to make a summary of this chapter. The missing words are on the right.

1 A _____ is a substance that dissolves in a _____ to form a _____.

2 When a substance _____, it is said to be soluble. Substances that do not dissolve are _____.

3 In a _____ (e.g. muddy water), the solid settles to the bottom when left standing. Solutions do not settle.

4 Many everyday substances are _____, with properties in between solutions and suspensions.

5 A _____ solution contains only a small amount of solute in a given volume of solvent. A _____ solution contains a larger amount of solute.

6 Separation techniques depend on differences in the _____ of the substances in the _____.

7 Suspensions can be separated by _____, using a centrifuge or by _____.

8 A dissolved solid can be separated from a solvent by evaporation or by _____.

9 A mixture of coloured substances can be separated by paper _____.

chromatography
decanting
solvent
distillation
filtering
concentrated
dissolves
insoluble
mixture
suspension
properties
solute
dilute
colloids
solution

CH•8 REVIEW

1 If you dissolve instant coffee in hot water, the water is the:
- **A** solvent.
- **B** solute.
- **C** solution.
- **D** suspension.

2 If more water is added to a coloured solution, it becomes:
- **A** more concentrated.
- **B** more dilute.
- **C** saturated.
- **D** a darker colour.

3 Water can be separated from alcohol by:
- **A** chromatography.
- **B** filtration.
- **C** evaporation.
- **D** distillation.

4 Four liquids—water, kerosene, alcohol and petrol—were used to test the solubility of three unknown solids, A, B and C.

Georgia did the tests and recorded the mass of solid that dissolved in equal volumes of the liquids.

Solvent	Grams of solid that dissolved		
	Solid A	Solid B	Solid C
water	5	6	0
kerosene	1	1	5
alcohol	4	3	4
petrol	1	0	6

- **a** Which liquid is the best solvent for solid B?
- **b** If solid A was accidentally mixed with solid C, which liquid could you use to separate them? Explain your answer.
- **c** Is there any way of separating a mixture of A and B?

ISBN 978 1 4202 3822 8

5 Look at the diagram below.
  a Name the pieces of equipment labelled A–E?
  b Label the filtrate and the residue.
  c There are two mistakes in the diagram. What are they?
  d Redraw the diagram correctly.

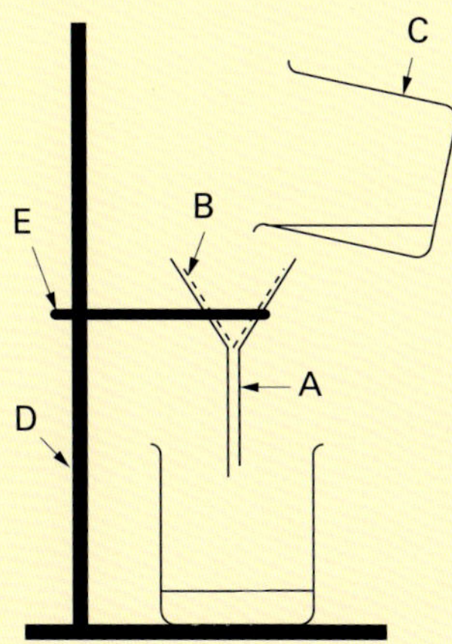

6 The apparatus below can be used to obtain pure water from salt water.
  a What is this separation method called?
  b Explain how the method works.
  c What is the purpose of the ice-cold water?

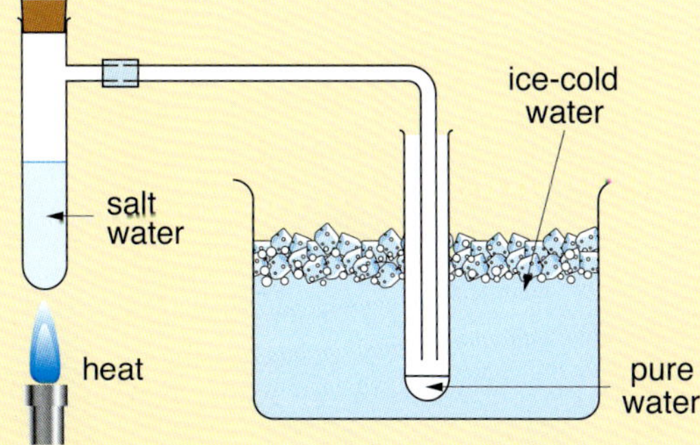

7 When a can of fruit juice is left to stand, a sediment forms on the bottom of the can. Is fruit juice a solution, a suspension, a colloid or a combination of these? Explain.

8 a Write one complete and scientifically correct sentence using these words:
colloid emulsion milk
  b Do the same for these words:
concentrated dilute solution

9 The police receive a ransom note written using a felt pen. They also have felt pens from three suspects. How could they use paper chromatography to find out who wrote the ransom note?

LAB REVIEW

You have a mixture of salt and dirt that Ken collected on his recent trip to Lake Eyre. Your task is to separate the salt by removing the dirt.
1 Work out a way of separating the salt.
2 Make a list of the equipment you will need.
3 Do the separation correctly and safely.
For step 3, work with a partner who will watch what you do and note any errors you make. They will discuss these with you when you have finished. Then swap jobs and check your partner's skills.

Check your answers on page 243.

ISBN 978 1 4202 3822 8

IN THIS CHAPTER

Science Understanding

> develop an understanding of how water is cycled through the environment
> explore ways of saving and recycling water
> explain how resources can be managed and used sustainably
> discuss the differences between renewable and non-renewable resources
> develop an appreciation of the hardships and successes of George Washington Carver

Science Inquiry Skills

> interpret graphs and use them to make inferences
> evaluate information in tables and draw conclusions from this information

CH•9 Sustainable Earth

ISBN 978 1 4202 3822 8

GET STARTED: *IMAGINE*

A sustainable Earth is one which has an ongoing capacity to sustain life.

Imagine you are living in the year 2120, and the Earth has survived, is healthy and has become more 'sustainable'. Humans have worked hard to make the Earth sustainable over the past 100 years. They have had to solve many environmental and other problems along the way to ensure human, plant and animal life can go on for future generations.

1 Work with a group to make a list of the problems that needed to be solved along the way to get to this point. Share your list with the class and make a class list.
2 Can you imagine ways of solving these problems?

ISBN 978 1 4202 3822 8

9.1 Water as a resource

All living things depend on water. Your body is about 70% water—so an average person contains about 45 litres of water. Fruit and vegetables contain a larger percentage of water—a tomato is 93% water and an orange is 86% water. Without water, living things cannot survive.

Over two-thirds of the Earth's surface is covered in water and about 96% of this is found in the oceans. The other 4% occurs as water vapour in the air, as water droplets in clouds, and as water in rivers and lakes and in the ground.

The water cycle

When you leave a bowl of water out in the sun for some time, the water disappears. It has turned into *water vapour*. This is the process of **evaporation** (e-VAP-o-RAY-shun). As the water evaporates, it leaves behind any dissolved substances, such as salt, so only pure water goes into the air. You can speed up the process by boiling the water. If the water vapour hits a cold surface, it changes back to liquid water. This is the process is of **condensation**.

The Earth's water is always moving from one place to another. The heat of the sun evaporates water from the soil and from the surfaces of lakes, rivers and oceans. Water also enters the air from trees and other plants by **transpiration**. This invisible water vapour is carried about by the wind. As it rises, it becomes much cooler and may condense to form clouds and perhaps rain, hail or snow.

Rain, hail and snow, also known as **precipitation**, return the water to the land or the ocean. The water that falls on land flows into rivers and lakes, and eventually reaches the oceans. Some water seeps into the soil and rocks to become **groundwater**. This process is known as *percolation*. Other water evaporates from the land. Once the water reaches streams, lakes and the oceans, the sun causes more evaporation, and the changes continue. These changes are known as the **water cycle** (see the diagram below).

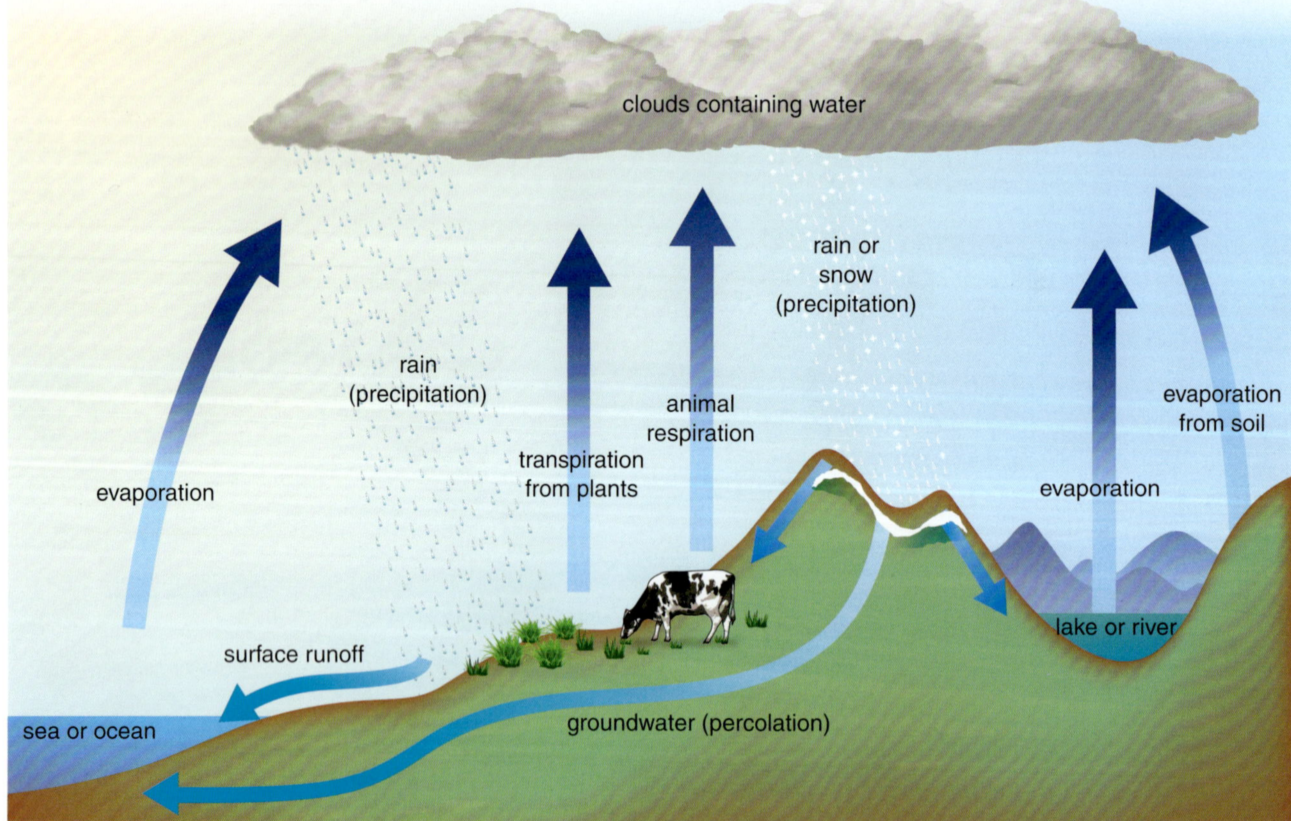

Figure 9.1 The water cycle

ISBN 978 1 4202 3822 8

What factors affect the water cycle?

Many factors can change how water moves through the water cycle.

Figure 9.2 Irrigation from groundwater or rivers and water locked in a glacier can change the flow of water in the water cycle. Which of these is a natural effect?

Factor	Effect on the water cycle
State of water	When water freezes to form ice, such as in a glacier or in the Antarctic, the solid water is trapped for long periods of time, even for thousands of years. This process locks water out of the water cycle for periods of time, slowing the movement of water.
Wind	Winds move clouds and evaporated water around the Earth. Moving air also causes faster evaporation of water. On a windy day a puddle will evaporate faster than on a still day.
Sunlight and temperature	Sunlight has a direct effect of heating up the ground, rivers, lakes and the ocean, and increasing evaporation of water into the atmosphere. Sunlight also increases the air temperature, which speeds up processes like transpiration from plants, releasing more water into the atmosphere.
Vegetation	Different environments affect the water cycle because of their different rates of water evaporation from the ground and/or plants. More water will enter the atmosphere from a rainforest which has more vegetation than from a desert with little vegetation.
Land formation	Steep land will produce more runoff water than a flat area that has more time for percolation to occur. Also, more rain tends to occur over mountains. This is due to the weather patterns that form over such areas.
Soil type	The type of soil and the structure of the ground affect the water cycle. A clay soil that is packed tightly with little space between the soil particles will hold a lot of water, while a sandy soil with lots of space between grains will allow water to percolate down faster.
El Niño	El Niño is an effect related to the warming of the ocean in the eastern and central Pacific. When this happens it affects the weather across the Pacific Ocean, including eastern and northern Australia, producing less rain. The opposite (La Niña) occurs when the Pacific cools, bringing more rain to eastern Australia.
Human intervention	Humans use water in many ways that affect the water cycle. Examples include: • building dams that stop runoff into rivers • pumping waste water into oceans • collecting water in tanks • industries release water into the atmosphere or rivers • removing forests and bushland which reduces evaporation and changes water flows • extracting groundwater for irrigation or farming • desalinating sea water to make it fresh, and using it for other purposes.

ISBN 978 1 4202 3822 8

Water use

Water is a very valuable resource, yet the average household water use per person is about 180 litres per day. That equates to drinking 720 glasses of water in a day! However, the water usage per person includes not only water for drinking, but for cooking, showers and baths, flushing the toilet, washing clothes and dishes, washing the dog and the car, and watering the garden. Figure 9.3 shows a breakdown of how we use water.

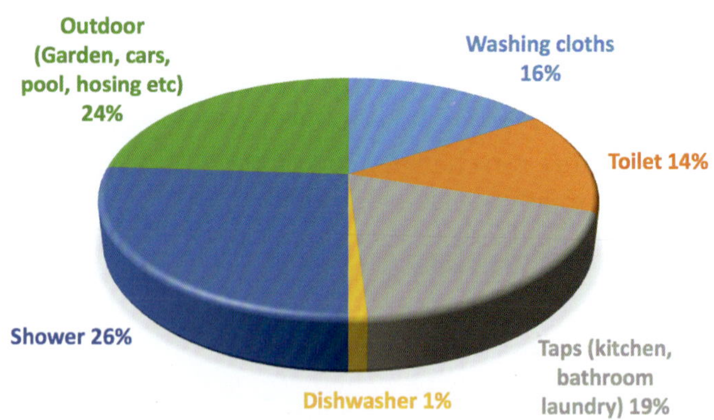

Figure 9.3 Typical household water use

PART A Your household water use

1 Find the water meter at your home. This is near the mains water tap and is usually just inside or just outside your front fence.

2 Read the meter and write down the reading. Read it again 10 days later.

Record your readings in a table and, by subtraction, find the volume of water your household used in 10 days.

Knowing the number of people in the house, calculate the volume of water used per person per day.

PART B How much water do you use?

For this part of the activity you are going to calculate the amount of water you use in a day. You will do this by estimating your water consumption over the two days on the weekend and find the average for one day.

1 Draw up a table like the one shown to record your water use.

How I used water	Volume of water used (L)
shower or bath	
toilet	
washing hands	
cleaning teeth	
drinking	

2 For each row in the *How I used water* column, design a method you will use to find the volume of water used.

For example, to find the volume of water used in the shower, you might use a bucket to find the volume of water used in a minute, then time your shower and calculate the water used over this time.

Note: You might design a clever way to find the volume of water used in your toilet. Alternatively, you could use these figures as an estimate of the flush volume:

- For modern toilets the full flush volume is 6 L and the half flush is 3 L.
- For older style toilets the full flush is about 9 L.

3 Record your water use and calculate the volume of water you used per day over the weekend.

How does your water usage compare with the average per person water usage of 180 litres? Suggest reasons for the difference.

Suppose your local council imposed a limit of 100 litres per person per day. How could you reduce your usage? What is the first thing you could do to reduce your usage? What is the last thing you would do?

Could some household water be used more than once? Suggest how.

Draw a graph of your water use, like the one in Figure 9.3.

ISBN 978 1 4202 3822 8

Managing water

Australia is the driest inhabited continent, with many areas often experiencing drought. Two desalination plants located on Western Australia's coast provide 47% of the state's water (see page 203). Another 46% is groundwater, with only 7% of WA's water coming from rainwater flowing into dams. The photo below shows the Mundaring Weir to the east of Perth, which has a capacity of 63.6 million kilolitres.

Figure 9.4 Mundaring Weir

Water restrictions

One way cities can save water is by enforcing water restrictions. Some of the water restrictions that have been implemented in Western Australia include:

- allocated days for using sprinkler systems in summer
- total sprinkler system bans in winter
- subsidies for bore water systems.

EXPLORE ONLINE

Go online and research the suggested methods of saving water in Western Australia.

Collecting water

Rain that falls will remain in the water cycle unless collected. Water for household use is collected in large dams, like the Mundaring Weir in Figure 9.4. Before the water is used, it is purified to make sure it is safe, then pumped into houses through a large network of pipes that bring the water to us. Governments and water authorities charge a fee for the water we use to pay for building and maintaining the dams supplying the water.

Figure 9.5 Rainwater can also be collected in tanks.

Rainwater can also be collected in tanks. However, there can be problems using water from a tank.

Health authorities in all states recommend that houses that are connected to mains water and have tanks, use the tank water for all their house needs, except for drinking and cooking. This is because the air in cities often contains dust and pollutants from traffic and industries. When this falls on a roof, it is washed into the tank when it rains. Animal droppings and dead animals sometimes also get caught in the roof guttering. The bacteria from these are carried by rainwater to the tank and could be a source of infection.

In rural areas where the air is generally cleaner, many households rely on rainwater tanks for all their water needs. In Western Australia, a subsidy is available to people who want to install a rainwater tank.

ISBN 978 1 4202 3822 8

Reusing water

The water that is used in showers, baths, washing machines and laundry tubs can be reused for garden irrigation. This water is called *greywater*. In times of water restrictions, greywater can be used to keep garden plants and lawns alive.

Greywater from the kitchen sink should not be used as the water contains food wastes and chemicals that are not easily broken down by the organisms in the soil. The safest greywater to use is the rinse water from washing machines, followed by the wash-cycle water, then the water from showers and baths.

Toilet water is called blackwater, and cannot be easily reused as it contains a high level of dangerous bacteria.

Purifying water

The water in reservoirs and dams is not usually suitable for drinking. It contains many dissolved minerals, some of which may be harmful. The water may also contain suspended materials, such as bacteria and other microscopic organisms as well as pieces of dead plants.

Water for households is treated (purified) in a water treatment plant. There are usually four steps in the treatment process.

Step 1: The first step is to get rid of the suspended materials. The easiest and cheapest way to do this is to add a chemical called alum to the water to make these suspended particles clump together.

Step 2: The clumped particles then settle out of the water by themselves as a sludge. The sludge is removed from the bottom of the large settling tanks and used as landfill.

Step 3: The water is then filtered to remove any suspended solids that are left.

Step 4: The water is finally disinfected using chlorine to remove any harmful bacteria.

In some cities, the water is fluoridated after step 4. A chemical containing fluoride is added to the water at the water treatment plant. The fluoride helps prevent tooth decay. Fluoride does not help purify water.

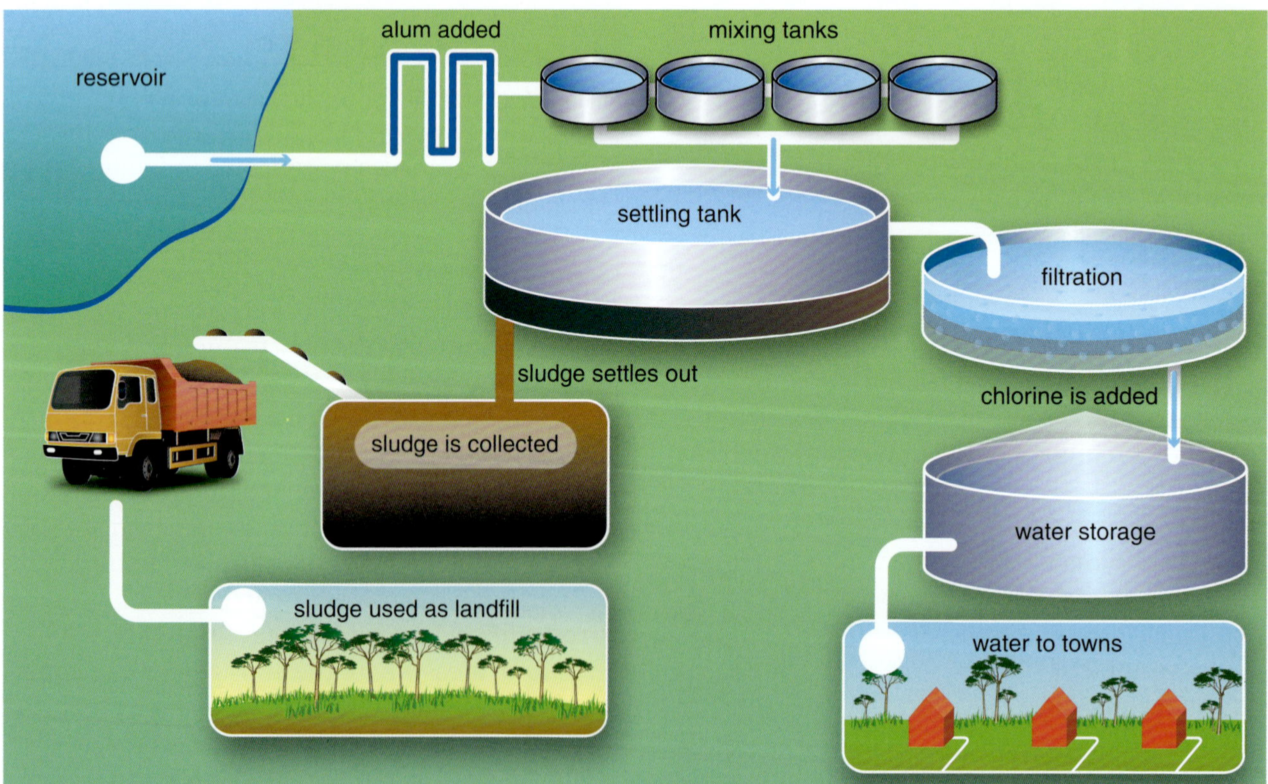

Figure 9.6 The water purification process

ISBN 978 1 4202 3822 8

1 Which of the following statements are true and which are false? Rewrite the false ones to make them correct.
   a Water disappearing from a bowl on a hot, windy day is an example of condensation.
   b Greywater is the water that flows into a water treatment plant from a reservoir.
   c The first step at a water treatment plant is to add chlorine to kill bacteria.
   d Groundwater is water that seeps into soil and rocks.

2 Hannah's mass is 60 kg. How much water does her body contain? (Assume 1 L of water has a mass of 1 kg.)

3 When you breathe out on a very cold morning, your breath forms little clouds.
   a Explain why this happens.
   b Which process in the water cycle does this demonstrate?

4 What causes the water in the oceans and in lakes to evaporate?

5 Predict what would happen if the water on the Earth's surface didn't evaporate for a year.

6 Describe some of the ways in which a city or town can reduce its water consumption.

7 a What is blackwater? Why can't it be recycled and used to water the garden?
   b The greywater from washing machines and showers contain soaps and bacteria. Suggest why health authorities recommend that the greywater is used straightaway and is not stored in tanks for more than 24 hours before use.

8 An older type shower head sprays out 14 L of water per minute. A new shower head uses 9 L per minute. If you have a 4-minute shower, how much water do you save by using the newer shower head?

9 What happens in the settling tank at a water treatment plant? Why is chlorine added?

10 a Draw a simple version of the water cycle using arrows to connect at least the following words: condensation, precipitation, percolation, groundwater, runoff, evaporation, clouds, transpiration.
   b Describe three natural factors that affect the water cycle.
   c Describe three ways in which humans can affect the water cycle.
   d Explain three ways that you can reduce your personal impact on the water cycle.

1 The table below shows how water is used in Australia. The total water use in a year is 15 million megalitres.

How water is used	Percentage use
Agriculture	62%
Households	11%
Industries	18%
Mining	4%
Manufacturing	5%

   a What volume of water is used for agriculture in a year?
   b About 90% of the water used in agriculture is used for irrigation. How much water is this?
   c What proportion of water is used in the cities and towns? What assumptions have you made in finding your answer?

2 Use Figure 9.6 on the previous page to draw a flow diagram showing how water is purified. The flow diagram is started for you below.

water from reservoir → alum is added →

3 Figure 9.3 on page 212 shows the water usage in an average household.
You have been asked to devise a plan to reduce household water consumption. Suggest ways to reduce the water consumption by 20%, and then by 50%.

ISBN 978 1 4202 3822 8

9.2 Sustainable resources

You just learnt about water, which is an important resource. Figure 9.7 below shows some other things people consider to be resources.

A resource is something that humans find useful in providing food, air, water, shelter, clothing, fuel, transport and other items needed for life. Resources can be grouped as follows:

- clean air
- clean water
- animals and plants
- soil
- minerals and fuels.

The time taken for a resource to be replaced naturally is called its *regeneration time*. Living resources, such as wood, and plants and animals used for food have short regeneration times. These resources are called **renewable**.

Non-living resources such as oil and coal have long regeneration times (measured in millions of years) and are therefore called **non-renewable**. The problem for humans is that some resources are being used faster than they can be renewed.

ACTIVITY

Read through all the comments in the cartoon below.

1. List all the resources referred to in the comments. Explain why each is a resource.
2. Suggest other example of resources that could be included in the cartoon.
3. Classify the resources into two groups—renewable and non-renewable. Discuss any that you have trouble classifying.
4. Are clean air and water renewable resources? Give reasons for your answer.

Figure 9.7 Different types of resources

 ISBN 978 1 4202 3822 8

Managing resources

To manage resources in a sustainable way is a challenge in a society where we use so much every day to meet our lifestyle needs. To be sustainable we need to find a balance between using resources and conserving the natural environment. **Sustainable** development means using resources to meet today's needs without endangering the needs of future generations. Some activities, such as clearing native forests, are not sustainable as the animals and plants will either not recover or take a very long time to do so. Other activities, such as growing wood in a plantation forest or making fuel from sugar cane that can be grown quickly, are more sustainable.

It is better to use renewable resources that can be regenerated at the rate at which we use them. In future we need to continue to find more alternatives to many of our resources. We will also need to reduce, reuse and recycle more, and find new ways to reuse and recover resources after their initial use. In the game below you will use strategies to manage a model resource.

Figure 9.8 Research and technology breakthroughs are crucial in finding solutions to how we make our use of resources sustainable.

Renewable resource game

The aim of this game is to devise a strategy to use a renewable resource (the marbles) over a 10-year period (the 10 rounds).

Work in groups of three with each group playing other groups in the class. Each group will need a shallow bowl containing 20 marbles (or plastic discs).

These are the rules of the game:

- The idea is to take as many marbles from the bowl as possible over 10 rounds.
- The game has to last for 10 rounds. If you take all the marbles before the last round, your group is out of the game (because the resource would not be sustainable).
- The number of marbles you take out can vary by only one from the previous round.
- At the end of each round, you must put back one marble for every four marbles left in the bowl. For example, if there were 11 marbles left (two groups of four and three remaining) you put back two marbles.

- Record the number of marbles the group takes out each round.
- Plot a graph of your results with the number of marbles taken each round on the vertical axis, and the number of rounds (years) on the horizontal axis.
- What is the most marbles you can take out of the bowl year after year and still have marbles left?
- Compare the strategy your group used with the strategies used by other groups.
- Explain how the game models the sustainable use of a renewable resource.

ISBN 978 1 4202 3822 8

SCIENCE AS A HUMAN ENDEAVOUR

Sustainable fishing

The world's fish numbers are in crisis. Scientists have estimated that humans have managed to remove 90% of the edible fish from the oceans.

One fish whose numbers have declined rapidly in the last 50 years is the orange roughy. You may not have heard of the orange roughy. In fish markets and shops it is called sea perch or deep sea perch.

Figure 9.9 A catch of orange roughy

The life cycle of the orange roughy

The scientists found that the orange roughy grows very slowly. It takes about 20 years to reach a length of 30 cm. The fish do not mature and start to breed until they are 25–30 years old. A female rarely carries more than 90 000 eggs, which is less than 10% of the eggs that other female fish carry.

Threats to the orange roughy

Commercial fishers use large trawling nets to catch orange roughy. Because the fish live in deep water near the sea floor, these nets are weighted to trawl over the bottom.

In the process of trawling, the nets catch other types of organisms that live on the sea floor. These organisms die and are thrown away when they are brought to the surface. The nets also damage cold water corals which the young orange roughies use for shelter and protection from predators.

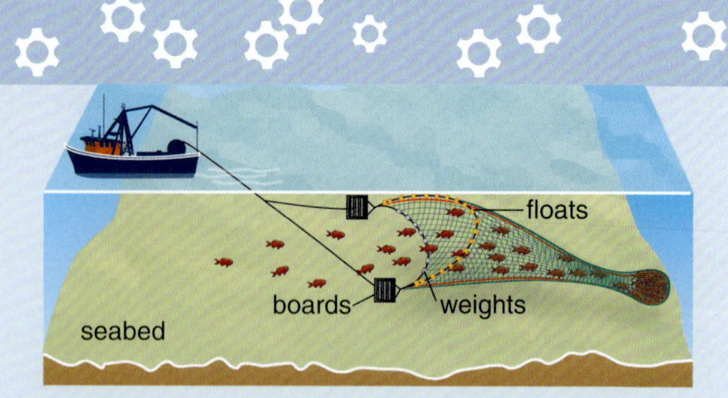

Figure 9.10 Trawling for orange roughy

Catching orange roughy

Orange roughy live in very deep water, up to 1000 m deep, off the continental shelf around the southern Australian mainland and Tasmania.

Fishing for the orange roughy started there in 1985. In that year, 400 tonnes were netted. In the next year, 4600 tonnes were netted. By 1992, more than 85 000 tonnes had been caught. Commercial fishers thought they had struck gold!

Surveys continued to show a decline in orange roughy numbers, down as low as 7% of biomass. In 2006, all fisheries but one were closed as the population was considered unsustainable.

Stocks were monitored and are now believed to be around 26% of biomass. This is now considered sustainable as long as the catch levels do not cause the stock to reduce further, and there's opportunity for the population to continue to grow. This means tight catch limits are set for each fishery.

Questions

1 If the average orange roughy weighs 1 kg, how many fish were caught in 1985? How many had been caught by 1992?

2 Was is the rate of growth (in cm per year) of a 20-year-old orange roughy? What assumptions did you make in arriving at your answer?

3 Why do marine biologists worry about the commercial netting of orange roughy more than they do about some other fish?

4 Do you think the government actions are sufficient to ensure sustainable fishing of the orange roughy? Discuss your opinions with the class.

ISBN 978 1 4202 3822 8

Resources from plants

Plants are a very important resource for us. They are a good source of sustainable resources with many uses.

Fibres

Plants are a source of fibres for making cloth and other materials. The coarse fibre from the hemp plant is used in making rope, mats and bags. Finer fibres such as cotton and rayon are used for making clothes. Wood pulp from plants is used to make paper and cardboard.

Building materials

From ancient times wood from trees has been used as a building material for houses and other buildings. It is also used for transportation—for building boats and carts. Other plants such as bamboo, grass and bark are also used as building materials.

Food

Green plants make their own food using sunlight as an energy source. Most animals rely on plants as their primary source of food. Almost every food web is dependent upon green plants.

Recreation, decoration and beauty

Humans have used plants, particularly flowers, as a symbol of happiness, love and wellbeing. People decorate their homes with plants in the garden and indoors. Plants are also used for shade. Parks and forests are used for bush walking, camping and picnics.

Fuel

Wood from trees has been the most important fuel for cooking and heating in all human societies from earliest times. Only in the last few hundred years has it been replaced by coal, oil and gas. Alcohol, made from sugar cane, is also used as a fuel. It is added to petrol for motor vehicles.

Medicines

All cultures throughout history have used plants for healing injuries and curing diseases. Indigenous Australians made a paste from the nut of the quandong that was rubbed on the skin to ease aches. The indigenous North Americans used the leaves and bark of the willow tree to ease pain. And indigenous people in Peru extracted the sap (quinine) from the bark of the cinchona tree to treat malaria. It is important to preserve natural forests, as future technologies will discover more beneficial substances from plants.

Figure 9.11 Plant resources

ISBN 978 1 4202 3822 8

SCIENCE AS A HUMAN ENDEAVOUR

George Washington Carver 1864–1943

George Washington Carver was born in Missouri, USA, to slave parents. His father was killed in a logging accident and he and his mother were bought by a couple, who set them free. Carver was kidnapped from his mother in his early teens.

Carver moved to Kansas when he was in his mid-teens and was accepted by a college in Kansas, but was rejected when they found out that he was black. After high school, he was accepted into the faculty of Agriculture at the Iowa State University. He was the first and only African American in this faculty.

During his time at the university, he developed a reputation and was known as the Plant Doctor because of his love of plants. He believed that plants were a valuable resource, and more than just food for animals. Carver loved peanuts and thought they could be made into many useful products. He was the first person to make peanut butter (Skippy brand), and also peanut oil.

Figure 9.12 George Washington Carver

George Washington Carver's inventions

During the 50 years he spent at university, Carver taught thousands of students, and invented hundreds of products from plants. His inventions include:

- about 300 peanut products including leather dyes, cloth dyes, wood stains, wall boards, face cream, soap, flooring, oil and foods
- plastics made from soybeans
- synthetic marble made from wood shavings
- foods such as starches, confectionary, flours and breakfast foods from sweet potato
- methods of rotating crops using legumes to give soil back the nitrogen taken from it by the crops.

Carver the conservationist

As a young man, Carver was very poor and had to be careful about having enough to eat. He wasted little and recycled almost everything. At university he made much of his equipment from discarded junk.

Carver taught students that discarded wastes could be recycled as soil fertilisers and mulch. He also taught farmers to rotate crops and recycle everything. His work was a lifetime lesson for students and farmers in sustainable development.

Throughout his life, Carver struggled against poverty, prejudice and racial injustice in the USA. When he died, his life savings went towards setting up the George Washington Carver Foundation for agricultural research.

Questions

1 Why were Carver and his mother 'bought' when he was young?

2 Why do you think Carver believed that plants were such a valuable resource?

3 What are some of the products Carver invented? Which do you think were his best inventions? Why?

4 Why was his knowledge of plants important to farmers?

5 What does the sentence 'Carver struggled against poverty and prejudice' mean?

ISBN 978 1 4202 3822 8

Forest resources

Australian forests are generally well managed. Strict government laws protect large areas of native forests. These areas are National Parks or State Parks. And for those forests that are logged, there is a quota on the number of trees that can be taken.

There are also many sustainable plantations that produce wood. If these are carefully managed we will have wood for the future.

What do we use forests for?

Forests are one of Australia's most important resources, and are used to:

- conserve the plants and animal species that live in the forests, and protect biodiversity
- supply high-quality timber for furniture making and for building.
- make medicines and cosmetic oils
- produce woodchip to make pulp for the paper industry.

Wood pulp and paper

To make paper, trees are cut down and then chipped to make woodchip. There are two sources of wood for woodchip—the trees in managed native forests, and trees that are grown in plantations.

Plantation trees can either be hardwoods like eucalypts or softwoods like pine trees. The trees are turned into chips which are then taken to paper mills.

Figure 9.13 Most of the woodchip produced from Australian forests and plantations is exported to Asia.

Your teacher will set up a microscope to look at different types of paper.

Collect samples of different types of paper, such as tissues, writing paper, newspaper, recycled paper, paper towel.

- **a** Place a small piece of paper on a microscope stage. Shine a lamp on the paper.
- **b** Observe the fibres in each type of paper.
- **c** If there is print on the paper, observe how the ink marks the surface.
  Record your observation in a table or list.

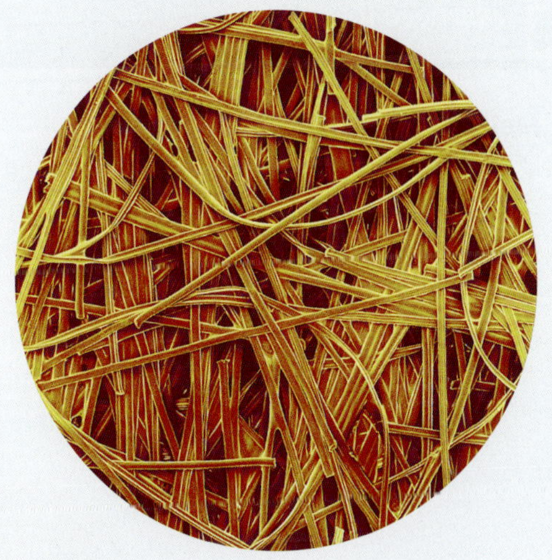

Figure 9.14 The fibres in a piece of writing paper

Search the internet for these websites for more information about paper production:

Making and recycling paper at home

History of paper

INVESTIGATION 9.1 Making recycled paper

Aim

To make recycled paper from old newspapers or scrap paper.

Materials

- 100 g of old newspapers or scrap paper
- a coathanger
- old pantyhose
- a blender (or electric drill with paint stirrer)
- square 20 L plastic laundry tub
- 10 L bucket
- PVA (white) glue
- iron (optional)

Risk assessment and planning

Carefully read through the Investigation. Prepare a simple flowchart showing what you have to do in each step.

Make a list of the safety precautions you will need for this investigation.

Method

1 Tear the paper into small pieces. Place them in a bucket with 2 L of warm water. (Warm water starts to break down the paper into fibres.)

2 While you are waiting for the paper to soak, prepare your paper frame. Bend a coathanger into a rectangular or square shape. Then stretch the pantyhose over the frame until it is tight.

3 Pour the soaked paper and water into a blender, and blend it until it turns into a soupy mass.

4 Add 4 L of water to a square plastic laundry tub, and then add about 30 mL of PVC glue. Mix the glue until it has dissolved.

5 Pour the soupy mass in the blender into the square plastic tub. Stir the paper pulp.

6 Scoop a layer of pulp onto the pantyhose screen, and allow the water to drain back into the tub. You should be left with a thin layer of wet paper.

7 Leave your paper to dry, preferably in the sun. You may have to leave it overnight.

8 When the paper is completely dry, carefully peel it off the pantyhose. You can iron out any moisture in the paper with an iron set to the cotton setting.

9 If you have another frame, you can make another sheet of paper.

Discussion

1 Test your paper by using various pens to write on it. How does it rate with commercial recycled paper or new paper?

2 Suggest ways to improve the quality of your recycled paper.

ISBN 978 1 4202 3822 8

CHECK

1 Which of the following statements are true and which are false? Rewrite the false ones to make them correct.
   a Oil is an example of a renewable resource.
   b The regeneration time for coal is many millions of years.
   c A female orange roughy matures at a very young age and produces hundreds of thousands of eggs.
   d Woodchip is an example of a resource obtained from plants.

2 Explain in your own words what sustainable development means.

3 List three uses of forests and describe how each use can be made sustainable.

4 What is wood pulp? What is it used for?

5 Some biologists suggest that when you buy fish to eat, you should buy fish other than sea perch or orange roughy. Why is this? Do you agree with this suggestion?

6 Plant fibres are an important renewable resource for making paper. Give examples of how other plant fibres are used.

7 The photo below shows a periwinkle plant. A substance extracted from this plant is used to treat cancers such as leukaemia.

   a What does the word 'extracted' mean?
   b What other plants are used to make medicines?
   c Why is it important to preserve natural bush and forests for the future?

8 Give an example of how we could live more sustainably by:
   a reusing
   b reducing
   c recycling
   d finding alternatives.

CHALLENGE

1 The table below shows how paper is used in Australia.

How paper is used	Percentage use of paper
Newsprint	16%
Printing and writing paper	41%
Household	7%
Packaging and paper for industry	36%
Total use	100%

   a Over half the paper consumed is used for people to read. What does this statement mean?
   b What type of paper products would be included in the household category?
   c Suggest how we could reduce our paper usage in each of the categories in the table.
   d Suppose that in a year the average office worker uses 10 000 sheets of copy paper (paper that has been printed on by an office printer or a photocopier). If there are 500 sheets of A4 paper in a ream, and a ream of paper weighs 2.5 kg, what mass of paper is used in a year?

2 Most of our forest woodchip is exported to pollution-producing paper mills in Asia.
   a Find out which countries buy our woodchip.
   b Would it be a better way to manage our resources if we didn't export woodchip, but built paper mills in Australia and supplied the paper to them? Give reasons for and against.

ISBN 978 1 4202 3822 8

9.3 Minerals and energy

Minerals are non-renewable resources and include metals such as iron and aluminium. **Fossil fuels** are also non-renewable and include coal, oil and gas. These fuels were formed from the remains of dead plants and animals over millions of years.

The Earth has a certain amount of non-renewable resources, and once they are used up they cannot be replaced. The use of well-managed renewable resources is like using the interest you earn in your bank account without reducing the balance. With a non-renewable resource, you are using the interest as well as the balance. And your bank account keeps getting smaller.

Australia's mineral reserves

The table below shows the reserves of some of the mineral resources mined in Australia. Australia has more reserves of brown coal, uranium and nickel than any other country in the world. It also has the world's greatest reserves of zinc, lead and silver.

How long will these non-renewable resources last? The activity on the right looks at some of the time scales for mining a metal ore.

Mineral	Reserves in Australia	Percentage of World resources 2014 (%)
Black coal	79 345 Mt	9
Brown coal	49 075 Mt	19
Iron ore	54 412 Mt	29
Copper	88.5 Mt	13
Nickel	19.0 Mt	23
Uranium	1.2 Mt	29
Tin	413 000 t	9
Gold	9112 t	17

Note: t = tonne Mt = million tonne
Source: Geosciences Australia

ACTIVITY

How long will it last?

Suppose ore X is used to make a metal that is used widely in the building industry.

Look at the graph below. Each curve represents the use of ore X at different rates. All lines start at the same point on the left of the graph. This point is the total amount of ore X that can be extracted from known reserves using our current technology.

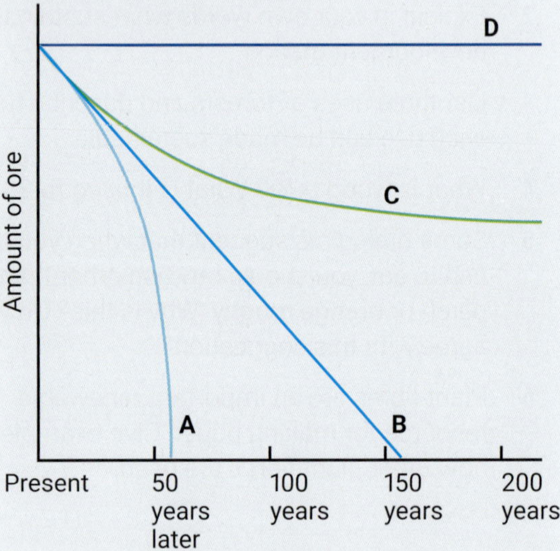

1 Curve B is a straight line. It predicts that we will continue to extract ore X at the same rate as we do now. How long will the reserves last? In which year will it run out?

2 Look at curve A. How is it different from curve B? What assumptions does it make about the future use of ore X?

3 Is curve D possible? How?

4 Suppose we find a new metal to replace the metal we make from ore X. Which curve would represent this?

5 Are there any other possibilities? If so, add them to the graph and explain them.

ISBN 978 1 4202 3822 8

The dangers of predicting

In the last activity you looked at some consequences of changing the rate at which a metal ore is extracted.

How accurate is it to predict how long a non-renewable resource will last? And what factors might change these predictions? You can investigate these questions in the activity below.

ACTIVITY

The Australian government gathers data on how much of a particular mineral resource is left. This is measured in 'years remaining'.

Data was recorded in 1999, 2009 and 2015. The table below shows this data.

For example, you can see that in 1999 there was 150 years of black coal left, and in 2009 there was 100 years left.

Resource	Years of mineral remaining		
	1999	2009	2015
Black coal	150	100	100
Copper	30	95	90
Iron ore	100	70	75
Nickel	80	145	75

1 Which of the minerals showed an overall decrease in their deposits between 1999 and 2015?

2 Which of the minerals showed an overall increase in their deposits between 1999 and 2015?

3 Chile, in South America, has the world's largest deposits of copper. Suppose Chile reduced its price for copper ore. How would this affect copper mining in Australia?

4 In 2008, it was estimated that there were 30 years of silver deposits left. In 2011, there were still 30 years of silver left. What can you infer from these figures?

Does recycling make a difference?

Metals can be reused over and over without losing any of their properties. For this reason they are ideal to recycle.

Aluminium is the world's most recycled metal. Using recycled aluminium saves a lot of energy. Recycling requires only 5% of the energy needed to make aluminium from its ore, bauxite. It also produces only 5% of the carbon dioxide (CO_2) produced when making the metal from its ore. Australia recycles around 68% of its aluminium cans!

The advantages of using recycled metals rather than making new metals are as follows:

- Their production requires much less energy, produces far less CO_2 and other pollution, and uses less than half the water that is used in making the metal from its ore.
- Their use saves thousands of tonnes of waste that would otherwise be dumped as landfill.
- Their use helps make the use of metals more sustainable by reducing the amount of minerals that need to be mined.

Glass is made from minerals obtained from sand. Australians use 850 000 tonnes of glass every year. We also recycle 350 000 tonnes. Like recycling metals, recycling glass saves energy, reduces water and air pollution, saves landfill space and conserves resources.

Figure 9.15 We have a choice to waste or recycle resources.

ISBN 978 1 4202 3822 8

Energy resources

Most of our energy needs, such as fuel for transport and factories, and electricity for homes and industries, currently comes from the fossil fuels coal, oil and gas. These are non-renewable resources.

The use of renewable energy such as solar, hydro-electricity, biomass (plant sources) and wind is small in comparison, but growing. The table below shows the use of renewable and non-renewable energy in Australia in 2014.

Energy resource		Percentage use
Non-renewables	Coal	31.7%
	Oil	38.4%
	Gas	24.0%
Renewables	Wind	6%
	Hydro-electricity	
	Solar	
	Biomass	

The amount of renewable energy overall is small at only 6% of total energy use. But in electricity production, renewable sources are becoming more important:

Electricity production source		Percentage produced	Growth in use over 10 years
Non-renewables	Coal	61.2%	-3.7%
	Oil	2.0%	6.5%
	Gas	21.9%	9.6%
Renewables	Wind	4.1%	1.9%
	Hydro	7.4%	31.3%
	Solar	2.0%	58.3%
	Biomass	1.4%	-1.0%

Another problem with burning fossil fuels is that they produce gases such as carbon dioxide. Excess carbon dioxide in the atmosphere is leading to global warming and climate change.

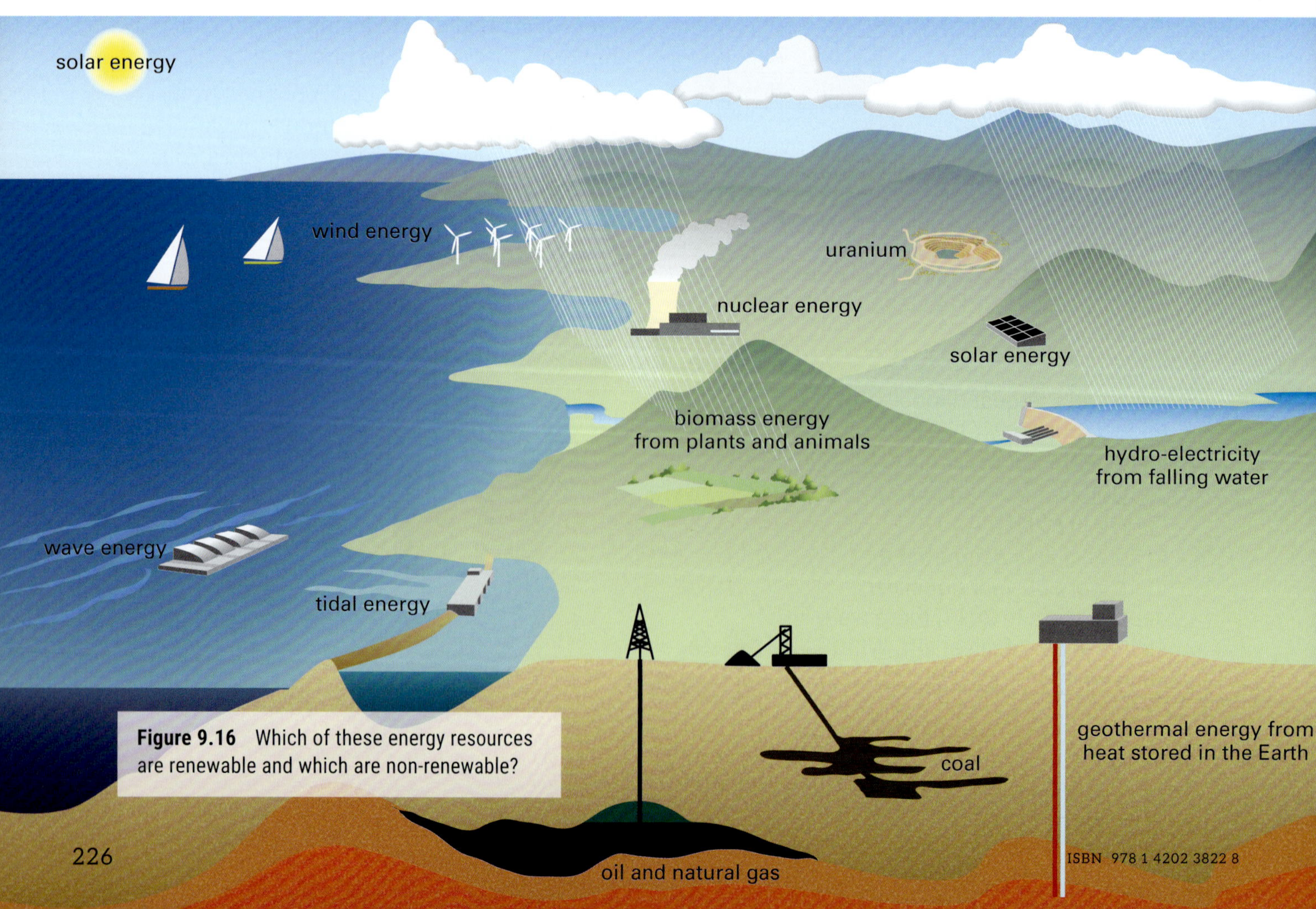

Figure 9.16 Which of these energy resources are renewable and which are non-renewable?

ISBN 978 1 4202 3822 8

Energy futures

Four possible energy futures for Australia are:

1 *Find new reserves of coal, oil and gas*

Continuing exploration will find more reserves of coal, oil and gas. However, when these reserves are used up the problem still remains. We would also need to work out how to use them more cleanly.

2 *Use nuclear energy*

Currently Australia has no nuclear power stations. All the uranium mined here is exported.

3 *Conserve energy*

Energy is often wasted and we can reduce the amount of energy we use.

4 *Increase the use of renewable energy*

At present we obtain about 94% of our energy from fossil fuels. However, some time in the near future these non-renewable resources will run out. We will then have to rely on renewable energy sources.

Renewable energy sources

Solar energy

The solar energy striking the Earth every minute is enough to supply the world's energy needs for a year! Solar energy can be used to heat water for household use in solar hot water systems, and for heating swimming pools. A well-designed house or building will also use solar energy to directly heat the building, reducing energy use and heating costs.

Solar cells

Solar cells can convert light energy into electricity. The first solar cells were only 6% efficient, but efficiencies of more than 40% are now possible.

The use of solar cells has become widespread in devices such as watches and calculators. They are also used to provide power to homes, to pump water and to operate roadside signs and street lighting systems.

Solar cells are expensive and do not operate at night. However, as the cost of solar cells decreases and their efficiency increases, they are becoming more widely used. They are already 100 times cheaper than they were 25 years ago!

SCIENCE AS A HUMAN ENDEAVOUR

Green Power

Each year the Australia Prize is awarded in an area of science and technology promoting human welfare. In 1999, it was awarded to Martin Green and Stuart Wenham from the University of New South Wales for their pioneering work with solar cells. For almost 20 years they held the world record for the most efficient solar cells. These cells were used by the winning car in the 1999 World Solar Challenge.

Green and Wenham worked with an Australian company called Pacific Solar (now CSG Solar). They worked out a way of depositing thin layers of silicon onto glass sheets, instead of using expensive silicon wafers. This cut the cost of solar power by two-thirds. Six modules of these solar cells mounted on your roof will generate 1.5 kilowatts, about a third of the electricity you need. In doing so, they will reduce greenhouse gases by almost 2 tonnes every year.

Figure 9.17 These solar panels generate electricity directly from sunlight.

ISBN 978 1 4202 3822 8

SCIENCE AS A HUMAN ENDEAVOUR

Solar power tower

A Melbourne-based renewable energy developer called EnviroMission has plans to develop Solar Tower power stations in Australia. Solar Towers will combine innovative design with the sun's energy in a unique way to make clean electricity. The design features a tall hollow tower up to 1000 metres tall positioned at the centre of a vast translucent circular canopy with a radius up to 2.5 km. Air inside the canopy is heated by the sun to make a thermal wind. As the hot wind rises up the hollow tower at 35–50 km/h, it passes through 32 turbines located around the base of the tower. These turbines will then generate up to 200 MW of pollution-free electricity. A 200 MW Solar Tower power station will generate enough clean electricity for up to 300 000 average households. No water is used in this electricity generation method.

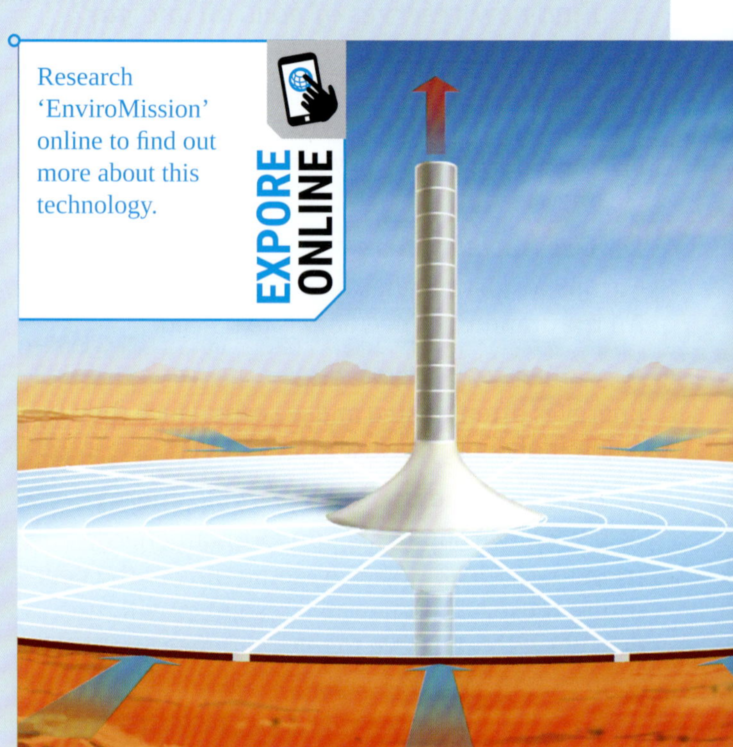

Wind energy

Windmills have been used in Australia for many years to pump water.

A small wind generator can produce 40–50 kilowatts, enough electricity for a single house. A huge generator whose blades sweep a circle 80 metres in diameter can produce 1.65 megawatts of electricity. A normal coal-burning power station generates about 2000 megawatts, so you need 1200 wind generators to produce the same amount of electricity as a single power station!

Australia's largest wind farm is at Macarthur, near Hamilton, 260 km west of Melbourne. It has 140 wind turbines and generates enough electricity to supply 220 000 homes. Each turbine is 85 m tall and the blades are 55 m long. The largest wind farm in Western Australia is at Collgar, about 230 km east of Perth.

Figure 9.18 Albany wind farm on the south coast of Wesern Australia

Biomass

Biomass is plant and animal material used as a source of renewable energy. It may be wood from forests, residues from agriculture and industry, or human and animal wastes.

Biogas

In Australia we throw away over 10 million tonnes of household rubbish every year. Much of this is biomass and contains about the same energy as 3 million tonnes of coal. When it is dumped and covered with soil, biodegradable materials break down to produce biogas, which is mainly methane. The gas is collected and burnt to generate electricity.

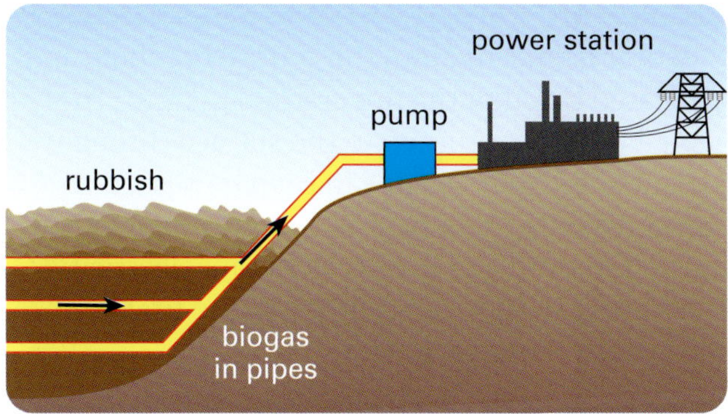

Figure 9.19 How a biogas power station works

Some processed-food manufacturers are using waste peelings to produce biogas. Some sewage treatment plants, such as the one at Woodman Point in Fremantle, use biogas to produce electricity to power the plant.

Biofuel

Fuels such as ethanol (alcohol) can also be produced from biomass such as corn and cane sugar waste after crops are processed. This liquid fuel can be used for many purposes and is already added to petrol to run cars.

Biodiesel

Waste vegetable oils such as those used in cooking, or oils extracted directly from plants such as palm oil or canola, can also be turned into a fuel called biodiesel.

Hydro-electricity

In Australia, about 7.4% of our total electricity is produced from falling water in hydro-electric power stations. However, hydro-electric power stations can only be built in mountainous areas, and there are few suitable sites remaining. Also, the building of storage dams can cause serious damage to the environment by flooding unique habitats.

An advantage of hydro-electric power generation is the ease with which generators can be started up and shut down. It thus provides a convenient and cheap way to supply the additional power needed during peak periods of use. Water from the upper reservoir can be used to produce power during the day when demand is high, and at night when demand is low, the water can be pumped back into the upper reservoir for re-use.

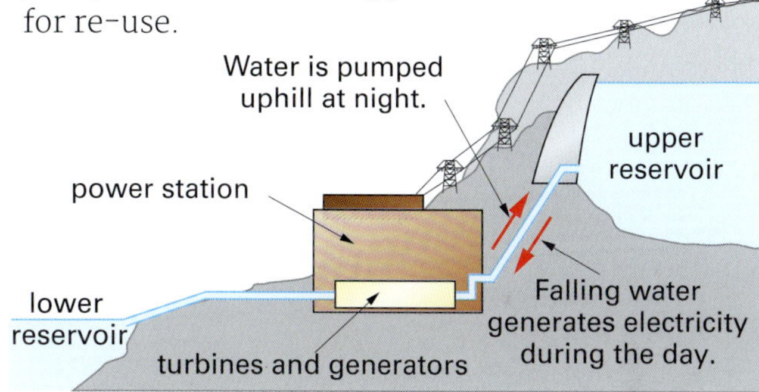

Figure 9.20 A pumped storage hydro-electric system

Tidal and wave energy

The tides of the oceans contain vast amounts of energy. As the tide comes in, water flows and drives turbines, which generate electricity. As the tide goes out, water flows through the turbines in the opposite direction, generating more electricity (see Figure 9.21). For this process to work effectively, you need large tides and these occur only in a few places.

The north-west coast of Australia is ideal for power generation because of the large tidal range there. A tidal power station is proposed for near Derby, with dams across two arms of a tidal creek and a canal between them. With this design it is possible to generate electricity continuously.

Ocean waves also have a large amount of energy, as you will know if you have been

ISBN 978 1 4202 3822 8

'dumped' by a wave. It is possible to use the up and down movement of waves to generate electricity, and scientists have suggested several different ways of doing this. One idea that has been shown to work is illustrated in Figure 9.22.

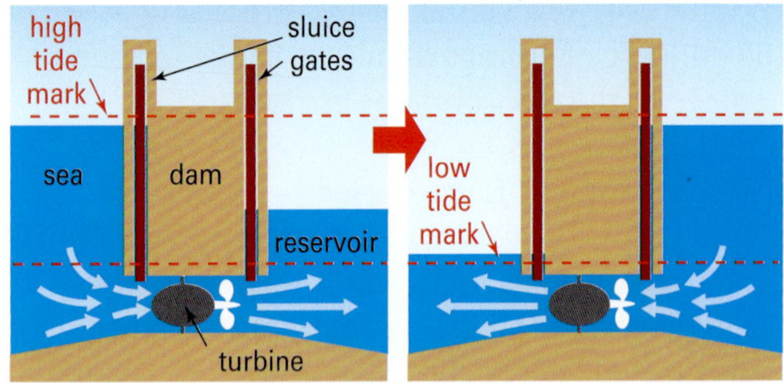

Figure 9.21 In a tidal power station the turbine can be driven by water flowing either way—from the sea to the reservoir as the tide comes in, and from the reservoir to the sea as the tide goes out.

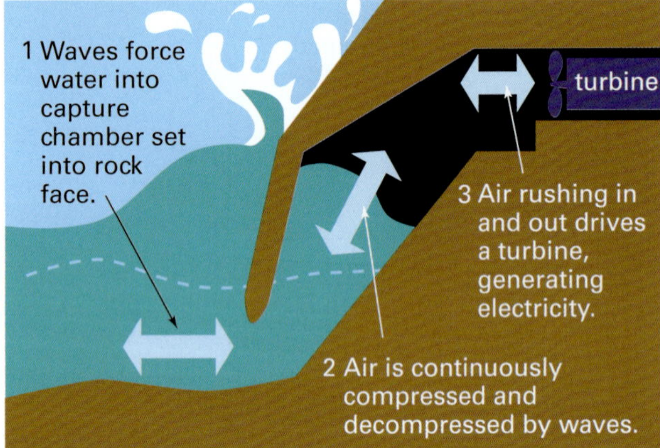

Figure 9.22 One design for a wave power station

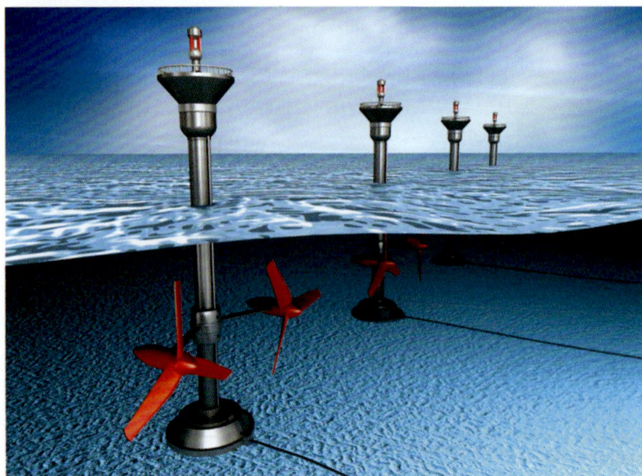

Figure 9.23 Tidal or wave energy could also power submerged propellers, like wind energy does.

Geothermal energy

Geothermal power stations make use of the heat inside the Earth. In some parts of the world, there is a great deal of volcanic activity and hot water and steam reaches the surface. New Zealand, Italy, Iceland and the United States have all built geothermal power stations that use this steam to turn turbines and generate electricity. Even in Australia, if holes are drilled down several kilometres, hot granite rocks are found.

This granite contains radioactive elements such as uranium and thorium, which release heat as they decay. Suitable hot rocks have been found near Innamincka in north-eastern South Australia. The rocks are at a temperature of around 250 °C.

Water is injected into the central borehole and circulated through the hot cracked granite. The heated water is returned to the surface through the outside boreholes. It can then be passed through a heat exchanger where most of its heat can be removed and used to generate electricity. The water can then be returned to the first borehole and used again.

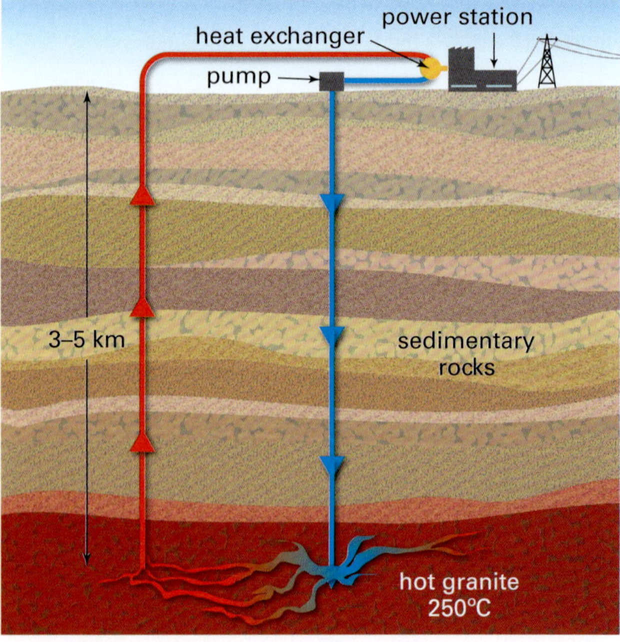

Figure 9.24 How electricity can be generated from hot rocks

ISBN 978 1 4202 3822 8

ACTIVITY

Energy supply proposal

Suppose a mine is to be developed on the Nullarbor Plain in Western Australia. Your task is to prepare a proposal for the supply of energy. You will work in small groups, each group representing a different company responding to the notice on the right.

Step 1: Select an energy system

Research all the energy alternatives available. Use the information on pages 227–30, but look for more detailed information by exploring online.

Future energy needs (NOVA)
Sustainability Western Australia
Renewable energy (NSW)

To summarise your findings, draw up a table giving advantages, disadvantages and notes for each source.

Step 2: Collect information

Collect detailed information on your chosen energy system. Identify the strengths and weaknesses of your energy system, and suggest ways of overcoming any problems (as part of your proposal).

Step 3: Prepare proposal

Decide on a name for an energy supply company, and a title for your proposal. Then prepare the proposal for presentation to the group.

Step 4: Presentation

Give a 5-minute presentation of your proposal for an energy supply system, including arguments for its use. Be prepared for other students to ask questions about your proposal and try to identify weaknesses in your arguments.

Step 5: Discussion

To finish, have a general discussion of the difficulties of energy supply systems on the Nullarbor Plain. Compare the Nullarbor with your local situation.

Nullarbor Mine

Tender for supply of energy system

Manufacturers and suppliers of energy systems are invited to register their interest in tendering for the supply of energy to the proposed uranium mine at Nullarbor. Potential suppliers will be required to provide a general description of their proposed method of energy supply. The proposed method should be cost-effective and suited to the area. The estimated energy requirement for the mine and settlement is 500 megawatts. All proposals must include an environmental impact study.

The Nullarbor mine is approximately 1000 km east of Perth. The settlement is close to the main highway and about 125 km south of the railway, but it is very expensive to bring in fuel by road, rail or air.

No coal, oil or gas has been found in the area. It hardly ever rains and there are few cloudy days. There are no trees and no streams, but water can be obtained from underground bores.

Nullarbor is close to the ocean, at the top of 80-metre-high cliffs. The average difference between high tide and low tide is 1.5 metres. The average wind speed is 8 km/h in summer and 25 km/h in winter, from the west.

Figure 9.25 Nullarbor Plain, Western Australia

ISBN 978 1 4202 3822 8

CHECK

1 Some of the following statements are false. Choose the false ones and rewrite them to make them correct.
   a Recycling aluminium uses more energy than making it from the ore.
   b Australia has more reserves of uranium than any other country.
   c Solar, wind, uranium and geothermal are all sources of renewable energy.
   d Glass is made from minerals in coal.

2 Look at the table on page 224. Which of the listed minerals are energy resources? Explain your answer.

3 What are the advantages of using recycled metals rather than producing metals from their ores?

4 Scientists calculated that this year we had 30 years of metal X left.
   a Suppose they predict we will have 10 years of metal X left in five years time. What reasons would they put forward to make this prediction?
   b After five years scientists find that there is slightly more than 40 years of metal X. Give reasons for the difference.

5 Look at the table on page 226 about electricity production in Australia.
   a Calculate the total percentage of electricity produced from renewables and non-renewables.
   b Identify which two sources of electricity have had the most growth in use over 10 years. Suggest a reason for this.
   c What has happened to the percentage of electricity produced from coal over the past 10 years? Suggest a reason for this.

6 Solar energy can be used in different ways. Describe three uses of solar energy.

7 Name two types of fuel that can be made from plants? How are these fuels used?

8 Suggest why solar is a better energy source for making electricity than burning coal. What is the disadvantage of solar? Why do we have so many coal-burning power stations in Australia?

9 Select four different renewable energy sources, and construct a table as follows to summarise the information about them.

Energy source	How it works	Benefits	Disadvantages

CHALLENGE

1 Look at page 225. What percentage of glass is recycled every year? How could we increase the amount of glass we recycle?

2 Some people predict that fossil fuels, particularly oil, will never run out. They will run down but never run out. Suggest what is meant by this.

3 Australia is self-sufficient in coal. What does this statement mean? Is Australia self-sufficient in oil?

4 The table below shows the annual energy consumption per person in various countries.

Annual energy consumption per person (energy units)	
Australia	277
Canada	427
Congo	7
Pakistan	14
Spain	161
USA	334

   a Which country has the highest yearly energy consumption per person? Which has the lowest?
   b Why is the energy consumption much lower in Pakistan than in Australia?
   c Canada and Australia have similar economies and lifestyles, Suggest why Canada has a higher energy consumption?
   d The world's average energy use per person per year is 72.4 energy units. What does this say about the energy use of the 'average person' on Earth?

5 Do you think it is possible to recycle 100% of a particular metal without having to produce more of it from its ore? Explain your answer.

6 Suggest where the best places for wind farms would be. Some people are against the construction of wind farms. Why do you think this would be?

ISBN 978 1 4202 3822 8

Copy and complete these statements to make a summary of this chapter. The missing words are on the right.

1 Water _____ from the surface of the Earth, _____, and then returns as rain, hail or snow. This is called the _____.

2 Homes can save water by restricting its use, installing _____ and using the water, called _____, from showers and laundries.

3 In a water treatment plant, alum is added to the water to remove the _____ material, then the water is _____, and lastly _____ is added to remove harmful bacteria.

4 Resources such as timber, plant and animals used for food are _____. Mineral resources such as oil and coal are _____.

5 _____ development occurs when there is a balance between using a resource and conserving the resource.

6 Australia has huge _____ deposits but they are all non-renewable and will run out one day.

7 Metals and glass can be _____. The advantage of this is it saves _____ and water, and reduces the amount of _____ that goes to landfill.

8 _______ and _______energy are more sustainable as they are renewable.

9 ______ electricity is made by water flowing downhill through turbines that spin.

sustainable
chlorine
hydro
mineral
greywater
energy
recycled
evaporates
water cycle
filtered
non-renewable
suspended
rainwater tanks
waste
renewable
solar
condenses
wind

CH•9 REVIEW

1 Match the term in the list with the correct description below.

greywater	groundwater	condensation
renewable	geothermal	sustainable

a the process in which water vapour turns in liquid water

b waste water from household bathrooms and laundries

c type of energy produced from hot rocks

d a resource that has a short replacement time

e development that uses a resource in a way that meets human needs but also conserves it for the future

f water that soaks into the soil or rocks from rain

2 Use the terms below to complete the flow diagram of a water treatment plant.

chlorine	sludge	settling tank
alum	filtering	

untreated water → [ ] (← [ ]) → [ ] → [ ] (← [ ]) → storage → drinking water

ISBN 978 1 4202 3822 8

3 What is the difference between a renewable resource and a non-renewable resource? Give two examples of each in your answer.

4 The pie chart below shows energy consumption in Australia.

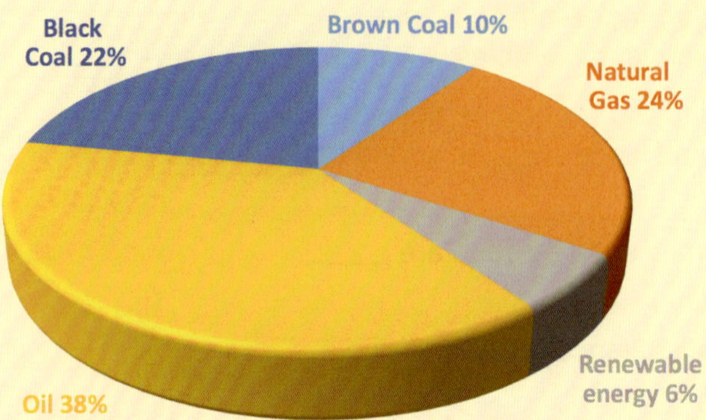

a How much of Australia's energy comes from coal?
b What is coal used for?
c What do we use oil for?
d Oil makes up slightly less than 1% of our total energy reserves but 38% of our energy consumption. What does this mean for our future?

5 Metals can be recycled over and over without losing any of their properties. Describe two advantages of using recycled metals over the extraction of metal ores to make metals.

6 Forests are living renewable resources.
a Describe some of the things we use forests for.
b Australians use a huge amount of paper and cardboard. Where do we get the raw materials for the paper industry? How do we make the paper industry sustainable?

7 Suppose you were designing power-generating stations for each of the five main renewable energy sources:

solar
wind
hydro-electricity
biomass
geothermal

Where would you locate each type of power station to get maximum power output? Give a reason for each answer.

8 The mudcrab is found in rivers throughout the northern parts of Australia. It is considered a delicacy and is expensive to buy. By law, all mudcrabs with a shell width less than 15 cm are protected. All female crabs are also protected. Suggest why we have these laws in Australia.

9 The table below shows how much biomass could be produced in Australia every year. The unit of energy is the petajoule (PJ).
a Which is the best potential source of biomass?

Biomass	Energy in PJ
cereal straw (mainly wheat)	200
bagasse (waste from sugar cane)	40
sawmill waste, sawdust	70
forest waste	140
urban garbage	170

b How much energy could be obtained from biomass in Australia each year?
c Australia uses 5500 PJ of energy each year. What proportion of the energy could be obtained from biomass?
d Which of the biomass sources do you think could be most easily collected to fuel city power plants? Give reasons for your choice.

ISBN 978 1 4202 3822 8

10 Label parts A to G on the diagram of the water cycle.

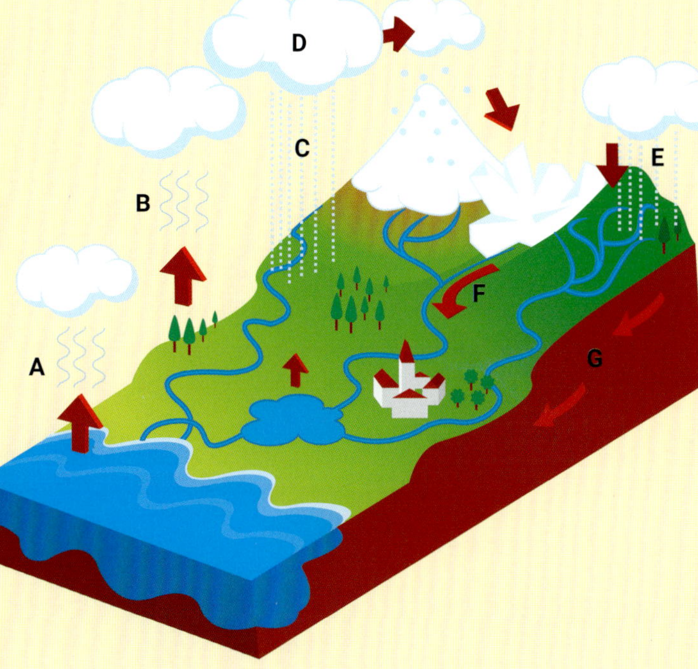

11 Look at each of the following photos. Explain why the activity shown has an effect on the water cycle, and what that effect might be.

B

C

Check your answers on page 244.

ISBN 978 1 4202 3822 8

Answers to Reviews

If your answer does not agree with the answer given here, go back to the chapter and read the relevant section again. Your answers may be slightly different from the answers given here. If in doubt, check with your teacher.

CH•1 Introduction to the lab

1 See the cartoon below.

2 **a** laboratory **b** apparatus
c Bunsen burner **d** metal tongs
e stirring rod **f** gauze mat
g aim

3 **a** Hold it under cold running water and tell your teacher.
b Flood the area immediately with lots of water and tell your teacher. Check whether any acid has splashed onto you or your clothing.
c Wash your eye immediately with lots of water and tell your teacher.
d Immediately extinguish it under a running tap and tell your teacher.

4 Safety glasses are to protect your eyes in the laboratory, especially from splashing liquids. You should wear them whenever you use a Bunsen burner to heat a liquid in a test tube or a beaker.

5 **a** true
b false. A spatula is used for picking up small amounts of solids.
c true
d false. If you spill acid on yourself, wash it off immediately with lots of water.
e false. You should turn on the gas after striking the match—see Investigation 1.1, step 3, page 13.
f false. You should always put a heatproof mat under a Bunsen burner.
g false. The hottest part of a Bunsen flame is just above the dark blue cone.
h true

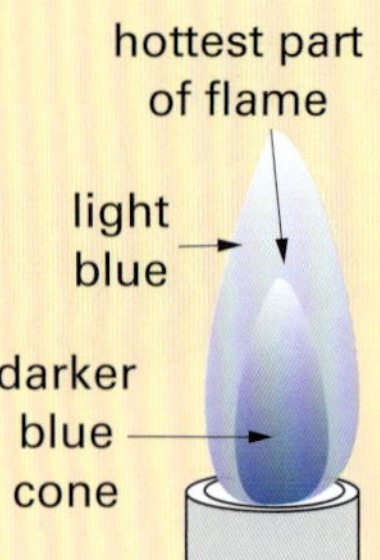

ISBN 978 1 4202 3822 8

6

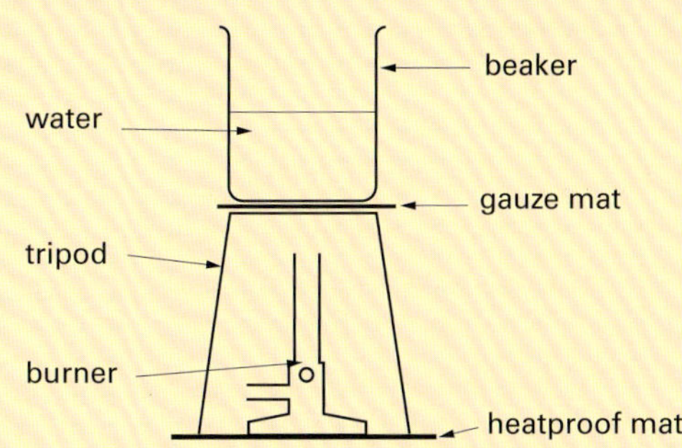

7 The air hole will be open to give a hotter flame. If the air hole is closed, the saucepans will become black with the soot from the yellow flame.

8 **a** Wash it down the sink with lots of water.
b Put it in a special container for proper disposal.
c Put them in a special container for waste solids.

9 **a** **A** **b** **C** **c** **B**

10 Organism D has:
- four legs
- back legs longer than front legs
- several toes on each leg
- large eyes on top of head
- lighter green patch behind eye .

LAB REVIEW

See page 13 Part A

CH•2 Working scientifically

1 **a** observation
b observation
c inference, as she didn't actually see the dog do it
d observation

2 **a** aim **b** prediction
c observation **d** conclusion

3 **a** 19.1 °C
b 20.9 °C
c 20.9 − 19.1 = 1.8 °C
d average temperature $= \frac{19.1 + 20.9}{2}$

$= \frac{40.0}{2} = 20\ °C$

4 **a** The pattern is more obvious if you reorganise the data table as shown below:

Temperature of water (°C)	Dissolving time (seconds)
20	27
25	24
30	22
35	20
40	18
45	16

As the temperature increases, the dissolving time decreases. Or, tablets dissolve more quickly at higher temperatures.

b A line graph shows the relationship between temperature and dissolving time.

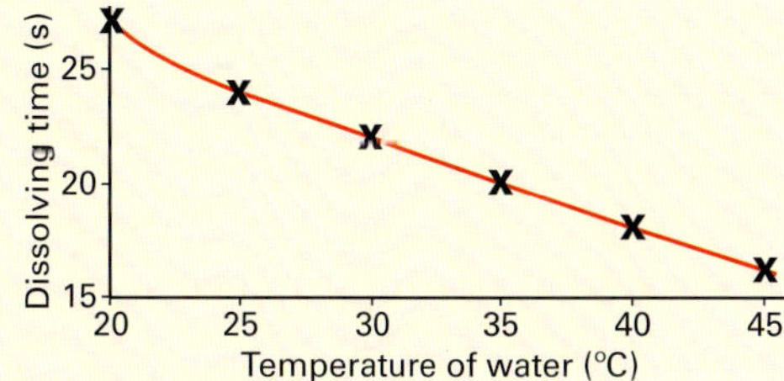

5 **a** Just over 50 cm—halfway between 35 cm (for a drop height of 50 cm) and 70 cm (for a drop height of 100 cm).
b 140 cm. (If you double the drop height, the bounce height will probably double also.)

ISBN 978 1 4202 3822 8

5 a Just over 50 cm—halfway between 35 cm (for a drop height of 50 cm) and 70 cm (for a drop height of 100 cm).
b 140 cm. (If you double the drop height, the bounce height will probably double also.)

6 a *Leaking can*

Aim: To find out what effect the depth of water above the hole has on how far the water shoots out from the can.

Materials: 4 cans, nail to punch holes, water, ruler.

Method: I used four identical cans, and

Distance of hole from top of can (cm)	Distance water shoots out of hole (cm)
2	3
4	7
6	9
9	13

Discussion: The results are about what you would expect.

Conclusion: As the distance of the hole from the top of the can increases, so does the distance the water shoots out.

b A line graph is better because you want to show the relationship between the distance of the hole from the top of the can and the distance the water shoots out.

c The water would shoot out about 17 cm. (If the water shoots out an extra 4 cm going from a 6 cm hole to a 9 cm one, then you might expect another 4 cm increase going from a 9 cm hole to a 12 cm one.)

LAB REVIEW

Temperature—see page 35 Part B

Volume of drop—use a small measuring cylinder to find the volume of 10 drops, then divide by 10 to find the volume of 1 drop.

CH•3 Forces

1
force	push or pull
carpet	surface with large friction
friction	force that exists when one surface rubs on another
newton	unit of force
gravity	a pull from a large body
spring balance	measures force
lubricant	reduces friction
heat	produced when there is friction

2 Gravity on the moon is only about 1/6 of gravity on Earth. The mass of the can is still 1 kilogram on the moon, but its weight (downwards pull of gravity) is only 9.8/6 = 1.6 newtons.

3 *Contact forces*—pushing something with your hand, pulling on a rope, wind blowing, waves crashing on beach
Non-contact forces—two magnets attracting without touching, gravitational and electrostatic forces

4 **A**, **B** and **D**
A—the more mass, the more friction
B—more friction between the wheels and the rubber mats than between the wheels and the mud
D—with less air in the tyres, there is a greater area of contact between the tyres and the mud, therefore there is more friction

5 **B**—600 N. In a lift descending rapidly, the floor tends to drop under you and you weigh slightly less.

6 a Advantage—friction helps slow you down
b Disadvantage—friction makes it hard to push
c Disadvantage—parts of engine rub together, causing wear
d Advantage—friction helps slow you down

7 **B**

8 When you are skating, friction can be a nuisance. It is therefore important that the friction between the ice and your skates is as low as possible. This is why you sharpen your skates. But when you are riding a mountain bike, you need as much friction as possible between the ice and the tyres. You would therefore use wide tyres with a thick, rough tread.

ISBN 978 1 4202 3822 8

9 a

	Average force (N)
concrete	20.3 $\left(\frac{20 + 23 + 20 + 18}{4}\right)$
newspaper	8.3
vinyl tiles	4.0
sandpaper	34.3
grass	17.5

b

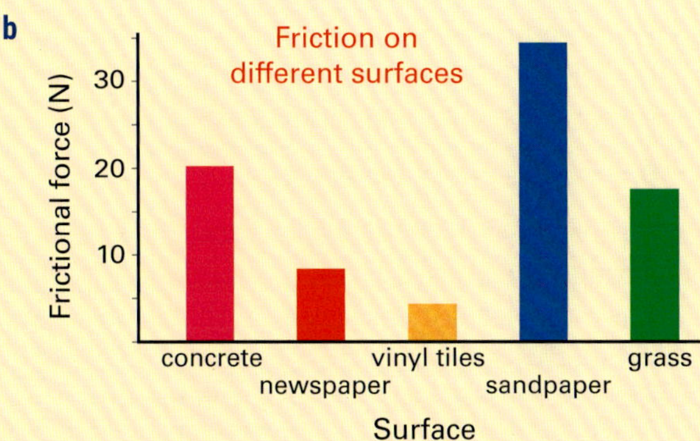

10 a C—opposite to the direction in which the van is moving

b D

c Yes—A and C are balanced, and so are B and D. (The van will therefore keep moving at a constant speed.)

11 A (gravity acts on all objects near the Earth) and **B** (there is air resistance between the ball and the air).

C is incorrect since the force of the hit acts only at the instant Lorna hits the ball, not as the ball moves across the net.

CH•4 Simple machine technology

1 A—screw (inclined plane), B—gear, C—wheel and axle

2 **C** and **E**

3 **B** The screwdriver will give the most leverage because it is the longest and strongest. The coin is not long enough to act as an effective lever, the fingernail file is too flexible, and the ice-cream stick would break.

4 Simple machines:

- magnify the force you use (A crowbar and a screwdriver do this.)
- change the direction of the force (A pulley and the claw of a hammer do this.)
- make things go faster. (A hand drill and bicycle gears do this.)

5 **D** The pivot is at one end of the lever and the effort is at the other end, with the load (cable) in between. This is like a wheelbarrow (page 75), not a crowbar.

6 a Gear wheel B has fewer teeth than gear wheel A, therefore gear B will turn more than one revolution. It will also turn faster than gear A.

b A hand drill, a kitchen beater and a non-geared bicycle have this gear wheel arrangement.

7 To reach the jetty, John has to push backwards with one foot so that he can move forwards. The boat will therefore probably move away from the jetty. For John to reach the jetty, the boat would have to be roped to the jetty, or his companion would have to hold the oars in the water to stop the boat moving.

8 The jet engines give enough thrust for the aircraft to travel forwards quickly. The shape of the wings gives the aircraft lift and the flaps are used to increase the lift.
The speed of the air moving over the wings creates lift. So the faster the plane travels down the runway, the greater the lift force.

ISBN 978 1 4202 3822 8

9

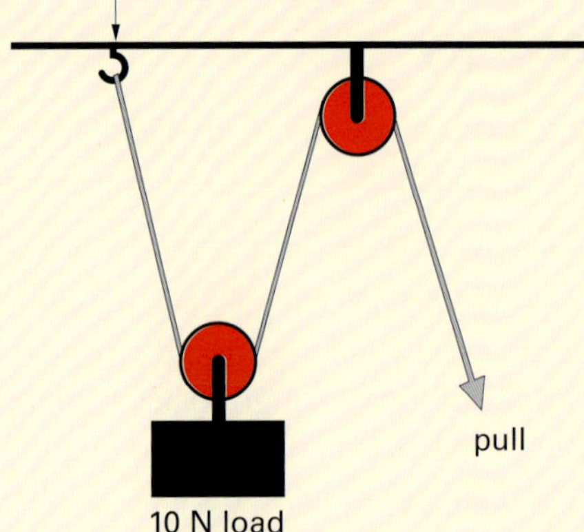

10 a 12 N

b wheelbarrow

c mechanical advantage $= \frac{\text{load}}{\text{effort}}$

$2.5 = \frac{20\text{ N}}{\text{effort}}$

$\text{effort} = \frac{20\text{ N}}{2.5}$

$= 8\text{ N}$ at **40 cm**

d If the ruler was 150 cm long, the load would be easier to lift. The effort using the 150 cm ruler would be 100/150 times the effort used with the 100 cm one, that is:

$$\frac{100}{150} \times 8\text{ N} = 5.3\text{ N}$$

CH•5 Classifying living things

1 **A** When classifying organisms, their structural or functional characteristics are generally used.

2 **D**

3 Zian was correct because the dolphin is a mammal and breathes oxygen from the air through lungs, while the shark is a fish that breathes through gills. (Lungs and gills are structural characteristics.)

4 **D**—see the plant key on page 121

5 A reptile has a dry, scaly skin and lays eggs with a tough, flexible covering. Amphibians, however, have a moist skin and lay eggs that do not have a protective covering. Amphibians also spend some of their life cycle breathing through gills.

6 The organism is green and has root-like structures, so it would be classified as a plant. It would then be placed in the moss group because it has no stem.

7 **B** Flowering plants and conifers produce seeds, but ferns produce spores.

8 a The organisms all have a segmented body, a number of legs, antennae and eyes.

b Organism D has four pairs of legs (8 legs), while the others have three pairs (6 legs).

c

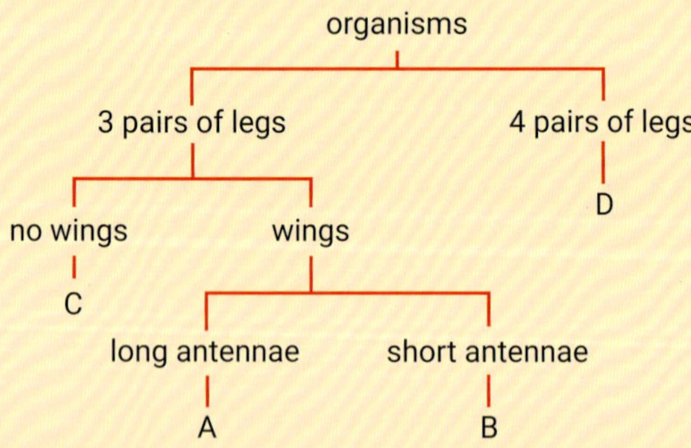

9 a Mites and water spiders are animals that have a jointed, hard covering (exoskeleton) and more than three pairs of legs.

b Flatworms are similar to leeches because they do not have a shell or hard body covering. However, flatworms do not have body segments as leeches do.

ISBN 978 1 4202 3822 8

c The animals in this key are all invertebrates, but frogs and fish are vertebrates (they have a backbone). Therefore the key cannot be used to classify them.

10 a *Pan troglodytes*

b The lion and chimpanzee are most similar, as they belong to the same class, whereas the bearded dragon is in a different class.

c Class Mammalia.
- give birth to live young
- have hair
- constant body temperature
- breath through lungs
- feed their young milk

d
- give birth to live young
- have hair
- constant body temperature
- breathe through lungs
- feed their young milk
- walk on two legs
- large brains.

e they all have a spinal cord.

CH•6 Ecosystems

1 **C**

2 **D**

3 a

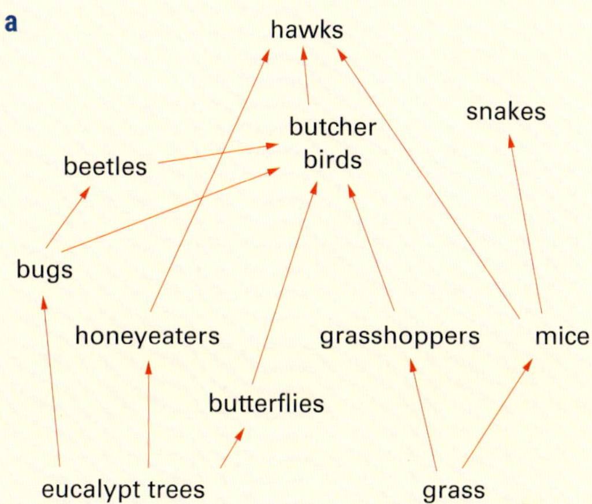

b The community could be called a eucalypt community or an open forest community.

4 Scavengers and decomposers are similar in that they both attack dead organisms. However, scavengers eat the dead organisms and digest them internally, whereas decomposers release chemicals which break down the organism's body externally.

5 a Algae and water plants are the producers.

b Small fish, lobsters, large fish and tadpoles are second-order consumers. (The large fish, lobster and tadpole can also be third-order consumers.)

c Tadpoles and lobsters are competitors of small fish because they eat the same type of food (tadpoles eat microscopic animals and lobsters eat insect larvae).

6 Decomposers are very important for communities because they:
- remove dead organisms by attacking and breaking down the bodies to simple substances
- recycle the simple substances into the soil and increase its fertility.

7 a Process 5 is respiration.

b When organisms die they are attacked by decomposers that break down the material in the dead organisms' bodies.

ISBN 978 1 4202 3822 8

- **c** The red arrows represent the food chain in which one organism is the food for the other.
- **d** Process 1 is photosynthesis, which is the start of all food webs.
- **e** Pathway 1 shows carbon dioxide taken out of the air. Pathways 5, 6 and 7 show carbon dioxide given off into the air.

8 In a natural ecosystem there is high biodiversity with many different varieties of plants and animals, such as mammals, birds, lizards and insects. When feral cats are introduced into this environment, these animals are not used to feral cats in their ecosystem and become easy prey. In this way, the feral cats can destroy the balance of the natural food web and reduce the biodiversity.

9 **a** To show that the oxygen produced by photosynthesis in plants is reduced by the amount of silt in the water.

b Tom changed the amount of silt in the tubes. He kept the amount of water in the tube, the size of the plant and the amount of light the same.

c The amount of oxygen given off by a green plant decreases as the amount of silt in the water increases.

10 **a** **A** sunlight
B oxygen
C carbon dioxide
D water

b chlorophyll

c glucose, which is stored as starch

d Plants contain chlorophyll in their leave which absorb sunlight. The energy from the sun is used by the plant to turn water and carbon dioxide into glucose which stores the energy for the plant to use later. Glucose can be converted to starch and stored if not required immediately. During the process of photosynthesis, oxygen is also produced and released into the air.

11 **A** Oxygen

B, C, D energy, water, carbon dioxide (in any order)

CH•7 Earth, moon and sun

1 **A**

2 **B**, **C**

3 **C** The planet rotates in the opposite direction to Earth, which rotates from west to east. Since daylight lasts 14 hours, the time for one rotation is twice this—28 hours.

4 **D** The relative positions of Frankenstein and Dracula are the same as the positions of the Earth and the moon for the crescent moon (position 2 in Figure 7.9 on page 169).

5 **a** **A**

b **D**

6 Rotation is the spinning or turning of a body on its axis. Revolution is the movement of one body around another body. The Earth rotates on its axis in 24 hours (1 day), but revolves around the sun in 365 days (1 year).

7 The moon is always in orbit around the Earth. It is just that we only notice it at night. If you are observant, you can sometimes see it during the day. Tides therefore occur during the day as well as at night.

8

Cause	Effect
a tilt of the Earth's axis	seasons
b Earth moves between sun and moon	lunar eclipse
c Earth rotates	night follows day
d Moon completely blocks the sun	total eclipse of the sun
e Moon revolves in same time that it rotates	same side of moon always faces Earth
f Earth orbits the sun	you see different stars at different times of the year

9 **B**

10 **D**

ISBN 978 1 4202 3822 8

CH•8 Separating mixtures

1 **A**

2 **B**—see page 189

3 **D**—see page 196

4 a **Water**—It dissolves more of solid B than the other liquids do.

b **Water**—If you add water to the mixture of A and C, only A will dissolve. You can then separate them by filtration and evaporation.

c A mixture of A and B could be separated using petrol. Some of A would dissolve, and this could be recovered by evaporation. However, this is not a good method, because A doesn't dissolve very well in petrol.

5 **a**, **b** and **d**

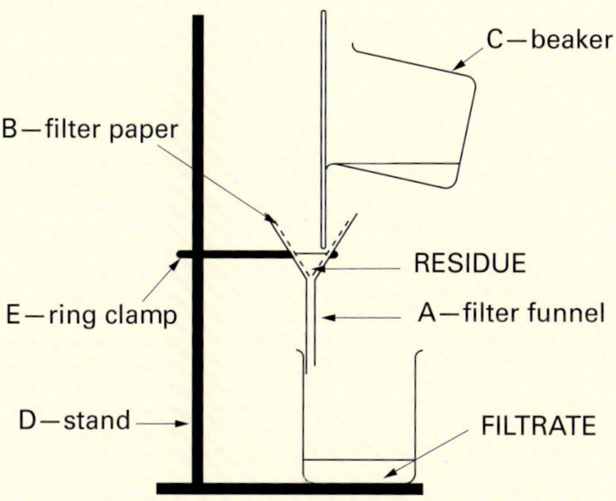

c The mixture should be poured from the beaker down a stirring rod—as shown in Figure 8.15 on page 194. The stem of the filter funnel should be touching the inside of the beaker.

6 a Distillation—see page 196

b Heating causes the water in the left-hand test tube to boil. Water vapour travels along the tube and condenses to form pure water in the right-hand test tube. The salt that was dissolved in the water is left behind in the left-hand test tube.

c The ice-cold water lowers the temperature inside the right-hand test tube. This causes the water vapour in the tube to condense back to liquid water.

7 The fact that a sediment settles out on standing indicates that some of the fruit juice is in suspension. However, some is also in solution since the liquid is coloured. And if it doesn't form a clear solution on settling, then the fruit juice is a colloid (see page 190). So fruit juice is a solution, a suspension and possibly a colloid as well.

8 There are many possible sentences using these words. For example:

a *Milk* is a liquid-in-liquid *colloid* called an *emulsion* (see page 190).

b A *concentrated solution* contains more solute than a *dilute* solution (see page 189).

9 The ink in a felt pen is a mixture of several different colours. Different felt pens contain different ink mixtures, which can be separated by paper chromatography.

To start with, the police would test a sample of ink from the ransom note and the ink from the felt pens of each of the three suspects. If they get the same pattern of colours as in the ransom note, then the owner of this pen is probably guilty. (It is of course possible that the note was not written using this particular felt pen, but with another pen of the same type.)

LAB REVIEW

1 Add water to the mixture and stir. The salt dissolves but the dirt does not.

2 Filter the mixture as in Investigation 8.2 on pages 194–5. The residue on the filter paper is the dirt, and the filtrate is the salt solution.

Equipment needed:

- piece of filter paper
- stand and ring clamp
- glass stirring rod
- filter funnel
- wash bottle
- 2 beakers

3 Evaporate the salt solution as in Part A of Investigation 8.3 on page 197.

Equipment needed:

- watch glass
- Bunsen burner
- heatproof mat
- gauze mat
- matches
- metal tongs
- boiling chips
- tripod

ISBN 978 1 4202 3822 8

CH•9 Sustainable Earth

1 a condensation
b greywater
c geothermal
d renewable
e sustainable
f groundwater

2

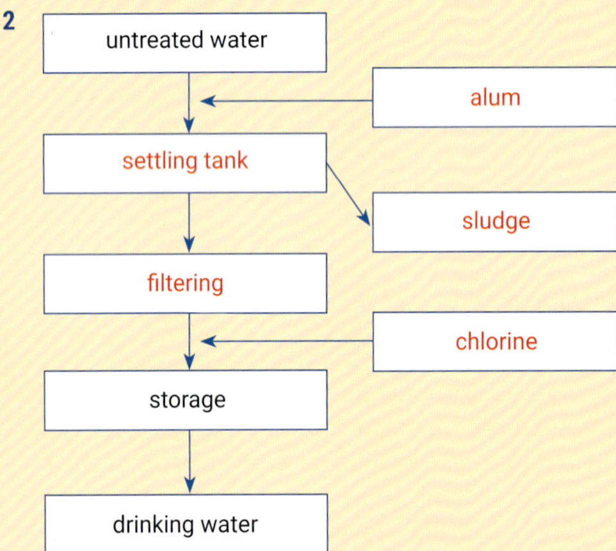

3 A renewable resource is one that can be replaced in a short period of time. Examples include animals bred for food, and timber used for furniture. A non-renewable resource is one that generally cannot be replaced. Examples include iron ore and coal.

4 a 32%
b Coal is mostly used in power stations to generate electricity.
c Oil is generally used to make fuels for transport (cars, trucks, planes and ships).
d We need to import oil from other countries as we are quickly using up our oil reserves.
e We urgently need to find alternatives for oil. We could use another fuel source such as ethanol or biodiesel from plants, or use electric cars.

5 Recycling metals:
- saves a lot of energy and water which is used when the metal ores are mined
- avoids a lot of waste (landfill) when the metal products are thrown away after use.

6 a Forests are used for making paper, building houses and furniture, and as a source of chemicals for medicines.
b Woodchip is the raw material for making paper. It is made from the timber in trees chopped down from plantations or native forests. The paper industry could be more sustainable if we recycled more used paper and used plantation trees instead of native forest trees.

7 *Solar*—would be located in a region where there is lots of sunshine and very little cloudy weather, such as the central parts of Australia.

Wind—the wind generators would be located on the coast in the southern parts of Australia where the wind blows strongly and for much of the year.

Hydro-electricity—the power station would be located on a large fast-flowing river, which falls down from a mountainous area.

Biomass—would be located where there is a large source of plants, such as sugar cane farms, for example, in northern NSW and Queensland.

Geothermal—would be located in an area where there are hot rocks beneath the surface.

8 The laws are designed to protect the young crabs so that they can grow, reach maturity and breed. Female crabs are protected so they can produce eggs and therefore more mudcrabs.

9 a Cereal straw is the best potential source.
b The total energy from biomass is 620 PJ.
c Biomass could contribute 11.3 % of our energy needs ($620 \div 5500 \times 100 = 11.3\%$).
d Urban garbage is the biomass source most easily collected from cities.

10 a evaporation
b transpiration.
c precipitation (rain or snow)
d water in clouds
e precipitation (rain or snow)
f surface run-off
g groundwater or percolation

11 **Photo A**: Clearing native forests reduces transpiration and less water enters the atmosphere. This means in such areas rainfall can decrease and weather patterns can change.

ISBN 978 1 4202 3822 8

This in turn can change the flow of water in that area. Also, there will be more run-off and erosion when it does rain.

Photo B: Building a dam traps water that would otherwise run off into rivers, and therefore reduces the flow of water in the rivers. Instead, the water trapped in the dam is used for farming animals or crops. So the dam has changed the natural flow of water in this area.

Photo C: Plantation crops absorb a lot of water from the soil as they grow, and reduce run-off. They also release water into the air by transpiration. But when the trees are cut down the opposite occurs—less transpiration and more run-off.

ISBN 978 1 4202 3822 8

Glossary

The words in this list occur in dark type throughout the book. The number after each entry gives the page where you will find more information. For some words the pronunciation is given. The syllable in capitals should be stressed; for example, laboratory (la-BOR-a-tree).

abiotic factors: physical or non-living factors that affect survival of organisms in ecosystems; for example, soil type, availability of clean water and temperature. 145

action and reaction: equal and opposite forces that occur together 89

apparatus: science equipment put together for an experiment 6

arthropod: an invertebrate animal that has a jointed exoskeleton covering its body, for example crabs and beetles 116

axis: imaginary line through the centre of an object in space; the object rotates around this axis 162

binomial name: a two part scientific name used to name each species 115

biodiversity: all the differences within and between living things 151

biologists: scientists who study living things 102

biomass: plant and animal material used as a source of renewable energy 229

biotic factors: biological or living factors that affect the survival of organisms in ecosystems; for example, predators and availability of food. 145

carnivore (CARN-e-vore): a consumer that eats other animals 138

cells: the building blocks of all living things; cells are usually microscopic 103

chlorophyll (KLOR-a-fill): the green substance in plants that is able to absorb the energy of sunlight 108

chromatography (CROW-ma-TOG-ra-fee): a technique used to separate small amounts of soluble substances in a mixture; for example, the coloured substances in ink can be separated using filter paper 201

class: a division for classifying animals and plants, between phylum and order 115

classification: a process of placing living things in groups based on their structural and functional characteristics 100

colloid (COL-oid): a mixture which has properties in between those of a solution and a suspension; the particles in the colloid may be tiny bits of solid, liquid droplets or gas bubbles 190

concentrated (CON-cen-TRAY-ted): describes a solution containing a large amount of solute compared with other solutions 189

concentration: the amount of solute dissolved in a certain volume of solution 191

condensation: changing from a vapour into a liquid; condensation is the opposite of evaporation 210

conifer: a type of plant that produces seeds in cones 122

consumer: an organism that eats other organisms 137

crescent moon: thin curved shape of the moon when only one part of its face is lit up, as seen from Earth 169

data: numerical results gathered by observation, experiment or research 7

data table: a table in which data are recorded 7

decomposers: organisms (such as some bacteria and fungi) that break down the bodies of dead organisms to simpler substances 139

decanting: gently pouring off a liquid, leaving the solid in the container 193

dilute (dye-LOOT): describes a solution containing a small amount of solute, compared with other solutions 189

dissolves: when two or more substances mix completely, so that they appear as one; for example, sugar dissolves in water 187

distillation: a separation technique that involves evaporating a liquid, then condensing the vapour in a separate container 196

ISBN 978 1 4202 3822 8

driven gear: the gear wheel that is forced to spin by a force from the driving gear 82

driving gear: the gear wheel that applies the driving force 82

ecosystem: a system of relationships among organisms and the way they interact with the non-living parts of their environment 145

effort: the force you apply to move a load 75

emulsion (ee-MULL-shun): a colloid with tiny droplets of one liquid spread through a second liquid; milk is an emulsion 190

evaporation (ee-VAP-o-RAY-shun): the process in which a liquid turns into a vapour and seems to disappear 210

exoskeleton: the protective, jointed covering on the outside of arthropods, such as insects, spiders and crabs 116

experiment: a well thought out scientific test, designed to answer a question or solve a problem 42

fair test: an experiment where you change something, measure something and keep everything else the same 130

family: a division for classifying animals and plants, between order and genus 115

fern: a type of plant that has a stem and reproduces by spores 121

filtering (filtration): a way of separating a solid from a liquid (or gas) using a filter 193

flowering plant: a type of plant that produces flowers and whose seeds are contained in a fruit 122

food chain: a diagram that shows a chain of organisms in which each organism is eaten by the next in the chain 134

food web: a number of food chains together showing what all organisms in a particular area eat 138

force: any push or pull, measured in newtons (N); it may act by contact or at a distance 50

fossil fuels: fuels obtained from material that was once living, for example, oil, coal and natural gas 224

friction: a force that exists when two things rub against each other; it slows down or prevents movement 59

full moon: the shape of the moon when its face is seen fully from Earth 169

fungi: plant-like organisms that do not contain chlorophyll; they obtain their food from dead or living organisms, for example mushrooms and moulds 109

generalisation: a statement or conclusion based on many observations that holds true in most cases; for example, most plants are green 15

genus: a division for classifying animals and plants, between family and species (plural = genera) 115

gibbous moon: the shape of the moon when about two-thirds of its face is lit up, as seen from Earth 169

gravity: the force of attraction between any two objects; for example, between a person and Earth 65

groundwater: water that has soaked underground through the soil and rocks 210

habitat: the living place of an organism 144

herbivore (HER-be-vore): a first-order consumer that eats only plants 138

hypothesis (high-POTH-e-sis): a generalisation that explains a set of observations or gives a possible answer to a question; it can be tested by experimenting 42

inclined plane: a slanting surface (ramp) used as a simple machine 77

inference: an explanation of an observation; it may or may not be correct 28

invertebrate (in-VER-te-brate): an animal without a bony skeleton, for example jellyfish or insects 103

key: a diagram showing the grouping of objects or living things and the characteristics used to group them 101

kingdoms: the major groups used in classifying living things; the five kingdoms contain all the living things on Earth 107

laboratory: (la-BOR-a-tree): a special room used for science experiments 3

ISBN 978 1 4202 3822 8

lever: a long bar that moves around a fixed point called the pivot or fulcrum 75

load: the force (usually the weight of an object) that you want to move with a simple machine 75

lubricant (LOO-bri-kint): a substance that reduces friction and allows surfaces to slip easily over each other 61

lunar eclipse: the darkening of the moon by the shadow of Earth when Earth is between the sun and the moon 172

mass: the amount of matter in an object; measured in kilograms 65

mechanical advantage: a measure of how useful a simple machine is (by dividing the load by the effort) 75

meniscus (men-NIS-kus): the curved upper surface of a liquid in a measuring cylinder or tube 34

mixture: two or more pure substances mixed together but not chemically combined 186

model: a way of representing something that cannot be observed directly because it is too small, too large or too complicated, for example a model of the solar system 164

mollusc: an invertebrate animal that has a soft body and lives in water or in moist surroundings; most molluscs have a shell, for example snails and periwinkles 117

moneran (MON-e-ran): an organism that belongs to the Monera kingdom; these organisms include bacteria and blue-green algae 111

moss: a type of small plant that has simple leaves and very simple roots and no stem 121

new moon: the phase of the moon when its face is almost invisible from Earth 169

newton: the unit used to measure force 55

non-renewable resources: resources that cannot be replaced as they are used, for example metals ores and oil 216

observation: information about objects and events collected by using one or more of your senses 14

orbit: the path followed by an object in space as it revolves around another object 163

order: a division for classifying animals and plants, between class and family 115

organism: a term used to mean any living thing 103

parallax error: an error that occurs when you don't look square-on to a measuring instrument 34

phases of the moon: the different shapes of the sunlit face of the moon as seen from Earth 169

photosynthesis (FOE-toe-SIN-thu-sis): the process in which the energy of sunlight is absorbed by chlorophyll in green plants and used to make food and oxygen 108

phylum: a division for classifying animals and plants, between kingdom and class (plural = phyla) 115

pivot (or fulcrum): the point or support on which a lever pivots or turns 75

precipitation: any form of water that falls from the clouds; for example, rain, hail, or snow 210

predicting: making a forecast of what a future observation will be, based on past observations 29

producer: an organism that makes its own food using the energy of sunlight 137

properties: the characteristics of materials; for example, strength, colour, hardness and flexibility 186

protist: an organism that belongs to the Protist kingdom; these organisms live in water and include mostly unicellular organisms and algae 110

pulley: a type of simple machine made up of a grooved wheel with a rope around it 80

pure substance: matter containing only one substance (either an element or a compound); it has a fixed composition and fixed properties 186

qualitative (KWOL-i-tate-ive): a type of observation using words without measurement 33

quantitative (KWON-ti-tate-ive): a type of observation that involves measurement 33

renewable resources: resources that can be replaced as they are used, for example timber and solar energy 216

respiration (RES-pe-RAY-shun): the process in living things of getting energy from foods 135

ISBN 978 1 4202 3822 8

revolution: the movement of one body around another body along a curved path (orbit) 163

risk assessment: a process to check that an experiment, investigation or activity is safe, or that risks can be reduced to make it safer 3

rotation: the spinning or turning of a body on its axis 162

saturated: describes a solution that contains the maximum amount of solute that will dissolve at that temperature 189

scavengers: animals that eat the flesh and organs of dead or dying organisms 139

simple machine: a simple machine that magnifies a force or changes its direction 74

solar eclipse: the shadow cast on Earth when the moon is between the sun and Earth 171

solubility: the amount of solute that will dissolve in a measured volume of solvent at a particular temperature 189

solute: a substance that dissolves in a solvent to form a solution 187

solution: a liquid (or solid) containing one or more solutes dissolved in a solvent; salt water is a solution 187

solvent: a substance that can dissolve other substances 187

space probe: a robotic spacecraft sent beyond Earth's orbit to explore space 176

species: a specific organism with a binomial name; a species has unique features and is different from all other species 115

spores: tiny reproductive cells in some groups of organisms, for example fungi, ferns and mosses 109

suspension: a mixture in which tiny bits of solid (or liquid) are evenly spread through a liquid (or gas) but are not dissolved; if allowed to stand, the suspended matter slowly settles out 187

sustainable: an activity is sustainable if it can be continued for a very long time without damaging the environment or using up our natural resources 217

system: something that has parts that work together as a whole; for example, the solar system or the many parts of a bicycle 83

telescope: a device that uses lenses and mirrors to focus light and create an image of objects far away 175

transpiration: the process in which water evaporates from the surface of a leaf 210

vertebrate (VER-te-brate): an animal that has a bony skeleton to support its body, for example a dog, bird or fish 103

water cycle: the cycling of water on Earth as it evaporates from the oceans, condenses into clouds, falls as rain and returns to the oceans 210

weight: the force that is exerted on an object by gravity; it is measured in newtons (N) 65

wheel-and-axle: a simple machine consisting of a central rod or axle with an attached handle or wheel 78

ISBN 978 1 4202 3822 8

Index

3D printing 95

abiotic factors 145
accuracy of measurements 34
action and reaction 89
aeroplanes
    flight controls 91
    wings and lift 89–90
Agenzia Spaziale Italiana (ASI) 177
algae 110
amphibians 118
angle of attack 90
animal kingdom 116–18
animals 108
    classification key (partial) 104
Antarctica
    food web example 140
    polar ecosystem 148
Apollo 11 mirror 173
arthropods 116–17
averaging 29
axes (tools) 77
axis of Earth 162

balanced and unbalanced forces 54
bar graphs 40
bicycles 82, 83–5
    forces 54
    gears 84–5
binomial naming system 115
biodiversity 151–2
biologists 102
biomass 229
biotic factors 145
birds (aves) 118
blackwater 214
block and tackle 80
blue whale 103, 108
blue-green algae 111
bread mould 109, 110
bumping (in test tubes) 197
Bunsen, Robert 12
Bunsen burner use 12–14
    flames and temperatures 13–14
buoyancy forces 52
burns 10

carbon dioxide, testing for 137
carnivores 138
carp, problem of 155
Carver, George Washington 220
Cassini probe 177–8
Cassini-Huygens probe 177
cells 103
    types of 111
centrifuges 193
chain wheels 83
chemicals, disposal of 11
chlorophyll 108
chromatography 201–2
classes (level of classification) 115
classification systems 115
classification process 100–1
colloids 190
competitors 146
concentrated solutions 189
concentration of solutions 189, 191
condensation 210
conifers 122
consumers 137
contact forces 53, 59
crescent moons 169
cuts 10
    data display 40–1
data tables 7
decanting 193, 195
decomposers 109, 139
dependent measurements 41
desalination plant 203
dilution 189
disposal of chemicals 11
dissolving of substances 187
distance advantage 74–5
distillation 196, 198
Double Helix Club 131
drag 89
driven gear 82
driving gear 82

Earth
    axis 162
    revolution 163
    rotation 162
Earth-based telescopes 175
ecosystems 144–5, 148
    changes in 150–1
effort 75
electrostatic forces 51–2
emulsion 190
energy futures 227
energy resources 226
energy supply proposals 231
errors in measurement 34
estimating readings 34
euglena 102–3
European Space Agency (ESA) 177
evaporation 196–7, 210
exoskeletons 116–17
experimenting 42
eye injuries 9

families (classification) 115
ferns 121–2
filter paper 15, 193
filters and filtering 15, 193–5
filtrate 193
fingerprints 20–1
fires 10
fish 118
flocculation 200
flowering plants 122
food chains 134–7
food webs 138–9
force advantage 74–5

ISBN 978 1 4202 3822 8

forces 50–3
balanced 54
bicycle 54
contact 53, 59
frictional 59–62
measuring them 55
non-contact 53
unbalanced 54
forensic entomologists 139
fossil fuels 224
friction 54, 59–62
everyday examples 62
measuring it 59–60
reducing it 61
fronds 121
froth flotation 199
full moon 169
functions of organisms 103
fungi 109

gas chromatography 201
gears 82–5
genera/genus 115
generalisation 15
geothermal energy 230
gibbous moons 169
goldfish (carp) 155
gold panning 199
graphs
drawing of 40–1
using 40
gravitational forces 51, 65
gravity 65
greywater 214
groundwater 210

habitats 144, 146
Haswell's frog 144
heating apparatus 5
herbivores 138
hovercrafts 64
Hubble Space telescope 176
Huygens probe 178
hydro-electricity 229
hypotheses 42

inclined planes 77
independent measurements 41
inferring 28
InMoov robot 94
insoluble substances 187–8
invertebrates 103
classifying them 116
investigative skills 17

jacks (tool) 77
James Webb Space Telescope (JWST) 176
jet engines 90
Juno probe 179

keys, for classifying 101
kingdoms 107–11, 115–16, 121
King's Lomatia 108

laboratories 3
laboratory equipment 4–5
containers 4
drawing of 6
heating apparatus 5
support devices 5
laboratory safety 9–11
leap years 163
levers 75
lift (upward force) 88
line of best fit, drawing 41, 67
line graphs 40–1
lipstick, components 186
living things, requirements 102
load 75
Lowell, Percival 29
lunar eclipses 172

magnetic forces 52
mammals 118
mangroves 124–5, 143
Mars 28, 179–81
mass 65
measuring of 37
measurement errors 34
measuring 33–4
mechanical advantage 75

meniscus 34
milk glue, making 16–17
mixtures 186
model wing, making 91
models 164–5
molluscs 117
monerans 111
Montgolfier, Jacques 88
Montgolfier, Joseph 88
mosses 121
multicellular 103
Mylne, Dr Josh 8

NASA 177
neap tides 173
new moon 169
Newton, Sir Isaac 55, 68
newtons (N) 55, 65
non-contact forces 53

observations 14
recording of 18
qualitative 33
qualitative 33
orange roughies (fish) 218
orbits 163
orders 115
organisms 103
oxygen, testing for 136

paper chromatography 201–2
paper gliders, making 91
parallax errors 34
parasites 109
pendulums 30
percolation 210
phases of the moon 169–70
photosynthesis 108, 135–6
phyla/phylum 115
pivots (fulcrums) 75
plant kingdom 121–22
plant medicines 123
plants 108
classifying them 121–22
plasma (blood) 193
Pluto 29

ISBN 978 1 4202 3822 8

poisoning 10
polar ecosystems, Antarctica 148
precipitation 210
predators 146
predicting 29
producers 137
propellers 90
properties of substances 186
proportions 186
protective clothing 10
protists 110
pulleys 80–1
pure substances 186
pushing forces 51

qualitative observations 33
qualitative observations 33
quokkas 1, 120

rainforest ecosystems 148
reaction time, measuring 44
recycled paper, making 222
recycling 225
regeneration time 216
renewable energy sources 227–30
renewable resource game 217
report writing 15–17, 43
reptiles 118
research projects 128–31
residue 193
resources
  from forests 221
  from plants 219
  minerals and energy 224–5
  recycling 225
  renewable and non-renewable 216, 224–30
respiration 135–6
rhizomes 121
risk assessment and planning 3
robots 93–4
rock plants 102
rockets 89
rocky shore ecosystems 148

safety glasses 9
safety in the laboratory 9–11
saturated solutions 189
scales, reading 33–4
scavengers 139
science, and future careers 45
science research projects 128–31
scientific investigations 14
screws 77
seasons 165, 167
separating colours 201–2
separating solids 199
sieves 193
simple machines 74
solar cells 227
solar eclipses 171
solar energy 226–7
solar power tower 228
solubility 189
soluble substances 188–9
solute 187
solutions 187
  separating them 196–8
solvents 187
space probes 176–8
space-based telescopes 176
species, naming them 115
speed advantage 74
spores 109
spring tides 173
sprockets 83–5
star trails 161, 162
STEM (Science, Technology, Engineering and Mathematics) 93
structure of organisms 103
survival mechanisms 146–7
suspensions 187, 190
  separating them 193–5
sustainable development 217
sustainable fishing 218
systems 83

telescopes 175–6
temperature, and activity 146
terrariums 145
thermometers, hints for using 35
thrust 89, 90
tidal energy 229–30
tides 173
  neap 173
  spring 173
Tombaugh, Clyde 29
transpiration 210
turbidity 153, 155

unicellular organisms 103
units of measurement 33
unsaturated solutions 189

vertebra 103
vertebrates 103
  classifying them 118
viruses 112
volume, measuring 36

water, purification of 200, 214
water cycle 210, 211
water management 213–4
water use 212
  restrictions on use 213
  reusing water 214
  saving and collecting water 213
wave energy 229–30
weight 65
wheel-and-axle 78
wind energy 228
woodchip 221

ISBN 978 1 4202 3822 8

Acknowledgements

The author and publisher are grateful to the following for permission to reproduce copyright material:

PHOTOGRAPHS

Alamy Stock Photo/Hilke Maunder, **190**, /Martin Shields, **37** (bottom), Andrew Brookes, National Physical Laboratory/Science Photo Library, **55**; Anteater Publications, **12**, **13**, **15**, **20**, (dog left) **22**, **22** (dog right), **44** (right), **84**, **97**, **110** (left), **120** (top), **122**, **139**, **141**, **169**, **192**, **193**, **199** (left); Auscape/Ben Cropp, **150** (bottom), /Karen Gowlett-Holmes-Oxford Scientific Films, **234**; Sovereign Hill Media Moguls, **199** (right); CSIRO Marine Research, **218**; Descent Imager/Spectral Radiometer, **178**; Dr Josh Mylne, **8**, **14**; Dreamstime/Elenathewise, **223**, /Enjoylife, **231**, /Kruwt, **196**, /Maystra, **122**, /Mopic, (right) **95**, /Nivi, **122**, /Peterorrell, **121**, /Thewhiteview, **148**; ESA/Hubble & NASA and S. Smartt (Queen's University Belfast)/Robert Gendler, **176** (left); Getty Images/Bettmann, **220**, /CDC, **111**, /D Roberts, **103** (left), /Grant Faint, **161** (Bottom), /Guillermo Gonzalez, **174**, Image © Threatened Species Section, Department of Primary Industries, Parks, Water and Environment, **108** (bottom), /John Fielding, **124** (B), /Kanua Black/EyeEm, 171, /Lynn Funkhouser, **151** (right), /Matt Meadows, **190**, /Neil McIntyre, **149** (bottom), /P. Steeger, **189**, /Reinhard Dirscherl, **151** (left), /Richard Cummins **221**, /Roger Ressmeyer/Corbis/VCG, **175** (top), /Roland Birke, 107 (D), /Visuals Unlimited, Inc./Kjell Sandved, **102** (left); iStockphoto/1001nights, **62** (top), /Aksonov, **73**, /andrearenata, **219** (cotton fibres), /Andrey Danilovich, **82**, /clearviewimages, **147**, /Dennis Donohue, **142**, /desnik, **122** (top), /djedzura, **134**(C), /Jacek Chabraszewski, **133** (top), /lauriek, **21**, /magaliB, **235**, /Martin Auldist, **155**, /Martin Strmko, **150** (top), /Michal Krakowiak, **91**, /Miguel Malo, **36**, /mik38, **64**, /Mike Morley, **137**, /Nicolas Loran, 62 (middle), /Oktay Ortakcioglu, **43** (left), /Richinpit, **61**, /Roberto Lo Savio, **209**, /Sandra Layne, **88**, /sportpoint, **53**, /ultramarinfoto, **59**, /Vladimir Pomortsev, **149** (top), /Wesley Tolhurst, **99** (snake), /YakobchukOlena, **186**; NASA, **28** (right), **65** (bottom), **180**, **173**; NASA/Chris Gunn, **176** (right), **177** (top), **177** (bottom); /ESA, **160**; /JPL-Caltech, **179**; /Lockheed Martin, **63**; /The Viking Project, **28** (left); Newspix /Derek Moore, **213** (right); Peter Saffin, **5**; Photolibrary/Christian Darkin, **108** (left), /Dennis Kunkel, **107** (E), /Royal Institution Of Great Britain, **68**; Science Photo Library, **184**, **205**; /Amelie-Benoist/BSIP, **95** (left), /Martyn F. Chillmaid, 37 (middle), /Russell Kightley, **102** (right); /Sheila Terry **175** (left), SUSUMU NISHINAGA, **221**; Shutterstock/Doloves, **25**, /Dragon Images, **135**, /Eric Isselée, **43** (right); /kuehdi, **113**, /zummolo, **143** (left), /Aaron Amat, **35**, /Adam Ke, **99** (starfish), /Adwo, **120** (bottom), /Africa Studio, **4**, /Africa Studio, **194** (left), /Air Images, **112**, /Alex Veresovich, **110** (right), /Alister G Jupp, **213** (left), /alphaspirit, **45**, /alybaba, **144**, /Andrea Danti, **159**, /Andrei Kholmov, **94**, /Andrey Armyagov, **66** (D), **132**, /antoni halim, **143** (right), /AustralianCamera, **107** (B), /Beat Bieler, **203** (top), /Bernhard Staehli, **211**, (bottom), /bikeriderlondon, **2**, /Bildagentur Zoonar GmbH, **99** (platypus), **219** (fuel), /brizmaker, **219** (building materials), /cbpix, **175** (bottom), /Couperfield, **17**, /CroMary, **44** (left), /CruZeWizard, **9**, /David Hyde, **235** (A), /Deenida, **123**, /DenisNata, **134** (B), /dezi, **99** (horse), /Dinda Yulianto, **99** (frog), /Dirk Ercken, **152** (left), /divedog, **99**, /Dmitri Melnik, **124** (A), /Dmitry Kalinovsky, **66** (C), /Doug Gordon, **107** (C), /Elena Elisseeva, **227**, /Elena Srubina, **99** (fern), /Eric Isselee, **107** (A), /Eric Isselee, **126** (B), /Eugene Onischenko, **66** (E), /George Burba, **192** (bottom), /goodluz, **217**, /Grigorii Pisotsckii, **138**, /GTS Productions, **133** (left), /Gualberto Becerra, **152** (right), /Ipatov, **48**, /Ivan Kuzmin, **99** (bat), /Jacob Lund, **23**, /Joey Chung, **42**, /John Carnemolla, **158**, /Kathryn Willmott, **144**, /kitti suwattanakoon, **66** (B), /ktsdesign, **193** (middle), /Kummeleon, **208**, /Kuznetcov_Konstantin, **71**, /kwest, **235** (B), /Labrador Photo Video, **99** (chicken), /Leah-Anne Thompson, **99** (magpie), /Leena Robinson, **172**, /Lemberg Vector studio, **25**, /Leonid Ikan, **11**, /Macrovector, **25**, /MarcelClemens, **162**, /Maria Bell, **65** (middle), /masik0553, **85** (left), /masik0553, **85** (right), /Mathee saengkaew, **25**, /Mauricio Graiki, **66** (A), /Meryll, **235**, /Mitch Gunn, **85**, /Monkey Business Images, **3**, /NaibankFotos, **99** (mosquito), /Neirfy, **126** (C), /nexus 7, **187**, /Nithid Memanee, **99** (pine), /Nixx Photography, **187**, /NorGal, **194** (right), /Patila, **99** (worm), /Paulphin Photography, **103** (right), /petovarga, **225**, /Praisaeng, **99** (sugar glider), /Pressmaster, **26**, /ProDesign studio, **99** (gum tree), /Rawpixel.com, **44** (middle), /Rob Bayer, **228**, /Sari Oneal, **99** (spider), /science photo, **72**, /sdecoret, **49**, /Serg64, **74**, /Serj Malomuzh, **219** (plants for medicine), /Shane Gross, **99** (shark), /Sihasakprachum, **66** (F), /smereka, **134** (A), /Sofiaworld, **219** (plants), /studiovin, **187**, /Subbotina Anna, **219** (food source), /Susan Schmitz, **99** (lizard), /Sylvana Rega, **220** (small), /Tory Kallman, **99** (dolphin), /Tosaphon C, **99** (fish), /Victor Maschek, Cover/1 /Vitaly Korovin, **126** (A) /Vladyslav Starozhylov, **83**, /volkova natalia, **99** (rose) /Volodymyr Goinyk, **133** (right), /Vudhikrai, **187**, /Wally Stemberger, **211** (top), /warayut, **124** (bottom), /Winston Link, **187**, /yayakizawa, **99** (moss), /Zern Liew, **203** (bottom).

OTHER MATERIAL

© School Curriculum and Standards Authority, extracts from the Year 7 Science Syllabus, 2016, pp 21–22, **viii – xi**.

While every care has been taken to trace and acknowledge copyright, the publisher tenders their apologies for any accidental infringement where copyright has proved untraceable. They would be pleased to come to a suitable arrangement with the rightful owner in each case.